W9-ANB-800

Lectures in Applied Mathematics
Volumes in This Series

Proceedings of the Summer Seminar, Boulder, Colorado, 1960

1 **Lectures in Statistical Mechanics,** *G. E. Uhlenbeck and G. W. Ford with E. W. Montroll*

2 **Mathematical Problems of Relativistic Physics,** *I. E. Segal with G. W. Mackey*

3 **Perturbation of Spectra in Hilbert Space,** *K. O. Friedrichs*

4 **Quantum Mechanics,** *R. Jost*

Proceedings of the Summer Seminar, Ithaca, New York, 1963

5 **Space Mathematics. Part 1,** *J. Barkley Rosser, Editor*

6 **Space Mathematics. Part 2,** *J. Barkley Rosser, Editor*

7 **Space Mathematics. Part 3,** *J. Barkley Rosser, Editor*

Proceedings of the Summer Seminar, Ithaca, New York, 1965

8 **Relativity Theory and Astrophysics. 1. Relativity and Cosmology,** *Jürgen Ehlers, Editor*

9 **Relativity Theory and Astrophysics. 2. Galactic Structure,** *Jürgen Ehlers, Editor*

10 **Relativity Theory and Astrophysics. 3. Stellar Structure,** *Jürgen Ehlers, Editor*

Proceedings of the Summer Seminar, Stanford, California, 1967

11 **Mathematics of the Decision Sciences, Part 1,** *George B. Dantzig and Arthur F. Veinott, Jr., Editors*

12 **Mathematics of the Decision Sciences, Part 2,** *George B. Dantzig and Arthur F. Veinott, Jr., Editors*

Proceedings of the Summer Seminar, Troy, New York, 1970

13 **Mathematical Problems in the Geophysical Sciences. 1. Geophysical Fluid Dynamics,** *William H. Reid, Editor*

14 **Mathematical Problems in the Geophysical Sciences. 2. Inverse Problems, Dynamo Theory, and Tides,** *William H. Reid, Editor*

Proceedings of the Summer Seminar, Potsdam, New York, 1972

15 Nonlinear Wave Motion, *Alan C. Newell, Editor*

Proceedings of the Summer Seminar, Troy, New York, 1975

16 Modern Modeling of Continuum Phenomena, *Richard C. DiPrima, Editor*

Proceedings of the Summer Seminar, Salt Lake City, Utah, 1978

17 Nonlinear Oscillations in Biology, *Frank C. Hoppensteadt, Editor*

Proceedings of the Summer Seminar, Cambridge, Massachusetts, 1979

18 Algebraic and Geometric Methods in Linear Systems Theory, *Christopher I. Brynes and Clyde F. Martin, Editors*

Proceedings of the Summer Seminar, Salt Lake City, Utah, 1980

19 Mathematical Aspects of Physiology, *Frank C. Hoppensteadt, Editor*

Proceedings of the Summer Seminar, Chicago, Illinois, 1981

20 Fluid Dynamics in Astrophysics and Geophysics, *Norman R. Lebovitz, Editor*

Proceedings of the Summer Seminar, Chicago, Illinois, 1982

21 Applications of Group Theory in Physics and Mathematical Physics, *Moshe Flato, Paul Sally and Gregg Zuckerman, Editors*

22 Large-Scale Computations in Fluid Mechanics, Parts 1 and 2, *B. Engquist, S. Osher and R. C. J. Somerville, Editors*

23 Nonlinear Systems of Partial Differential Equations in Applied Mathematics, Parts 1 and 2, *Basil Nicolaenko, Darryl D. Holm and James M. Hyman, Editors*

Nonlinear Systems of Partial Differential Equations in Applied Mathematics

Part 1

Volume 23 - Part 1
Lectures in Applied Mathematics

Nonlinear Systems of Partial Differential Equations in Applied Mathematics

Edited by
Basil Nicolaenko
Darryl D. Holm
James M. Hyman

1986
American Mathematical Society, Providence, Rhode Island

The proceedings of the SIAM-AMS Summer Seminar on Systems of Nonlinear Partial Differential Equations in Applied Mathematics were prepared by the American Mathematical Society with partial support from the following sources: National Science Foundation Grant DMS8402204 U. S. Army Research Office Contract DAAG 29-84-M-0343. (The views, opinions, and/or findings contained in this report are those of the authors and should not be construed as an official Department of the Army position, policy, or decision, unless so designated by other documentation.) The Center for Nonlinear Studies at Los Alamos National Laboratory Contract 5-LT4-N6232-1, through a grant from the Department of Energy.

1980 *Mathematics Subject Classification* (1985 Revision). Primary 35L65, 35L67, 35Q20; Secondary 58-XX, 76-XX, 80-XX.

Library of Congress Cataloging-in-Publication Data

SIAM-AMS Summer Seminar on Systems of Nonlinear Partial Differential Equations and Applications (1984 : College of Santa Fe)
Nonlinear systems of partial differential equations in applied mathematics.
(Lectures in applied mathematics, ISSN 0075-8485; v. 23)
"Proceedings of the SIAM-AMS Summer Seminar on Systems of Nonlinear Partial Differential Equations and Applications"—Verso t.p.
Hosted by the Los Alamos Center for Nonlinear Studies and held at the College of Santa Fe, 1984.
Includes bibliographies.
1. Differential equations, Partial—Congresses. I. Nicolaenko, Basil, 1943– II. Holm, Darryl D. III. Hyman, James M. IV. Society for Industrial and Applied Mathematics. V. American Mathematical Society. VI. Center for Nonlinear Studies (Los Alamos National Laboratory) VII. Title. VIII. Series.
QA377.S4917 1984 515.3'53 85-15107
ISBN 0-8218-1123-1 (alk. paper)
ISBN 0-8218-1125-8—Part I

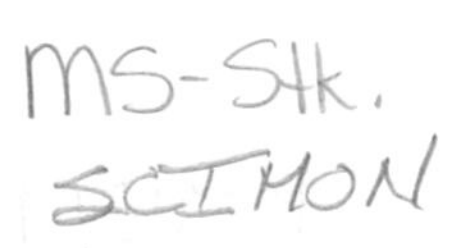

Contents

Part 1

Variational Problems

Evolutionary Systems

PREFACE

The AMS-SIAM Summer Seminar on "Nonlinear Systems of Partial Differential Equations in Applied Mathematics," hosted by the Los Alamos Center for Nonlinear Studies (CNLS), surveyed rapid recent developments, future research trends, and increased interplay of theoretical advances in nonlinear hyperbolic systems, completely integrable systems, and evolutionary systems of nonlinear partial differential equations (PDEs).

The variety of subjects discussed at the meeting reflects the breadth of the scientific interest and offers abundant and clear examples of the interwoven trends among these research areas. For example, techniques in compensated compactness and homogenization of oscillations have spurred a renaissance in the traditional field of nonlinear hyperbolic PDEs. These developments have, in turn, cross-fertilized research in near-integrable systems. The Hamiltonian structure in many integrable systems, dispersive systems, and hyperbolic systems is forming the basis for new nonlinear stability results. Milestones in nonlinear stability were presented for shock and diffusion waves, systems of conservation laws, and nonlinear parabolic equations with chaotic behavior.

These and many other interfield interactions were discussed at the conference and are reflected in the proceedings. We have loosely grouped the articles in the collection into five sections: Integrable Systems, Variational Problems, Dispersive Systems, Evolutionary Systems, and Hyperbolic Systems. In the spirit of the seminar, we urge our dedicated readers to seek additional conceptual links among the articles. We hope that this volume conveys some of the excitement and interactive vitality felt by those of us who participated in the conference.

Finally, we acknowledge, with pleasure, the tremendous personal and organizational support extended by Betty Verducci of the AMS and Frankie Gomez and Marian Martinez of the CNLS. We are most grateful to the Center for Nonlinear Studies, the US Department of Energy Office of Scientific Computing, the US Air Force Office of Scientific Research, and the US Army Research Office for their enabling financial support. The hospitality of the College of Santa Fe is also much appreciated.

Basil Nicolaenko

Darryl D. Holm

James M. Hyman

List of Participants

Mark J. Ablowitz, University of Rochester, Rochester, New York
Alejandro B. Aceves, Tucson, Arizona
Shafique Ahmed, University of Southern Mississippi, Hattiesburg, Mississippi
Darrell E. Allgaier, Richardson, Texas
Ian M. Anderson, Utah State University, Logan, Utah
Judith M. Armes, University of Washington, Seattle, Washington
Joel D. Avrin, University of North Carolina, Charlotte, North Carolina
Andre Bandrauk, Los Alamos National Laboratory, Los Alamos, New Mexico
Claude Bardos, Ecole Nationale Supérieure Ulm, Paris, France
Peter W. Bates, Brigham Young University, Provo, Utah
Patricia E. Bauman, Purdue University, West Lafayette, Indiana
Thomas J. Beale, Duke University, Durham, North Carolina
Melvin S. Berger, University of Massachusetts, Amherst, Massachusetts
Hemant Bhate, Ohio State University, Columbus, Ohio
Alan R. Bishop, Los Alamos National Laboratory, Los Alamos, New Mexico
Anthony M. Bloch, Harvard University, Cambridge, Massachusetts
Luis L. Bonilla, Stanford University, Stanford, California
Claude M. Brauner, Ecole Centrale de Lyon, Ecully, France
Mutiara Buys, University of New Mexico, Albuquerque, New Mexico
Russel E. Caflisch, Courant Institute, New York, New York
Gunduz R. Caginalp, University of Pittsburgh, Pittsburgh, Pennsylvania
David Campbell, Los Alamos National Laboratory, Los Alamos, New Mexico
Gregory Canavan, Los Alamos, New Mexico
Margaret Cheney, Duke University, Durham, North Carolina
I-Liang, Chern, Academia Sinica, Taipei, Taiwan, Republic of China
Donald S. Cohen, California Institute of Technology, Pasadena, California
Peter Constantin, Courant Institute of Mathematical Sciences, New York, New York
Charles A. Coppin, University of Dallas, Irving, Texas
Daniel David, Universite de Montreal, Montreal, Quebec, Canada
Pierre Degond, Ecole Polytechnique, Palaiseau, France
Francoise Demengel, Paris, France
Erik Deumens, Theoretische Natuurkunde Ryksuniversitair Centrum, Antwerpen, Belgium
Ronald J. DiPerna, Princeton University, Princeton, New Jersey
Kai J. Druhl, Albuquerque, New Mexico
W. Eckhaus, University of Utrecht, Utrecht, The Netherlands
J. C. Eilbeck, Heriot-Watt University, Riccarton, Edinburgh, United Kingdom
Nick Ercolani, Ohio State University, Columbus, Ohio
Paul C. Fife, University of Arizona, Tucson, Arizona
Allan Finkel, IBM T. J. Watson Research Center, Yorktown Heights, New York

W. H. Fleming, Brown University, Providence, Rhode Island
Ciprian I. Foias, Indiana University, Bloomington, Indiana
A. S. Fokas, University of Rochester, Rochester, New York
Thanasis S. Fokas, Clarkson University, Potsdam, New York
M. Forest, Ohio State University, Columbus, Ohio
Barry A. Friedman, University of Illinois, Urbana, Illinois
M. Garbey, Ecole Centrale de Lyon, Ecully, France
Robert Gardner, University of Heidelberg, Heidelberg, Federal Republic of Germany
P. Dean Gerber, IBM Thomas Watson Research Center, Yorktown Heights, New York
Jean-Michel Ghidaglia, Rueil-Malraison, France
Yoshikazu Giga, Nagoya University, Nagoya, Japan
Jerome A. Goldstein, Tulane University, New Orleans, Louisiana
Francois Golse, Ecole Nationale Supérieure Ulm, Paris, France
Martin A. Golubitsky, University of Houston, Houston, Texas
Jonathon B. Goodman, Courant Institute of Mathematical Sciences, New York, New York
James M. Greenberg, Ohio State University, Columbus, Ohio
Wilfred M. Greenlee, Tucson, Arizona
F. Alberto Grunbaum, University of California, Berkeley, California
John M. Guckenheimer, University of California, Santa Cruz, California
Richard Haberman, Southern Methodist University, Dallas, Texas
Paul J. Hansen, Potsdam, New York
B. Kent Harrison, Brigham Young University, Provo, Utah
Harumi Hattori, West Virginia University, Morgantown, West Virginia
Russell L. Herman, Clarkson University, Potsdam, New York
Reuben Hersh, University of New Mexico, Albuquerque, New Mexico
David C. Hoff, Indiana University, Bloomington, Indiana
Leonard I. Holder, Gettysburg College, Gettysburg, Pennsylvania
Kent Holing, STATOIL, The Norwegian Oil Company, Stavanger, Norway
Darryl D. Holm, Los Alamos National Laboratory, Los Alamos, New Mexico
Frederick A. Howes, University of California, Davis, California
John K. Hunter, Colorado State University, Fort Collins, Colorado
James M. Hyman, Los Alamos National Laboratory, Los Alamos, New Mexico
A. Evan Iverson, University of Arizona, Tucson, Arizona
Rainer Uwe Jettmar, Naval Surface Weapons Center, Silver Spring, Maryland
Per Jogi, Los Angeles, California
C. K. R. T. Jones, University of Arizona, Tucson, Arizona
Maria C. Jorge, University of New Mexico, Albuquerque, New Mexico
David J. Kaup, Clarkson University, Potsdam, New York
James D. Keeler, Los Alamos National Laboratories, Los Alamos, New Mexico
Barbara L. Keyfitz, University of Houston, Houston, Texas
Mohammad Essam Khalifa, New York, New York
Sergiu Klainerman, Courant Institute of Mathematical Sciences, New York, New York
Yuji Kodama, Nagoya University, Chikusa-Ku, Nagoya, Japan
Robert V. Kohn, Courant Institute of Mathematical Sciences, New York, New York

Philip Korman, University of Cincinnati, Cincinnati, Ohio
Martin D. Kruskal, Princeton University, Princeton, New Jersey
Boris A. Kupershmict, University of Tennessee Space Institute, Tullahoma, Tennessee
John E. Lagnese, Georgetown University, Silver Spring, Maryland
Horst R. Lange, Universitat Koln, Koln, Federal Republic of Germany
Peter Lax, Courant Institute of Mathematical Sciences, New York, New York
Jon Lee, Wright-Patterson Air Force Base, Ohio
Jyh-Hao Lee, Academica Sinica, Taipei, Taiwan, Republic of China
Kotik K. Lee, TRW Space and Technology, Redondo Beach, California
Randy LeVeque, University of California, Los Angeles, California
C. David Levermore, Lawrence Livermore National Laboratory, Livermore, California
Elliott H. Lieb, Princeton University, Princeton, New Jersey
Jacques Louis Lions, College de France, Paris, France
Pierre Louis Lions, Universite de Paris-IX, Paris, France
Tai-Ping Liu, University of Maryland, College Park, Maryland
Geoffrey S. Ludford, Cornell University, Ithaca, New York
Andrew J. Majda, Mountain View, California
Jerrold D. Marsden, Los Alamos National Laboratory, Los Alamos, New Mexico
David W. McLaughlin, University of Arizona, Tucson, Arizona
William R. Melvin, Los Alamos, New Mexico
Michael J. Miksis, Courant Institute, New York, New York
Adrian I. Nachman, Clarkson University, Potsdam, New York
K. Wayne Nagata, Colorado State University, Fort Collins, Colorado
Richard T. Newcomb II, University of Wisconsin, Madison, Wisconsin
Basil Nicolaenko, Los Alamos National Laboratory, Los Alamos, New Mexico
Louis Nirenberg, Courant Institute of Mathematical Sciences, New York, New York
Peter J. Olver, University of Minnesota, Minneapolis, Minnesota
Nils Ovrelid, Nesuddtg, Norway
Robert Palais, University of California, Berkeley, California
George C. Papanicolaou, Courant Institute of Mathematical Sciences, New York, New York
Robert L. Pego, University of Michigan, Ann Arbor, Michigan
Benoit Perthame, Courbeuvoie, France
George H. Pimbley, Los Alamos National Laboratory, Los Alamos, New Mexico
Stephanos Pnevmatiicos, Los Alamos National Laboratory, Los Alamos, New Mexico
Jean-Pierre Puel, Paris, France
Michel Rascle, Universite de St. Etienne, St. Etienne, France
Tudor Ratiu, University of California, Berkeley, California
Rodolfo R. Rosales, Massachusetts Institute of Technology, Cambridge, Massachusetts
Philip Rosenau, Technion, Haifa, Israel
Victor Roytburd, Rensselaer Polytechnic Institute, Troy, New York
John V. Ryef, Bethesda, Maryland
Richard Sanders, Northridge, California

I. Satija, Bartol Research Institute, University of Delaware, Newark, Delaware
Kasia Saxton, University of California, Los Angeles, California
Ralph A. Saxton, University of California, Los Angeles, California
Michelle Schatzman, Universite Claude-Bernard, Villeurbanne, France
B. Scheurer, Centre d'Etudes de Limeil-Valenton, Villeneuve, St. Georges, France
Johanna S. Schruben, Akron, Ohio
Alwyn C. Scott, Los Alamos National Laboratory, Los Alamos, New Mexico
Harvey Segur, University of California, Santa Barbara, California
George R. Sell, University of Minnesota, Minneapolis, Minnesota
Remi J. Sentis, Centre de Limeil-Valenton, Villeneuve St. Georges, France
Jalal M. Ithsan Shatah, Courant Institute of Mathematical Sciences, New York, New York
Michael J. Shelley, University of Arizona, Tucson, Arizona
Thomas C. Sideris, University of California, Santa Barbara, California
Eric Siggia, Cornell University, Ithaca, New York
I. M. Singer, University of California, Berkeley, California
Joel Smoller, University of Michigan, Ann Arbor, Michigan
Eduardo A. Socolovsky, Carnegie Mellon University, Pittsburgh, Pennsylvania
Panagiotis E. Souganidis, Brown University, Providence, Rhode Island
Ren-Ji Sun, Southern Methodist University, Dallas, Texas
Blair K. Swartz, Los Alamos National Laboratory, Los Alamos, New Mexico
Luc C. Tartar, Centre d'Etudes de Limeil-Valenton, Ecole Polytechnique, Paris, France
Roger Temam, CNRS and Université Paris Sud, Orsay, France
Enrique A. Thomann, Berkeley, California
Lu Ting, Duke University, Durham, North Carolina
Peter J. Tonellato, University of Arizona, Tucson, Arizona
Hans Troger, Technische Universitat Wien, Vienna, Austria
Kaising Tso, SUNY at Stony Brook, Stony Brook, New Yokr
A. Van Harten, University of Utrecht, Utrecht, The Netherlands
Stephanos Venakides, Stanford University, Stanford, California
Pierre A. Vuillermot, University of Texas, Arlington, Texas
David H. Wagner, University of Houston, Houston, Texas
Henry A. Warchall, University of Texas, Austin, Texas
Michael I. Weinstein, Princeton University, Princeton, New Jersey
Burton B. Wendroff, Los Angeles National Laboratory, Los Alamos, New Mexico
Weldon J. Wilson, University of Central Florida, Orlando, Florida
James W. Wiskin, University of Utah, Salt Lake City, Utah
J. Wysin, Cornell University, Ithaca, New York

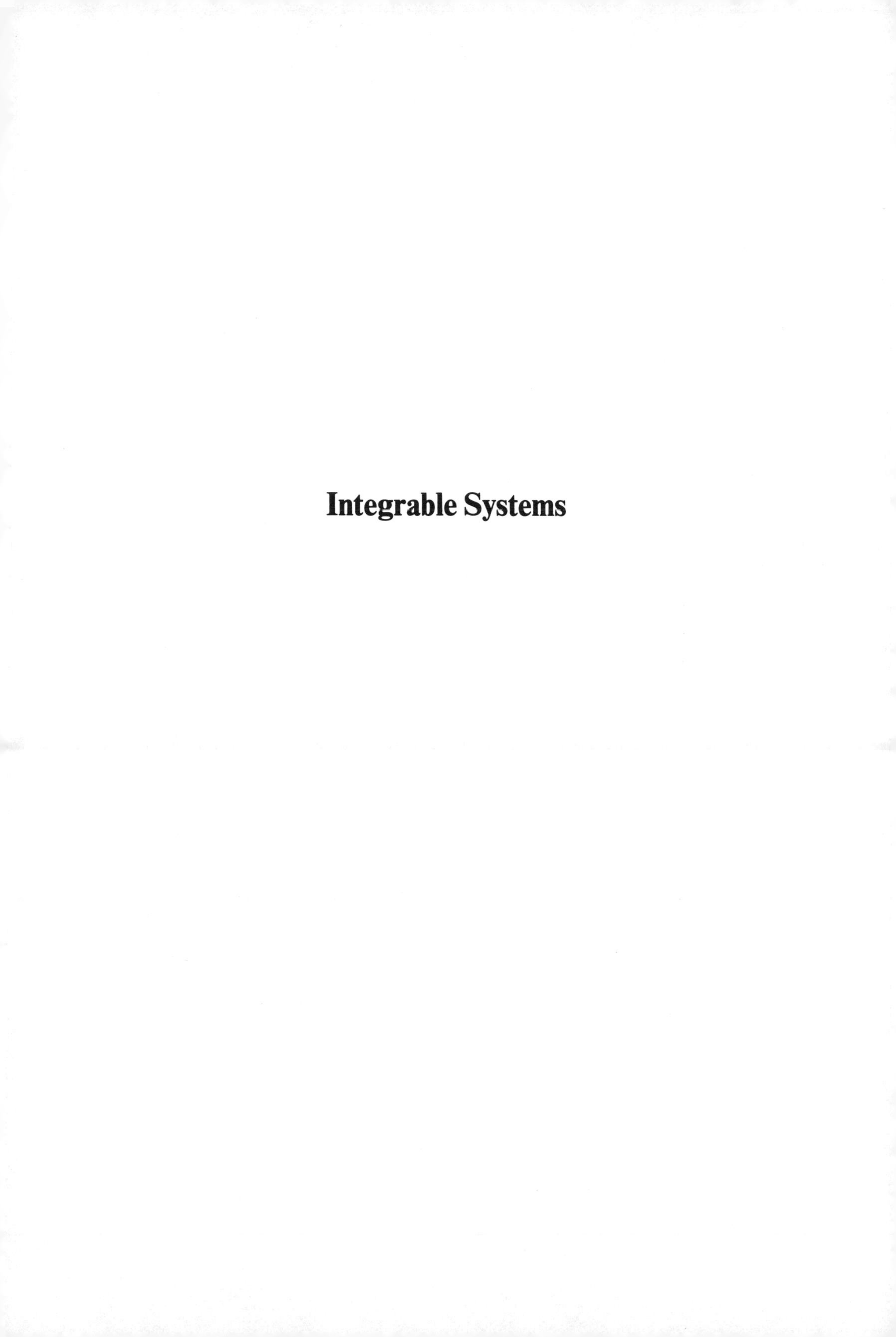

Integrable Systems

Lectures in Applied Mathematics
Volume 23, 1986

OSCILLATIONS AND INSTABILITIES IN NEAR INTEGRABLE PDE'S

by

N. Ercolani[1]
M. Forest[1]
D. W. McLaughlin[2]

Abstract

In this article we describe general qualitative information about solutions of nonlinear evolutionary PDE's which can be obtained from integrable methods. We emphasize, through examples, typical behavior of nonlinear waves which can be discovered and described by these methods. Because we believe this qualitative behavior is more general than the integrable methods themselves, we do not describe in any detail the specific mathematical techniques. For the most part, the examples are taken from our own work, although references to related studies are given. The qualitative properties which we illustrate fall into three classes: (i) the propagation of oscillations, (ii) coherent structures, and (iii) the generation of more complicated spatial structures

1980 Mathematics Subject Classification: 35B10,35B20,35Q20,58G25
[1]Supported in part by NSF MCS-8202288, DMS 8411002
[2]Supported in part by AFOSR 830227, NSF DMS 8403187

I. Introduction

Any list of the major theoretical problems of modern, nonlinear, evolutionary PDE's will certainly include (i) the propagation of oscillations; (ii) the formation, propagation, and interaction of spatially coherent structures (for example, solitary waves, vortices, shocks, and singularities); (iii) the chaotic evolution of coherent structures; and (iv) the breakdown of these structures and the subsequent generation of more complicated states. Integrable models provide exact solutions and complete representations with which to study such general problems. Of course, the solutions of integrable equations are simpler and more regular than their realistic counterparts; however, behavior in nonlinear PDE's is so complex that even simple integrable examples can provide substantial insight.

Our own work has concentrated on (i) the propagation of oscillations in integrable wavetrains; (ii) the behavior of solitary waves and wavetrains in the presence of perturbations; (iii) the chaotic evolution of solitary waves; and (iv) instabilities of integrable waves which generate more complicated spatial states.

In this article we will illustrate, primarily with examples from our own work, that general qualitative properties of nonlinear waves can be extracted with integrable methods. Specifically, in section III we describe the propagation of integrable oscillations, in section IV we identify coherent structures with a (numerical) spectral transform, and in Section V we discuss examples of the breakdown of coherence.

II. Integrable Preliminaries

In this section we present a minimal description of those objects from integrable theory which we need. In an Appendix we provide a more detailed account of this integrable machinery.

This section is organized in two parts: in Part A we describe explicit formulas which characterize robust families of solutions; in Part B, we describe the inverse spectral transform which provides a complete representation.

II.A. Robust Families of Solutions

One striking feature of integrable wave equations is that explicit, robust families of solutions are known. Each family is indexed by N = the number of dynamical degrees of freedom, $N = 1,2,\ldots$; moreover, one can list the free parameters for each member of the family. We describe two such families.

The structure of these families is similar for a class of integrable equations which includes Korteweg-deVries (KdV), nonlinear Schrodinger (NLS), and Sine-Gordon (SG). Therefore, it will suffice to present families for KdV,

$$u_t - 6uu_x + u_{xxx} = 0.$$

The simplest example of such a family is single solitons,

$$u(x,t;k,y) = -2k^2 \operatorname{sech}^2(k(x-y-k^2 t)).$$

The 2 parameters in this family are k^2(speed) and y(location). In this article we will be primarily concerned with a larger family of solutions: N phase, quasiperiodic wavetrains. This means a solution of the form

$$u(x,t) = W_N(\theta_1,\ldots,\theta_N;\ \lambda_0,\ldots,\lambda_{2N}) \qquad \text{(II.1)}$$

where W_N is C^∞, 2π periodic in each θ_i. The phase θ_i is linear in x and t,

$$\theta_i = \kappa_i x + \omega_i t + \theta_i^0 ; \qquad \text{(II.2)}$$

also, W_N is smooth in the real parameters $\lambda_0, \lambda_1, \ldots, \lambda_{2N}$. The function W_N can be explicitly constructed in terms of Θ functions. These well known functions are built from the hyperelliptic Riemann surface **R** of $[\Pi_j (\lambda - \lambda_j)]^{1/2}$, whose branch points are precisely the parameters $\{\lambda_i\}$. In addition to defining W_N, this Riemann surface also fixes the wave numbers κ_i and the frequencies ω_i by the following formulas:

$$\kappa_i = \oint_{a_i} \Omega_1$$

$$\omega_i = -12 \oint_{a_i} \Omega_2 .$$

Here Ω_1 and Ω_2 are normalized meromorphic differentials on **R** and $\{b_i\}$ are canonical closed paths on **R** . For explict formulas, see the Appendix.

Note Ω_1, Ω_2, κ_i, ω_i are uniquely determined by the parameters $\{\lambda_i\}$. It follows that the general N-phase wave for KdV depends upon $N + (2N + 1)$ parameters: N phases $\vec{\theta}_0 = (\theta_1^{(0)}, \ldots, \theta_N^{(0)})$ and 2N+1 parameters $\vec{\lambda} = (\lambda_0, \ldots, \lambda_{2N})$ which fix the N spatial wave numbers, the N temporal frequencies, and the mean of u. This parameter dependence is explicitly indicated in

$$u(x,t;\vec{\theta}_0, \vec{\lambda}) = W_N(\theta_1, \ldots, \theta_N; \vec{\lambda}) \qquad \text{(II.1')}$$

II.b. Inverse Spectral Representations

A second ramification of integrability is a complete representation of the solution to the initial value problem for soliton equations. Once again it will suffice to consider KdV since the other soliton equations have similar representations.

The integration of KdV under periodic boundary conditions is accomplished through Hill's operator $\hat{L}$,

$$\hat{L} = -\partial_{xx} + u. \tag{II.3}$$

The fundamental object in the spectral theory of $\hat{L}$ is the Floquet discriminant $\Delta(u;\lambda)$, which we now define. First compute, from a basis of solutions of $\hat{L}\psi = \lambda\psi$, the 2×2 fundamental matrix solution $M(x,x';\lambda;u)$, normalized to equal the identity when $x = x'$. The transfer matrix $M(x'+L, x';\lambda,u)$ maps solutions across one period of the potential u. The Floquet discriminant is defined in terms of this transfer matrix by

$$\Delta(\lambda;u) = \text{trace } [M(L;0;\lambda,u)]. \tag{II.4}$$

$\Delta^2(\lambda;u)-4$ is an entire function of λ, whose roots $\{\lambda_i\}$ determine the spectrum of $\hat{L}$.

Under the KdV flow, $\Delta(\lambda;u(t))$ is invariant. In fact, the Riemann surface $\mathbf{R}$, used to construct the families described in II.A, is just the Riemann surface of the function $\sqrt{\Delta^2(\lambda;u)-4}$. The roots $\{\lambda_i\}$, which are the branch points of $\mathbf{R}$, fix the physical characteristics of the wave. For instance, (i) there are exactly $2N+1$ simple roots of $(\Delta^2(\lambda;u)-4)$ iff u is an N phase wave; (ii) both solitons and radiation have a characterization through these simple roots; (iii) the wave numbers κ_i are given in terms of $\{\lambda_i\}$ by the formula

$$\kappa_i = \oint_{a_i} \frac{\Delta'}{\sqrt{\Delta^2-4}}\, d\lambda.$$

This exemplifies the general principle discussed further in section IV, that physical characteristics of the wave $u(x,t)$ can be efficiently measured by the invariant $\Delta(\lambda;u)$.

III. Propagation of Rapid, Integrable Oscillations.

In this section we apply the robust oscillatory families

(Section II.A) to the first general problem mentioned in the introduction - the propagation of oscillations. Consider a solution u of the Korteweg-deVries equation

$$u_t = 6uu_x - \varepsilon^2 u_{xxx}, \quad 0 < \varepsilon \ll 1, \qquad \text{(III.1)}$$

which at some time t is rapidly oscillating in the form

$$u \sim W_N \left(\frac{\vec{\theta}(x,t)}{\varepsilon}; \vec{\lambda}(x,t)\right) + O(\varepsilon), \qquad \text{(III.2)}$$

where

$$\vec{\theta}_t = \vec{\omega}(\vec{\lambda}) = -12 \oint_{\vec{a}} \Omega_2 \qquad \text{(III.3a)}$$

$$\vec{\theta}_x = \vec{\kappa}(\vec{\lambda}) = - \oint_{\vec{a}} \Omega_1, \qquad \text{(III.3b)}$$

and where the existence of $\vec{\theta}$ satisfying (III.3a,b) demands

$$\vec{\kappa}_t = \vec{\omega}_x. \qquad \text{(III.4)}$$

Here W_N denotes the N-phase KdV wave form described in (II.1). It depends upon $(2N+1)+N$ free parameters. Physically, $2N+1$ of these, $\lambda_0,\lambda_1,\ldots,\lambda_{2N}$, define the temporal frequencies $\vec{\omega}$ (III.3a), spatial wave numbers $\vec{\kappa}$ (III.3b), and the mean of the wave form. The remaining N variables fix the phases. Our goal is to pick the time evolution of the $\vec{\lambda}(x,t)$ variables so as to guarantee that there is a solution u of form (III.2).

As one attempts to construct a solution of (III.1) in the rapidly oscillating form (III.2), a necessary condition emerges.

THEOREM [1,2,3]. If a rapidly oscillating KdV solution of form (III.2) exists, then the $\vec{\lambda}(x,t)$ must satisfy modulation equations which can be represented in terms of the differentials $\Omega_j = \Omega_j(\lambda; \vec{\lambda}(x,t))$ as

$$\partial_t \Omega_1 - 12\, \partial_x \Omega_2 = 0. \qquad \text{(III.5)}$$

Proofs of this invariant representation may be found in reference [1]. Here we make several remarks about it.

Remark 1 (Level of Mathematical Arguments) The mathematical arguments which establish the validity of the rapidly oscillating wave form (III.2) are formal. For the single phase case $(N = 1)$, the modulation equations (III.5) can be shown necessary and sufficient [3,4]. In the multiphase $(N > 1)$ case, only necessity has been established [1,3,5]. The most detailed mathematical results are those of Lax, Levermore, and Venakides [6,7,8] who show that the weak limit of u^ε is characterized, for various types of initial data, by (III.5) for variable N.

Remark 2 (Invariant Representation as a Natural Deformation of a Riemann Surface) One must understand the invariant representation (III.5) as an equation for the variables $\vec{\lambda}(x,t)$. Expression (III.5) is a particularly nice way to write the equations for these variables. (III.5) may be viewed as defining a natural deformation of the underlying Riemann surface. Exploiting the λ dependence of $\Omega_j(\lambda;\vec{\lambda}(x,t))$, other equivalent forms of the modulation equations may be derived. In the next few remarks we describe three equivalent forms - (i) averaged conservation laws, (ii) a Hamiltonian form, and (iii) a Riemann invariant form.

Remark 3 (Averaged Conservation Laws) The most physically intuitive derivation of modulation equations results by averaging the conservation laws of the KdV equation. The first of these is the equation itself,

$$\partial_t[u] = \partial_x[3u^2 - \varepsilon^2 u_{xx}]. \qquad \text{(III.6a)}$$

Multiplying the equation by u generates the second,

$$\partial_t[\frac{u^2}{2}] = \partial_x[2u^3 + \frac{3}{2}\varepsilon^2 u_x^2 - \frac{1}{2}\varepsilon^2(u^2)_{xx}]. \quad \text{(III.6b)}$$

Continuing in a similar, although more complicated, fashion one can generate an infinite sequence of conservation laws of the form

$$\partial_t[\tau_j(u,\varepsilon u_x,\ldots)] = \partial_x[\chi_j(u,\varepsilon u_x,\ldots)]. \quad \text{(III.7)}$$

Inserting the rapidly oscillating wave form (III.2) yields

$$(\frac{1}{\varepsilon}\vec{\omega}\cdot\vec{\nabla} + \partial_t)\,\tau_j = (\frac{1}{\varepsilon}\vec{\kappa}\cdot\vec{\nabla} + \partial_x)\chi_j \quad \text{(III.7')}$$

where $\vec{\nabla} = \partial/\partial\vec{\theta}$. Averaging over the 2π periodic $\vec{\theta}$ yields

$$\partial_t \langle\tau_j(W_N,\ldots)\rangle = \partial_x \langle\chi_j(W_N,\ldots)\rangle,\ j = 1,2,\ldots \quad \text{(III.8)}$$

The averages $\langle\tau_j(W_N,\vec{\kappa}\cdot\vec{\nabla}W_N,\ldots,)\rangle$ and $\langle\chi_j(W_N,\vec{\kappa}\cdot\vec{\nabla}W_N,\ldots)\rangle$ depend only on the $\vec{\lambda}$ variables, or equivalently on the wave numbers $\vec{\kappa}$, the frequencies $\vec{\omega}$, and the mean. Thus, $(2N+1)$ conservation laws provide $(2N+1)$ equations for $\vec{\lambda}(x,t)$. Physically, these equations state that the densities are conserved on the average. Mathematically, there are difficulties. Is the system over determined? After all, we have an infinite number of equations for $(2N+1)$ unknowns. Does one get equivalent equations no matter which $(2N+1)$ conservation laws are averaged?

These averaged conservation laws may be derived [1] directly from representation (III.5). One expands $\Omega_j(\lambda;\vec{\lambda}(x,t))$ near $\lambda \simeq \infty$. The coefficients of (III.5) in this expansion are the averaged conservation laws. In addition, the procedure shows that if the first $(2N+1)$ averaged conservation laws are satisfied, (that is if the first $(2N+1)$ coefficients in the expansion of (III.5) near $\lambda = \infty$ are zero), then all averaged conservation laws are satisfied, (that is (III.5) is satisfied for all λ). This is a partial answer to the consistency question.

Remark 4 (Connection with Closure Problems) Averaged conservation laws are closely related to a "closure problem" in the engineering approach to the propagation of rapid fluctuations. The averaged conservation laws do not close in terms of powers of u; the equation of $\langle u \rangle$ depends upon $\langle u^2 \rangle$, the one for $\langle u^2 \rangle$ depends upon $\langle u^3 \rangle$, and so forth. Here they do close, however, on the parameters of a robust family of rapidly oscillating waves. These parameters, $\vec{\lambda}$, are the appropriate dynamical variables for the averaged system. They satisfy the modulation equations which are quite different from the original KdV equations. This successful closure should be constrasted with the unsuccessful, naive closures based on powers of u which are often used in the far more difficult problem of turbulence.

Remark 5 (Riemann Invariant Form of Modulation Equations) The most useful mathematical form of the modulation equations is the Riemann invariant form which is diagonal in the derivatives of $\vec{\lambda}$. Since the number of unknowns $(2N+1)$ exceeds 2, the existence of a Riemann invariant form is not guaranteed; indeed, it is rare for a hyperbolic system of size greater than two which arises in a physical problem to admit a diagonal form. This form of the modulational equations was discovered by Whitham [9,10] in the case $N = 1$ through clever explicit calculations.

This Riemann form is an immediate consequence [1] of the modulation representation (III.5). No calculations are required, even for arbitrary N. One expands $\Omega_k(\lambda,\vec{\lambda})$ near the ramification points $\lambda = \lambda_j$, which (besides ∞) are the only distinguished points on the Riemann surface. The invariant representation (III.5) is singular at $\lambda = \lambda_j$; the coefficient of the singular term must be zero, which implies, for each $j = 0,1,\ldots,2N$,

$$\frac{\partial}{\partial t}\lambda_j - s_j(\vec{\lambda})\frac{\partial}{\partial x}\lambda_j = 0. \qquad \text{(III.9)}$$

This is the Riemann invariant form of the modulation equations. It is diagonal in the derivatives $\lambda_{j,t}$ and $\lambda_{j,x}$. The $\vec{\lambda}$ are the Riemann variables, with the characteristic speeds explicitly given by the formula

$$s_j(\vec{\lambda}) = \frac{12\tilde{\Omega}_2(\lambda_j;\vec{\lambda})}{\tilde{\Omega}_1(\lambda_j;\vec{\lambda})}, \qquad \text{(III.10)}$$

where $\tilde{\Omega}_k$ is defined by the behavior of Ω_k near $\lambda = \lambda_j$:

$$\Omega_k(\lambda;\vec{\lambda}) \simeq \frac{\tilde{\Omega}_k(\lambda_j;\vec{\lambda})}{\lambda-\lambda_j}\,d\lambda \qquad \text{near } \lambda \simeq \lambda_j\,. \qquad \text{(III.11)}$$

Remark 6 (Generation of Additional Phases or Degrees of Freedom) The Riemann invariant form (III.9) has important physical consequences. Generally, the (2N+1) speeds s_j are real and distinct. In this situation, the modulation equations are strictly hyperbolic and the N-phase KdV wavetrain is modulationally stable. These distinct characteristics speeds replace the single group velocity in the theory of dispersive waves. For other integrable equations such as the sine-Gordon equation and the focusing nonlinear Schrodinger equation [10,11,12,13] these characteristic speeds can be complex. The modulation equations resemble elliptic equations and the wavetrain is modulationally unstable.

Returning to the KdV case, the modulation equations (III.9)

as solved by the method of characteristics can develop multiple-valued solutions. This indicates a breakdown of the N phase rapidly oscillating wave form. In retrospect, the resolution of this situation was implicitly indicated by [14]; however, the complete, thorough, and systematic resolution was given by Lax and Levermore [7].

In this context, a consequence of the work of Lax and Levermore is that the correct way to think about modulation equations (III.9) is not for fixed N. Rather, the entire family of modulation equations as indexed by N should be treated simultaneously. N is integer valued and changes values occasionally. The multivalued nature of solutions is resolved by increasing N. At the points in (x,t) at which the solution becomes multivalued, an additional phase is generated and an N phase solution becomes an $(N+1)$ phase solution. Alternatively, the $(N+1)$ phase solution continued backwards in time t has a point at which two of the distinct variables λ_i become equal. For KdV, this establishes a mechanism for "smoothing shocks" - namely, the generation of additional phases or degrees of freedom. As yet the generality of this mechanism for "smoothing shocks" in unknown; however, it is certainly suggestive that shocks in a conservative system are smoothed by increasing the number of active degrees of freedom in the front. See [15,16] for related work.

<u>Remark 7</u> (<u>Results of Lax-Levermore-Venakides</u>) The scenario described in the preceding remarks is not yet completely understood. First, Lax and Levermore do not use "modulation theory". Rather, they study the [illegible] value problem for u,

$$\partial_t u^\varepsilon = 6u^\varepsilon u^\varepsilon_x - \varepsilon^2 u^\varepsilon_{xxx}$$

$$u^\varepsilon(x,t=0) = u_0(x) \to 0 \quad \text{as} \quad |x| \to \infty.$$

for special classes of (ε independent) initial data, such as $u_0(x)$ negative with a single minimum. They characterize the weak limit of u^ε as $\varepsilon \to 0$. Their argument begins with the inverse scattering representation of u^ε. They argue that as $\varepsilon \to 0$, the radiation (or continuous spectrum) components of u^ε are negligible and that u^ε can be well approximated by a collection of solitons, the number of which goes to infinity as $\varepsilon \to 0$. Using a pure soliton representation of u, they characterize the weak limit of u by a variational problem whose solution is given in terms of the entire family (indexed by N) of equations (III.9). No interpretation of the physical meaning of the parameters $\vec{\lambda}$ is given, nor is the oscillatory structure of the N-phase wavetrain captured. However, the weak limit of u^ε is completely characterized, tied to the initial data u_0, and the necessity and procedure for changing N is established.

It is often asked if the weak limit of (III.12), as characterized by Lax and Levermore, is the same as the weak limit of Burger's equation,

$$u^\varepsilon_t = 6u^\varepsilon u^\varepsilon_x + \varepsilon^2 u^\varepsilon_{xx}$$
$$u^\varepsilon(x,t=0) = u_0(x), \qquad \text{(III.13)}$$

as characterized by discontinuous shocks whose speed and stability are fixed by the Rankine-Hugoniot jump conditions and the entropy conditions. The answer is that the weak limits of (III.12) and (III.13) are different, although earlier analysis [17,18] of the limit in Burgers' case through the Cole-Hopf transformation certainly motivated the Lax-Levermore analysis of the KdV case through the inverse scattering transformation.

Before the first breaking time, the limits of the two problems agree and are actually strong limits. After breaking, oscillations persist in the KdV case and the weak limit is characterized by (III.9), which plays no role in the description of the weak limit for Burgers' equation. Of course, the jump condition, which is a consequence only of the conservation law from the equations of motion, still fixes the "speed of the shock region" [19]; however, the local structure of the shock front is changed from a discontinuity in Burger's case to a fan (with increasing support) of rapid oscillations in the KdV case.

Venakides [8] has improved the arguments of Lax and Levermore and has captured the local oscillatory nature of the field u^{ε} for small, but positive, ε. He replaces the inverse scattering representation with the $\vec{\mu}$ representation (see the Appendix) from periodic spectral theory. In the small ε limit, this $\vec{\mu}$ representation is dominated by solitons and is quite tractable. Using this representation, Venakides has (i) recovered the Lax-Levermore characterization of the weak limit, together with an interpretation of its ingredients, (ii) shown that, for small but positive ε, the oscillations are N-phase wavetrains, and (iii) accomplished the same for KdV with period 1 boundary condition [20]. As yet, he is unable to control the phases of the wavetrains.

<u>Remark 8</u> <u>(Young's Measure of Tartar and DiPerna)</u> A very interesting theoretical description of weak limits of solutions of nonlinear pde's is being developed by Tartar and DiPerna [21,22]. In their framework, the weak limit is characterized by a measure in "u space"- the Young measure. In the case of Burger's equation (III.13), this measure consists of point masses with support on the (discontinuous) shocks. In the KdV case, the support of the measure is not point-like; the spread

in the support captures the oscillatory nature of the weak limit. In the KdV case, we have formally derived an explicit representation of Young's measure as it is characterized by the modulation equations (III.5). For the explicit formulas, we refer the reader to [3].

Remark 9 (A Hamiltonian Form of the Modulation Equations) The structure of the modulation equations is captured by their Hamiltonian form [10,23,24]. This form is easier to display for the sinh-Gordon equation

$$\varepsilon^2(u_{tt}-u_{xx}) = -\sinh u \tag{III.14}$$

than for KdV. The modulation theory for (III.14) is worked out in [11]. There $u = u^{\varepsilon}$ is represented in rapidly oscillating form

$$u^{\varepsilon} \sim W_N(\frac{\vec{\theta}}{\varepsilon}\,;\,\vec{E}) \tag{III.15}$$

where the parameters $\vec{E} = (E_1,\dots,E_{2N})$ (i) are the ramification points of the appropriate Riemann surface; (ii) fix the spatial wave number and frequencies $\vec{\kappa}$ and $\vec{\omega}$; (iii) are the Riemann variables. Again, the modulation equations can be expressed in terms of differentials $\Omega^{(\pm)}(E;\vec{E})$,

$$\Omega_t^{(-)} = \Omega_x^{(+)} \quad , \tag{III.16}$$

and modulational stability is predicted. The Hamiltonian form of these equations results by integrating the E dependence of $\Omega^{\pm}(E;\vec{E})$ around suitably chosen cycles on the Riemann surface. The result is [11]

$$\partial_t \vec{\kappa} = \partial_x \frac{\partial H}{\partial \vec{J}} , \tag{III.17}$$

$$\partial_t \vec{J} = \partial_x \frac{\partial H}{\partial \vec{\kappa}} ,$$

where

$$\vec{\kappa} \equiv - \oint_{\vec{a}} \Omega^{(-)}$$

$$\vec{J} \equiv \frac{2i}{\pi} \oint_{\vec{a}} \ell n E \ \Omega^{(-)} . \qquad \text{(III.18)}$$

(Equation (III.18a) for $\vec{\kappa}$ is just the consistency equation (III.4).) For modulation theory, the wave numbers $\vec{\kappa}$ and the action variables $\vec{J}$ are canonically conjugate variables, with respect to the symplectic structure generated by $\begin{pmatrix} 0 & 1 \\ 1 & 0 \end{pmatrix} \partial_x$. For more details, we refer the reader to [24,25].

Remark 10 (Nearly Integrable Equations) Modulation theory is certainly not restricted to integrable equations; moreover, invariant representations such as (III.5) can still be derived for nearly integrable equations provided the perturbation is sufficiently weak. As an example in [26], we study a perturbation of the KdV equation,

$$u_t^\varepsilon - 6u^\varepsilon u_x^\varepsilon + \varepsilon^2 u_{xxx}^\varepsilon = \delta^\varepsilon u_{xx}^\varepsilon . \qquad \text{(III.19)}$$

When $\delta^\varepsilon > \beta\varepsilon^2$, then the weak limit of (III.19) consists of Burger shocks [27]. When $\delta^\varepsilon < \beta\varepsilon^2$, then the weak limit of (III.19) is described by (III.5). When $\delta^\varepsilon = \beta\varepsilon^2$, the weak limit is characterized by new [26] modulation equations of the form

$$\partial_t \Omega_1 - 12 \ \partial_x \Omega_2 = \beta df, \qquad \text{(III.20)}$$

where the differential df is derived from u_{xx}^ε and depends only on $\vec{\lambda}$, not on any derivatives of $\vec{\lambda}$. In some sense, $\delta = \beta\varepsilon^2$ is on the boundary between KdV oscillations and Burgers' shocks. On this boundary the role of the term βdf

in shock smoothing is yet to be investigated. However, see the related work [28].

IV. Coherent Structures in Nearly Integrable Systems

In this section we use the inverse spectral representation (Section II.B) to study spatially coherent structures which persist even in temporally chaotic states. The theory of integrable waves, together with numerical experiments, have certainly taught us that the large time behavior of a dispersive wave is frequently dominated by solitary waves. The qualitative theory of dispersive waves without integrable methods is still in its infancy - the nonlinear stability of some solitary waves has been established [28-34]; the response of solitary waves to external perturbations has been formally computed by many workers, including ourselves; some estimates of the number of solitary waves in the asymptotic state have been given [35]; nonlinear scattering theory is being developed [36,37,38]; the development of singularities in finite time has been established in certain situations [39,40,41,42].

At present it is currently fashionable to study "spatially coherent structures" in nonlinear fields, even though such structures are never defined. Indeed, the prevalent attitude is that "one knows a coherent structure when one sees it". In realistic problems such as high Reynolds number fluid flows this attitude may be founded in necessity; however, it is not required when studying perturbations, even large perturbations, of integrable waves. In this situation solitons or soliton wave trains are the natural candidates for these coherent structures.

If a few solitons or solitary wave trains accurately describe the field, then soliton perturbation theory could be used to reduce the infinite dimensional field to a low dimensional approximation which is accurate enough to capture the essential features, even including chaotic responses. Recently, several authors have assumed the validity of a representation of the field in terms of a few solitons and concentrated on the derivation and consequences of the reduced equations [43,44]. Alternatively, we have emphasized that integrable machinery should first be used numerically to check the validity of the representation [45-50]. Is the field well described by a few solitons? If so, perturbation analysis is useful.

Explicitly, we have studied two problems with these techniques: (i) the damped, driven sine-Gordon equation under periodic boundary conditions [50] and (ii) a sequence of nonlinear Schrodinger equations which arise in a laser cavity [51]. Here we will concentrate on our work on the sine-Gordon equation. The laser cavity work is summarized elsewhere [51], although one of its features will be discussed in Section V.

In the remainder of this section we will describe: (a) numerical experiments on the damped, driven SG equation with periodic boundary conditions,

$$u_{tt} - u_{xx} + \sin u = \Gamma \cos(\omega t) - \alpha u_t, \qquad \text{(IV.1a)}$$

$$u(x+L,t) = u(x,t) \pmod{2\pi}; \qquad \text{(IV.1b)}$$

$$u_t(x+L,t) = u_t(x,t);$$

(b) the numerical identification of coherent spatial structures in these experiments; and (c) the status of analytical methods aimed at predicting the observed phenomena.

A. Numerical Experiments on (IV.1)

Extensive numerical experiments [52,53,54] have been performed on the perturbed periodic SG system (IV.1). The x-period L, the damping coefficient α, and the driving frequency ω are fixed; for given Cauchy data $\vec{u}(x,0)$, a bifurcation sequence of long time $(t \gg 1)$ behavior is computed as Γ, the driving strength, is varied. Here we describe the numerical results for Cauchy data of a periodic breather train with envelope frequency initially distinct from the driver. We will represent the asymptotic $(t \gg 1)$ response in each parameter range of Γ by a pair (spatial structure; temporal structure). We will abbreviate as follows. SPATIAL DESCRIPTION: period 1/N means N localized states within the basic period L; $k = 0$ means an x-independent, flat background. TEMPORAL DESCRIPTION: locked means locked temporally to the driver; chaotic means temporally chaotic.

Fix $\vec{u}_0$ an exact, period 1, breather train. In order of increasing Γ, the typical bifurcation sequence is:

(1) ($k = 0$; locked) (i.e., the breather decays to a flat state which is periodic at frequency ω);

(2) (period 1 plus $k = 0$; locked) (i.e., one local state per period, riding on a flat background);

(3) (period 1/2 plus $k = 0$; locked) (i.e., two local states per period, plus a flat background);

(4) (period 1, with several local states, plus $k = 0$; chaotic).

B. Data Analysis of the Numerical Experiments

We want to quantify what "coherent" or low dimensional means in this numerical experiment. Is it correct to represent the field by a few sine-Gordon breathers? To answer this, take the numerically computed spatial data, $\vec{u}(x;t_k)$, at each time step t_k in the integration. Then, numerically calculate the discriminant $\Delta(E;\vec{u}(\cdot;t_k))$ by integrating the ode system (A.10). Now analyze Δ as a function of E, and compute the simple roots $\{E_j\}_1^{2N}$ of $\Delta = \pm 2$, $0 \leqslant N \leqslant \infty$. These carry the required information.

A crucial point to be emphasized here is that any spatial

profile, periodic in x of period L, can be resolved in the SG basis, or a Fourier basis, or any other basis. This data analysis, using $\Delta(E;\vec{u}(\cdot,t_k))$, measures, in terms of the variables $\{E_j\}_1^{2N}$, the amplitudes of each SG basis mode. It measures how many modes are excited and the degree of excitation of each.

Either the spatial structure $\vec{u}(x,t_k)$ is well approximated by an N-dimensional SG basis element, defined by $\{E_j\}_1^{2N}$, for N small, or not. The integrable machinery, as just described, can resolve the answer. The results of the numerical experiments tested in [50] are that even when Γ is large enough for temporal chaos, the spatial structure is accurately resolved by a low dimensional SG wavetrain. For details, the reader is referred to [50].

C. Current Status of Perturbation Analysis on (IV.1)

The numerical data, and spectral analysis of it, for the perturbed SG system (IV.1) shows that coherent spatial structures which persist admit a low-dimensional SG approximation. This strongly indicates that soliton perturbation theory can be used to reduce the infinite-dimensional field to an approximate description in terms of a finite number of degrees of freedom, and yet capture the essential dynamical features (even with temporally chaotic responses).

The results stated in Section III systematically lead to reduced equations for the N-phase locked states. They take the following form:

$$\Omega_t^{(-)} = dF, \qquad \text{(IV.2a)}$$

$$(\theta_j)_t = \omega \,, \; j = 1,\ldots,N. \qquad \text{(IV.2b)}$$

The first equation is directly from section III; but F now depends explicitly on the phases by the frequency locking criterion, so (IV.2b) must be adjoined.

The nonlinear pde (IV.1), has been systematically reduced to this system of ode's, (IV.2). Nonetheless, these are quite complicated dynamical equations. So far, our only new result is the system (IV.2) itself. Several problems are currently being studied:

(1) Derive a practical condition for existence of locked states.

Equations (IV.2) give the correct predictions for the simplest locked states, but these results are derivable by classical methods as well. For example, we can reproduce the following results from (IV.2): (i) the classical locked pendulum state, (ii) the result of Kaup-Newell [55] on the frequency locked soliton, and (iii) its periodic extension.

(2) Derive a stability theory for the locked states.

As yet, only the stability of very simple locked states has been investigated. Two approaches are in progress - one using (IV.2) and a second using a basis of "squared eigenfunctions".

(3) Does this reduced dynamical system (IV.2) go chaotic according to the right scenario? A numerical study will be required to answer this question.

It remains to be seen if the structure (IV.2) yields results beyond the limits of classical perturbation methods. We are optimistic since we can at least systematically derive a low-dimensional closed, reduced system of equations.

V. The Breakup of Coherent Structures

Integrable machinery can also be used to describe instabilities which destroy coherent states by generating more complicated spatial structures. One example which we have already discussed is the conservative mechanism for smoothing shocks through the generation of additional phases at breaking points.

In the companion paper [56] we discuss a second example, the modulational or Benjamin-Feir [57] instability for the periodic sine-Gordon equation. There we use the inverse spectral transform to identify and to characterize the instabilities of a fixed coherent state. More importantly, we describe the state which finally emerges from the instability. This final state, although still coherent, has considerably more spatial structure than the initial state, and its mathematical description involves additional degrees of freedom. See also [58].

Here we briefly describe a third example - the propagation of a laser beam through a ring cavity [51]. Mathematically this problem reduces to a sequence of initial value problems for a nonlinear Schrodinger equation:

$$2i\partial_z\psi_n + \partial_{xx}\psi_n - N(\psi_n\psi_n^*)\,\psi_n = 0$$

$$\psi_n(x,z=0) = a(x) + Re^{i\gamma}\,\psi_{n-1}(x,z=L). \qquad \text{(V.1 a,b)}$$

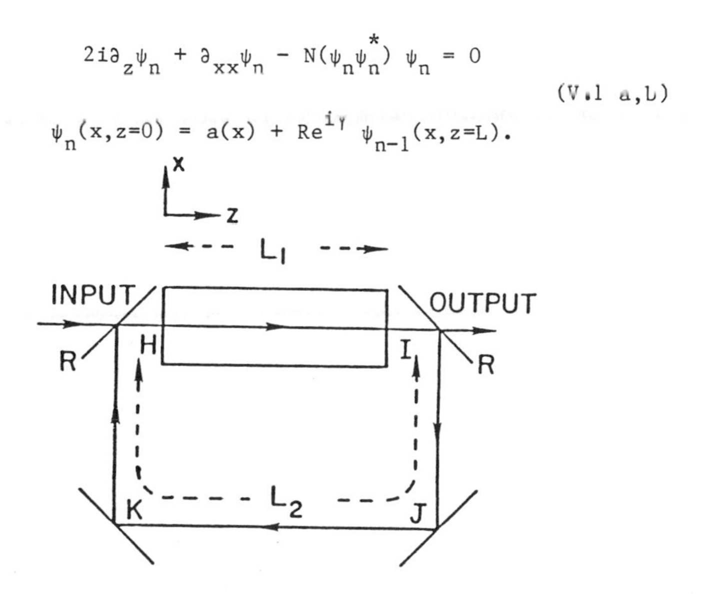

Figure 1. Schematic of a Ring Cavity

Physically ψ_n denotes the envelope of the electromagnetic field on the nth pass through the cavity; z is the coordinate in the direction of propagation down the cavity; x is a coordinate transverse to the direction of propagation; the nonlinearity

$$N(\psi\psi^*) = \frac{1}{1+2\psi\psi^*} \tag{V.2}$$

describes the nonlinear index of refraction inside the cavity; $a(x)$ is a Gaussian input laser field

$$a(x) = Ae^{-x^2}; \tag{V.3}$$

L is the length of the cavity. The initial data (V.1b) arises as follows: The field on the $(n-1)^{st}$ pass reaches the end of the cavity $(z=L)$, $\psi_{n-1}(x,z=L)$, and is brought back to the entry point $(z=0)$ by a system of mirrors. In this process, the field experiences a phase shift $e^{i\gamma}$ and mirror losses $(R < 1)$. This damaged field,

$$Re^{i\gamma}\,\psi_n\,(x,z=L),$$

is added to the incident field $a(x)$, and together they comprise the initial data for the nth pass.

The sequence of initial value problems (V.1) may be viewed as an infinite dimensional map

$$\psi_{n-1}(\cdot,L) \to \psi_n(\cdot,L). \tag{V.4}$$

The stable fixed points of this map in function space are solitary waves [59] with their parameters set at certain specific values. Thus, in some parameter regions the output field tends as $n \to \infty$ to a particular "sech-like" profile with respect to the transverse variable x.

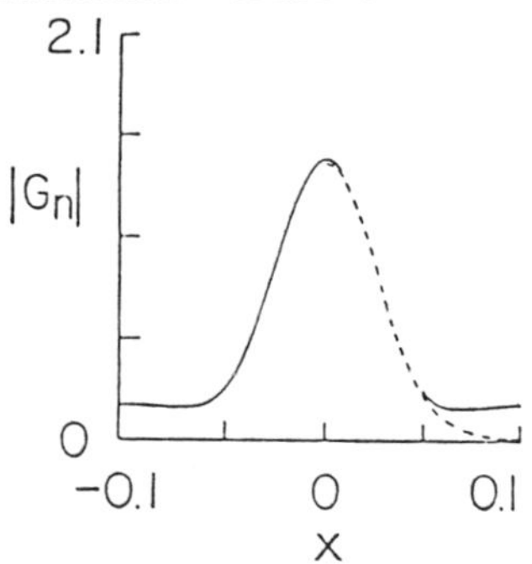

Figure 2. Solitary wave fixed points-the solid line is numerical; dotted line theoretical.

As the stress parameter, A in (V.3), is increased, this solitary wave fixed point becomes unstable to a period 2 state in n. More importantly this period 2 state has a more complicated transverse spatial structure than the solitary wave fixed point. We have observed two different types of period 2 states, depending upon the value of parameters other than A. One of these is a solitary wave together with short wavelength spatial oscillations of a very definite spatial period. The other seems to consist of several solitary waves. Which type appears can be predicted by a linearized stability analysis [60]. Both types consist of more degrees of freedom than the solitary wave fixed point.

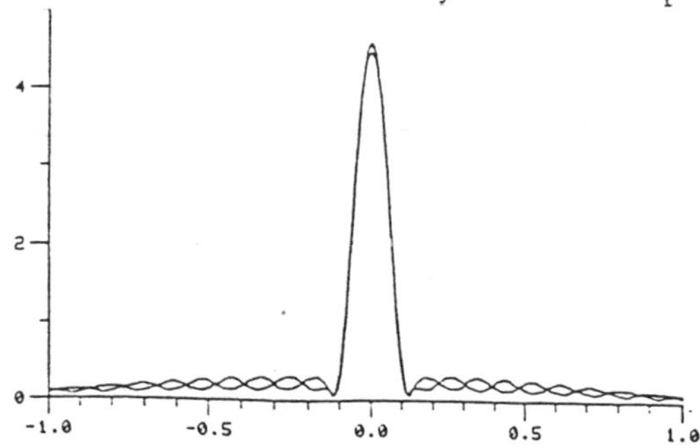

Figure 3. Period 2 and chaotic states. Note the regular, high frequency transverse structure.

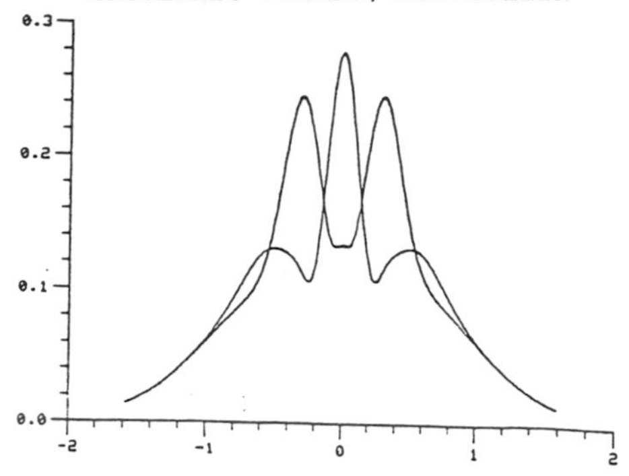

Figure 4. A state which seems to consist of several solitary waves. This transverse structure should be contrasted with the regular structure of Figure 3. The nature of the instabilities differ in the two cases because of different parameter values.

As A is further increased, a temporally chaotic state occurs. This state seems to have a very similar spatial structure to the period 2 state. As yet, we have not observed the generation of spatial chaos, which has been found in model solid state calculations [61].

Studies of the breakup of coherent structures by the generation of additional degrees of freedom, together with the properties of the more complicated spatial states which ensue, are just beginning. In these studies, near integrable machinery will provide theoretical information which will complement numerical experiments. The three examples mentioned in this section illustrate an approach which uses near integrable theory together with numerical experiments.

VI. Conclusion

In this article we have indicated, through examples ranging from a theoretical study of the propagation of oscillations to the development of a numerical measuring device, general qualitative information which can be obtained from integrable methods. In particular we have emphasized that integrable representations provide analytical descriptions of the breakdown of a certain level of coherence and the subsequent generation of more complicated structures. These analytical results, together with those of numerical experiments, provide the theoretical mathematician with new information about the qualitative properties of nonlinear evolution equations which should guide his further studies.

Appendix

Background Material From the Mathematical Theory of Soliton Wave Equations

In this review we focus on three illustrative equations, listed in increasing order of difficulty:

Korteweg-deVries (KdV)

$$u_t - 6uu_x + u_{xxx} = 0,$$

Nonlinear Schrodinger (NLS)

$$iu_t + u_{xx} \pm 2|u|^2 u = 0, \quad + = \text{focusing}, \ - = \text{defocusing}$$

Sine Gordon (S.G.)

$$u_{tt} - u_{xx} + \sin u = 0.$$

We will concentrate on KdV and SG with comments on NLS. The references for this section are [62-66,1,24,12].

A. Robust Families of Solutions

One striking feature of integrable wave equations is that explicit, robust families of solutions are known. Each family is indexed by the number N of dynamical degrees of freedom, N = 1,2,...; moreover, for each N one can list the number of free parameters in the solution. We describe two such families next.

KdV (i) N = 1, Traveling Wave Solutions

The traveling wave ansatz, $u(x,t) = u(\theta) = u(\kappa x+\omega t)$, reduces KdV to a third order o.d.e. Two integrations yield

$$u_x^2 = 2u^3 - Uu^2 + Au + B = 2(u-e_1)(u-e_2)(u-e_3), \tag{A.1}$$

where $U = \omega/\kappa$ and A,B are integration constants.

Under vanishing boundary conditions at $|x| = \infty$, A = B = 0, or equivalently, $e_1 = e_2 = 0$. It is then elementary to integrate (A.1) and obtain the

<u>Single Soliton</u> $u = -2\kappa^2 \operatorname{sech}^2(\kappa(x-x_0-\kappa^2 t))$.

This is a 1-parameter family, where $\kappa \in \mathbb{R}$ fixes the soliton amplitude, width and speed; x_0 merely centers the pulse at $t = 0$.

Under periodic, C^∞ boundary conditions, $A, B, U \in \mathbb{R}$ are such that the cubic in (A.1) has distinct real roots, $e_1 > e_2 > e_3$. In this case, (A.1) is the familiar o.d.e. for the Weierstrass elliptic function $\wp$, and we find the

<u>Elliptic Function Solution</u> $u = 2\wp(\frac{\theta}{2\pi} + ib/2 + \theta_0) + \text{constant}$,

where $\wp$ has periods 1 and ib, $i = \sqrt{-1}$. This is a 3-parameter family, indexed by (A,B,U) or (e_1,e_2,e_3).

This KdV cnoidal wave, $u = 2\wp$, in the limit of infinite x-period $(b \to \infty)$, converges to the single soliton.

<u>KdV</u> (ii) <u>N Degree of Freedom Solutions</u>

The above exact solutions generalize to N degrees of freedom. There are the well-known N soliton formulae, which are also special limits of periodic or quasiperiodic solutions. These are the N phase wavetrains, where $(\prod_0^{2N} (\lambda-\lambda_k))^{1/2}$ replaces the square root of a cubic in (A.1), and $u(x,t) = u(\theta_1,\ldots,\theta_N)$ is 2π-periodic in N independent phases, $\theta_j = \kappa_j x + \omega_j t$. These solutions are defined on the Riemann surface $\mathbf{R}$ $:(\lambda,R(\lambda))$, of

$R(\lambda) = (\prod_0^{2N} (\lambda-\lambda_k))^{1/2}$, and form a $(2N+1)$-parameter family, indexed by $\lambda_0,\ldots,\lambda_{2N}$.

We want to write down these explicit solutions in terms of the Riemann theta function, to define this theta function, and to give formulas for the wave numbers κ_j and frequencies ω_j in terms of the parameters $\{\lambda_k\}_0^{2N}$. This requires some fundamental ingredients of **R** [67].

First, one must choose a canonical homology basis on **R**, i.e. a basis for all closed paths of integration. There are 2N of these, a_j, b_j, $j = 1,\dots,N$, as chosen below in Figure 1. (This selection of branch points, λ_j, is useful for other discussions.) The differentials $\lambda^{j-1}d\lambda/R(\lambda)$, $j = 1,\dots,N$, are a basis for holomorphic differentials on **R**. We choose a normalized holomorphic basis, $\psi_i = \sum_{j=1}^{N} C_{ij}\,\lambda^{j-1}d\lambda/R(\lambda)$, $i = 1,\dots,N$, by imposing N^2 conditions on C_{ij}, $\oint_{a_j}\psi_i = \delta_{ij}$, the Kronecker symbol. The C_{ij}, and therefore ψ_i^j are uniquely specified in terms of $\{\lambda_j\}_0^{2N}$. The periods of ψ_i around the remaining cycles are now also uniquely determined, yielding the period matrix, $(B)_{ij}$, of **R**, $B_{ij} = \oint_{b_i}\psi_j$, which is always symmetric and with positive definite imaginary part.

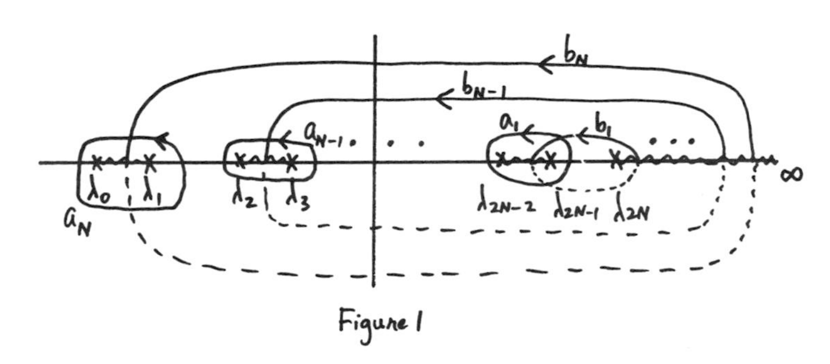

Figure 1

There are two additional, but meromorphic differentials, Ω_1, Ω_2, which are central to the KdV theory. These are unique differentials of the second kind (i.e., zero residues at the pole(s)), whose only pole is at $\lambda = \infty$, of order $2j$ for Ω_j, with Laurent expansion in the local parameter $\lambda = \xi^{-2}$ near ∞ on R specified by

$$\Omega_j \sim (1\cdot\xi^{-2j}+\dots+ 0\cdot\xi^{-1} + \text{holomorphic part})d\xi. \tag{A.2a}$$

This condition together with the constraints of vanishing b_k

cycles,

$$\oint_{b_k} \Omega_{1,2} = 0 \qquad j = 1,\dots,N, \tag{A.2b}$$

uniquely specifies $\Omega_{1,2}$. We are now in a position to state the N phase KdV Solutions

$$u(x,t) = u(\vec{\ell}(x,t)) = \text{constant} - 2\,\frac{d^2}{dx^2}\log\left(\Theta(\vec{\ell};B)\right), \tag{A.3a}$$

where

$$\underset{N\times 1}{\vec{\ell}}\,(x,t) = \vec{\ell}(\theta_1,\dots,\theta_N) = \sum_{j=1}^{N} \theta_j/2\pi\,\vec{B}_j + \vec{\ell}(0,0), \tag{A.3b}$$

$$\theta_j = \kappa_j x + \omega_j t, \quad \vec{B}_j \text{ is the } j^{th} \text{ column of } B, \tag{A.3c}$$

$$\kappa_j = \oint_{a_j}\Omega_1, \quad \omega_j = -12\oint_{a_j}\Omega_2,\ j = 1,\dots,N. \tag{A.3d}$$

The Riemann theta function in (A.3a) is defined by

$$((\vec{a},\vec{b}) = \sum_{i=1}^{N} a_i b_i)$$

$$\Theta(\vec{\ell};B) = \sum_{\vec{m}\in \mathbb{Z}^N} \exp\{2\pi i(\vec{m},\vec{\ell}) + \pi i(B\vec{m},\vec{m})\}. \tag{A.4}$$

It follows now from the properties of Θ that

$$u(\theta_1,\dots,\theta_j + 2\pi,\dots,\theta_N) = u(\theta_1,\dots,\theta_j,\dots,\theta_N), \text{ for all } j.$$

KdV (iii) Reality Constraints

The description above for the N phase KdV solutions, $u(\theta_1,\dots,\theta_N)$, is valid for general complex-valued $u(\vec{\theta})$ and for complex phases $\vec{\theta}(x,t)$. However, the real-valued, C^∞ solutions of this class are completely determined by the following reality constraints.

The branch points λ_j, $j = 0,\dots,2N$ are constrained to be real, with $-\infty < \lambda_0 < \lambda_1 < \ldots < \lambda_{2N} < \infty$, as in Figure 1. This completely determines the Riemann surface **R**, and thereby the unique objects defined above on **R**. In particular, it follows [1] that κ_j, ω_j as defined in (II.3d) are real, so the phases θ_j are real. Also, the period matrix B_{ij} is purely imaginary. The final ingredient is the integration constant $\vec{\ell}(0,0)$, and for all $\mathrm{Re}(\vec{\ell}(0,0)) = 0$, the theta function solution (A.3a) is real-valued and C^∞. Moreover, the set of all such solutions for given $\{\lambda_j\}_0^{2N} \in \mathbb{R}$ is a smooth manifold, in fact, a real N-torus, parametrized by the N phases

$$(\theta_1,\ldots,\theta_N) \in [0,2\pi)^N.$$

KdV (iv) Periodicity Constraints

As constructed above, this real N phase solution, $u(\vec{\theta})$, is quasi-periodic in x (i.e., N incommensurate periods in x, $\frac{2\pi}{\kappa_j} = L_j$). This family depends on $(2N+1)$ real parameters λ_j. One must impose (N-1) additional constraints on $\{\lambda_j\}$ to enforce periodicity:

$$n_1/\kappa_1 = n_2/\kappa_2 = \ldots = n_N/\kappa_N, \; n_j \in \mathbb{Z}-\{0\}.$$

If one fixes the period, $u(x+L) = u(x)$, this imposes one more constraint. Thus, the set of real, N phase, periodic in x of fixed period L, solutions of KdV is an (N+1)-parameter family.

KdV (v) Phase Space Complexity

We simply remark there are no separatrices in the periodic KdV phase space. This turns out to render the solution space particularly tame, in contrast to sine-Gordon and focusing NLS, which we discuss next.

Sine-Gordon (i) N=1, Traveling Wave Solutions

The traveling wave ansatz, $u(x,t) = u(\kappa x + \omega t)$, reduces (S.G.) to the familiar pendulum: $(\omega^2 - \kappa^2)u_{\theta\theta} + \sin u = 0$. One integration yields $\frac{1}{2}(U^2-1)u_x^2 - \cos u = E$, where $U = \omega/\kappa$ and E is constant. Recall the pendulum phase portrait, and we see that there are separatrices for (S.G.) already in the N=1 case.

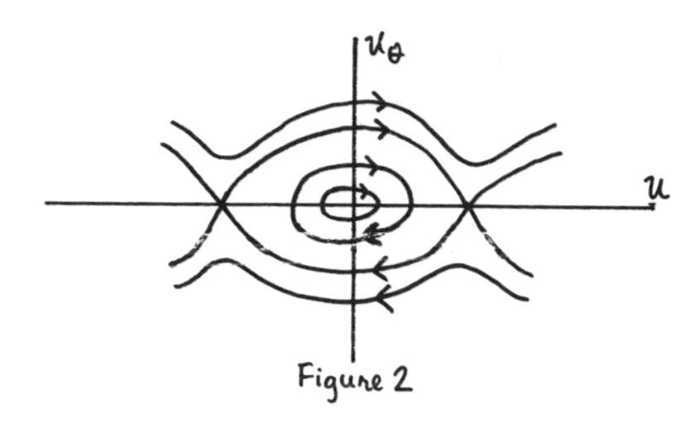

Figure 2

There are two types of 1-soliton solutions, the kink and antikink, which appear as the upper and lower branch of the separatrix in Figure 2. In the ode, these arise for the special value $E = 1$, for which an elementary integration yields $u = 4 \arctan(\exp(\pm\phi))$, $\phi = (x-x_0-vt)/\sqrt{1-v^2}$, where $+$ is for the kink soliton, $(-)$ for the antikink. This is clearly a 1-parameter family of each type, indexed by v, while x_0 centers the initial profile.

As Figure 2 also indicates, the presence of separatrices allows three disconnected regions of periodic solutions. There are the low energy $(-1 < E < 1)$, purely oscillatory solutions bounded inside the separatrix, plus two types of periodic mod 2π solutions $(E > 1)$, the kink trains lying above the kink separatrix and the antikink trains below their limiting soliton separatrix. These $N = 1$ periodic solutions are 2-parameter families, indexed by (E,U).

(S.G.) (ii) N degree of freedom solutions

The rich behavior already seen above for $N = 1$ extends to

higher N as well, due in essence to higher dimensional separatrices. There are four distinct two-soliton solutions; for example, in addition to the expected kink-antikink pair, there is a lower energy kink-antikink bound state, called a breather.

The N phase (S.G.) wavetrains, $u(\theta_1,\ldots,\theta_N)$, $\theta_j = \kappa_j x + \omega_j t$, which generalize the elliptic function solutions of the pendulum, are defined on the Riemann surface $\mathbf{R}$:(E, R(E)), of $R(E) = (E \prod_1^{2N} (E-E_j))^{1/2}$. These solutions form a 2N-parameter family, indexed by $\{E_j\}_1^{2N}$, or equivalently, $\{\kappa_j,\omega_j\}_1^N$. Again, we must define some ingredients on $\mathbf{R}$ to state the exact formulas. We choose canonical a_j, b_j cycles as indicated in Figure 3, where the E_j are picked for future relevance. The normalized basis of holomorphic differentials is uniquely defined by $\psi_j = \sum_{m=1}^N C_{jm} E^{N-m} dE/R(E)$, $j = 1,\ldots,N$, with $\oint_{a_k} \psi_j = \delta_{kj}$. The period matrix $B_{ij} = \oint_{b_i} \psi_j$, and again these elements C_{jm}, B_{ij} are fixed by $\{E_j\}_1^{2N}$. We define two unique meromorphic differentials of the second kind, $\Omega^{(+)}$ and $\Omega^{(-)}$: the only poles are second order poles at 0 and ∞, with expansions in the local parameters ($E = \xi^{-2}$ near ∞, $E = \xi^2$ near 0) of the form

$$\Omega^{\pm} \sim ((1 \pm \frac{1}{16})\xi^{-2} + \text{holomorphic part})d\xi;$$

$\Omega^{\pm}$ have vanishing a_j cycles, $\oint_{a_j} \Omega^{(\pm)} = 0$, $j = 1,\ldots,N$.

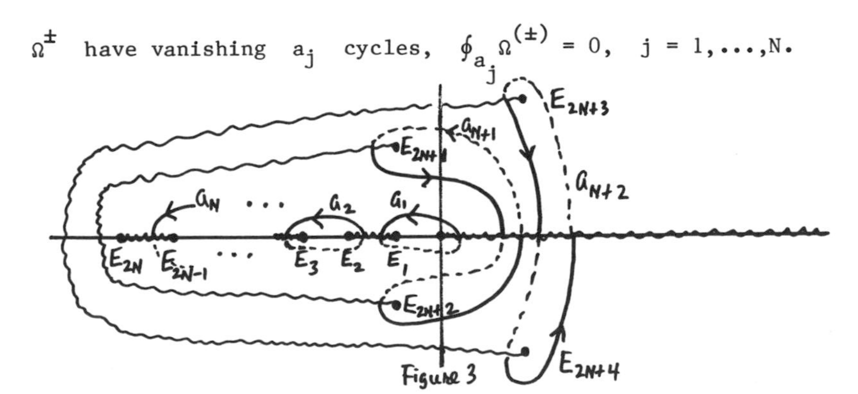

Figure 3

The N-phase sine-Gordon solutions are

$$u(x,t) = u(\vec{\ell}(x,t)) = 2i\ell n \left[\frac{\Theta(\vec{\ell} + \frac{1}{2}; B)}{\Theta(\vec{\ell};B)}\right], \tag{A.5a}$$

where the theta function is defined in (A.4),

$$\vec{\ell}(x,t) = \sum_{j=1}^{N} \theta_j/2\pi \, \vec{e}_j + \vec{\ell}(0,0), \tag{A.5b}$$

$$\frac{\vec{1}}{2} = (\frac{1}{2}, \frac{1}{2}, \ldots, \frac{1}{2}), \ (\vec{e}_j)_i = \delta_{ji}, \ \theta_j = \kappa_j x + \omega_j t, \tag{A.5c}$$

$$\kappa_j = \oint_{b_j} \Omega^{(-)}, \ \omega_j = \oint_{b_j} \Omega^{(+)}. \tag{A.5d}$$

The periodicity properties of $u(\theta_1, \ldots, \theta_N)$ depend on $\vec{\ell}(0,0)$; see below.

(S.G.) (iii) Reality Constraints

The solution u and phases θ_j in (A.5) are in general complex-valued. The reality constraints on $\{E_j\}^{2N}$, force E_j to come in pairs, with either $E_{2j} < E_{2j-1} < 0$ or $E_{2j}^* = E_{2j-1}$, as chosen in Figure 3. With these constraints in force, it follows that $\kappa_j, \omega_j \in \mathbb{R}$, so θ_j are real phases, $Re(C_{jk}) = 0$, and similar relations hold for B_{jk}. The final yet fundamental ingredient in this theta function representation of u is the integration constant $\vec{\ell}(0,0)$. We already know from the $N = 1$ case above that multiplicities may occur for the same choice of parameters ((E,U), $E > 1$, corresponds to two solutions). This situation, for general N, is rectified as follows. (See [66] for details.)

Each conjugate pair, $E_{2j}^* = E_{2j-1}$, corresponds to a strictly periodic degree of freedom, whereas each real pair $E_{2j} < E_{2j-1} < 0$ yields two (mod 2π) periodicity modes (kinktrain and antikink train). By systematic ordering, θ_j

corresponds to (E_{2j}, E_{2j-1}), which corresponds to ℓ_j, $j = 1,\dots,N$. (This diagonalization is apparent in (A.5b)). The result is, for construction of real, C^∞ $u(x,t)$, there are two distinct $\mathrm{Im}\,(\ell_j(0,0))$ for each real pair $E_{2j} < E_{2j-1} < 0$, and only one $\mathrm{Im}(\ell_j(0,0))$ for each conjugate pair. Moreover, the periodicity behavior of $u(\vec{\theta})$ is: $u(\theta_1,\dots,\theta_j + 2\pi,\dots,\theta_N) = u(\vec{\theta})$ if $E^*_{2j} = E_{2j-1} < 0$, whereas $u(\theta_1,\dots,\theta_j + 2\pi,\dots,\theta_N)$ $= u(\vec{\theta}) \pm 2\pi$ if $E_{2j} < E_{2j-1} < 0$, (+) sign for one choice of $\mathrm{Im}\,\ell_j(0,0)$, (−) sign for the other.

The total multiplicity of real, N-phase solutions goes as follows: Given $\{E_j\}_1^{2N}$, where $E^*_{2j} = E_{2j-1}$ or $E_{2j} < E_{2j-1} < 0$, let <u>k = the number of real pairs</u>. Then, there are 2^k distinct classes of real, C^∞, N phase solutions. The set of all such solutions for such given $\{E_j\}_1^{2N}$ is a smooth manifold consisting of 2^k connected components, each of which is a real N-torus parametrized by the N real phases $(\theta_1,\dots,\theta_N) \in [0,2\pi)^N$.

<u>(S.G.)</u> (iv) <u>Periodicity Constraints</u>

The real N phase solutions (A.5), parametrized by the 2N real parameters in $\{E_j\}_1^{2N}$, are quasiperiodic in x. To restrict the family to be periodic in x of fixed period L imposes N constraints on the E_j, $\frac{1}{2\pi} L = \frac{n_1}{\kappa_1} = \frac{n_2}{\kappa_2} = \dots = \frac{n_N}{\kappa_N}$. This yields a real N-parameter family.

B. <u>PERIODIC SPECTRAL THEORY FOR SOLITON EQUATIONS</u>

One more ramification of integrability is a complete representation of the solution to the initial-value problem (IVP) for soliton equations. Inverse scattering on the line is the most familiar success story, and leads to an explicit action-angle representation for the solution of the IVP. Most of this review focuses on periodic theory, so we briefly indicate the ingredients of the associated periodic spectral theory (or Floquet theory).

(i) KdV Periodic Theory

The integration of KdV is associated with the spectrum of Hill's operator (with spectral parameter λ),

$$(\mathbf{L} - \lambda)\psi = 0, \quad \mathbf{L} = \frac{-d^2}{dx^2} + u(x,\cdot),$$

or in two-component form,

$$\begin{pmatrix} -\partial_x & 1 \\ (u-\lambda) & -\partial_x \end{pmatrix} \begin{pmatrix} \psi_1 \\ \psi_2 \end{pmatrix} = \vec{0}. \tag{A.6}$$

The fundamental object of Floquet theory is the discriminant $\Delta(u;\lambda)$, which we now define. First, compute the normalized fundamental matrix of (A.4), $M(x;x';\lambda;u)$, a (2×2) matrix-valued solution of (A.4) satisfying $M(x';x';\lambda;u) = \begin{pmatrix} 1 & 0 \\ 0 & 1 \end{pmatrix}$. Next one defines the transfer matrix, $M(x'+L;\, x';\, \lambda;u)$, which maps a solution across one period (L) of the potential u. The Floquet discriminant is simply

$$\Delta(\lambda;u) = \text{trace } (M(L;0;\lambda;u)). \tag{A.7}$$

The continuous spectrum of Hill's operator, $\mathbf{L}$, is the closure of all λ for which $-2 \leqslant \Delta(\lambda;u) \leqslant 2$. Δ and M contain the spectral data for the 1:1 map from u to action-angle coordinates, as follows.

The simple roots, λ_j, of $\Delta(\lambda;u) = \pm 2$ form the simple periodic and antiperiodic spectrum, $\sum(u) = \{\lambda_j\}_0^{2N}$, $0 \leqslant N \leqslant \infty$. These λ_j are invariant in t (isospectral) as u(x) evolves by translation, KdV, or any of the higher flows in the KdV hierarchy. They provide the action variables of the periodic KdV action-angle transformation. The corresponding angle variables arise from the zeros, $\lambda = \mu_j$, of the M_{12} entry of the transfer matrix: $M_{12}(L;0;\lambda=\mu_j;u) = 0$. These μ_j are Dirichlet eigenvalues of (A.6), and are not isospectral. The KdV inversion formula in terms of these variables is

$$u = \sum_0^{2N} \lambda_j - 2\sum_1^N \mu_j, \quad 1 \leq N \leq \infty. \tag{A.8}$$

KdV Reality Constraints For real periodic potentials u, Hill's operator is self-adjoint under both periodic and Dirichlet boundary conditions. Thus, λ_j and μ_j are real, and can be shown by classical Sturm-Liouville oscillation theorems to interlace according to

$$\lambda_0 < \lambda_1 \leq \mu_1 \leq \lambda_2 < \lambda_3 \leq \mu_2 \leq \lambda_4 < \lambda_5 < \cdots \ . \tag{A.9}$$

So, on each interval $(\lambda_{2j-1}, \lambda_{2j})$, there is one μ_j trapped inside. If $\lambda_{2j-1} = \lambda_{2j}$, corresponding to a double periodic (anti) eigenvalue, hereafter called a double point, then the corresponding μ_j is tied at λ_{2j}. If $\lambda_{2j-1} \neq \lambda_{2j}$, there is a gap (instability band) in the continuous spectrum, and μ_j oscillates in that band under the flows of the KdV hierarchy. The N phase solutions (A.3) are in 1:1 correspondence with N gap potentials of (A.6); the formulas (A.3a), (A.8) coincide. Moreover, the (2N+1) branch points $\{\lambda_j\}_0^{2N}$ of **R** in Figure 1 are precisely the elements of $\sum(u)$. The N untied μ_j variables also parametrize the isospectral manifold of real periodic KdV potentials with prescribed $\sum(u) = \{\lambda_0 < \lambda_1 < \cdots < \lambda_N\}$. (These two equivalent realizations of the torus of real potentials are connected by the classical Abel-Jacobi map,

$$(\mu_1,\ldots,\mu_N) \in [\lambda_1,\lambda_2]\times\cdots\times[\lambda_{2N-1},\lambda_{2N}] \overset{1:1}{\leftrightarrow} (\theta_1,\ldots,\theta_N) \in [0,2\pi)^N.$$

In this way the $\vec{\mu}$-representation (A.5) of u leads to the theta function representation (A.2).)

When the constraints for periodicity in x are in force, we also find the important connection between the meromorphic differential Ω_1 (equations (A.2), (A.3)) and $\Delta(\lambda;u)$,

$$\Omega_1 = \frac{\Delta'(\lambda)d\lambda}{L\sqrt{4-\Delta^2}} .$$

We have yet to make such a connection for Ω_2.

(ii) Sine-Gordon Periodic Theory

The linear spectral problem associated to sine-Gordon is

$$\left[\begin{pmatrix} 0 & -1 \\ 1 & 0 \end{pmatrix} \frac{d}{dx} + \frac{i}{4}(u_t+u_x)\begin{pmatrix} 0 & 1 \\ 1 & 0 \end{pmatrix} + \frac{1}{16\sqrt{E}}\begin{pmatrix} e^{iu} & 0 \\ 0 & e^{-iu} \end{pmatrix} - \sqrt{E}\right]\begin{pmatrix} \psi_1 \\ \psi_2 \end{pmatrix} = \vec{0}, \quad \text{(A.10)}$$

with spectral parameter E. As before, one forms the normalized fundamental matrix of (A.7), $M(x;x';E;\vec{u})$, $\vec{u} \equiv (u,u_t+u_x)$, and the transfer matrix is $M(L;0;E;\vec{u})$. The Floquet discriminant is

$$\Delta(E;\vec{u}) = \text{trace } (M(L;0;E;\vec{u})). \quad \text{(A.11)}$$

The action variables again arise from the simple periodic, antiperiodic eigenvalues, which are the simple roots, E_j, of $\Delta(E=E_j; \vec{u}) = \pm 2$. The angle variables are determined (as before) from the simple zeros, μ_j, of M_{12}. The (S.G.) inversion formula in these variables is

$$u = i\ln \left(\prod_1^N \mu_j \Big/ \sqrt{\prod_1^{2N} E_j}\right). \quad \text{(A.12)}$$

Sine-Gordon Reality Constraints. There are profound differences in the constraints of real potentials for the (S.G.) linear system, (A.10), versus the KdV linear system, (A.6). For real u, $u_t + u_x$, (A.10) is non-selfadjoint. The spectral data $\{E_j\}_1^{2N}$, $\{\mu_j\}_1^N$ are no longer confined to the real axis, and the oscillation theorems which order the KdV data by (A.9) simply do not apply. New methods, based on algebraic geometry, were developed to understand the non-selfadjoint reality constraints pertinent to (S.G.) and focusing NLS. We will state the results of that work; see [66] for details.

The constraints on the action variables, E_j, are easy. The simple spectra $\{E_j\}_1^{2N}$ come in distinct pairs: either $E_{2j} < E_{2j-1} < 0$ or $E_{2j} = \overset{*}{E}_{2j-1}$, as depicted in Figure 3. The real flows of the dynamical angle variables, μ_j, are quite a bit richer.

As with KdV, for each pair E_{2j-1}, E_{2j} of periodic (anti) spectra, there is one angle variable μ_j. At the double points, $E_{2j} = E_{2j-1}$, μ_j is tied at E_{2j}. At the open degrees of freedom corresponding to simple spectra, $E_{2j-1} \neq E_{2j}$, the corresponding μ_j does still parametrize the open mode, but no longer lies on a fixed curve in the complex plane (except the N=1 case). The motion of the angle variables, μ_j, under the (S.G.) flows $(\theta_1,\ldots,\theta_N) \in [0,2\pi)^N$, has been topologically controlled, with the results that the cycles of μ_j (closed orbits as one or more $\theta_k: 0 \to 2\pi$) are invariant. I.e., the closed curves of μ_j deform as one varies over $\vec{\theta} \in [0,2\pi)^N$, but topologically the curves remain invariant. In the Figure 3, the result is: $\mu_j \sim a_j$, $j = 1,\ldots,N$, ($\sim$ means "is homologous to".)

When the N phase sine-Gordon solution is periodic in x, of fixed period L, we once again find an important connection between the meromorphic differential, $\Omega^{(-)}$, and the discriminant Δ:

$$\Omega^{(-)} = \frac{\Delta'(E)dE}{L\sqrt{4-\Delta^2(E)}} .$$

REMARKS: The focusing NLS theory lies somewhere between (S.G.) and KdV. The linear system (AKNS) is non-selfadjoint, but the multiplicity of solutions for (S.G.) does not appear. The action variables, λ_j, which define the NLS Riemann surface $\mathbf{R}$: $(\lambda, R(\lambda))$, where $R(\lambda) = \sqrt{\prod_{1}^{2N} (\lambda-\lambda_j)}$, always come in conjugate pairs, $\lambda_{2j}^{*} = \lambda_{2j-1}$. This constraint makes the NLS periodic theory quite parallel to (S.G.) theory restricted to purely periodic degrees of freedom. For example, the NLS N-phase isospectral manifold is a real N-torus. The angle variables, μ_j, once again do not lie on fixed curves, but as in (S.G.) are constrained to the same topological cycles.

The importance in our applied investigations of the exact periodic and quasiperiodic structure outlined above will become apparent in the rest of this survey and the ensuing article [56].

References

1. H. Flaschka, M. G. Forest, and D. W. McLaughlin, "Multiphase averaging and the inverse spectral solution of the Korteweg de Vries equation," Comm. Pure Appl. Math. 33, 1980, pp. 739-784.

2. D. W. McLaughlin, "Modulations of KdV Wavetrains," Physica D3, 1981, pp. 335-343.

3. D. W. McLaughlin, "On the Construction of Modulating, Multiphase Wavetrains," University of Arizona Preprint, 1982.

4. R. Miura and M, Kruskal, "Application of a nonlinear WKB method to the Korteweg de Vries equation," SIAM J. Appl. Math. 26, 1974, pp. 376-395, and J. C. Luke, "A perturbation method for nonlinear dispersive wave problems," Proc. Roy Soc. A 292, 1966 (403-412).

5. M. J. Ablowitz and D. J. Benney, "The evolution of multi-phase modes for nonlinear dispersive waves," Stud. Appl. Math. 49, 1970, pp. 225-238.

6. Lax, P. D. and C. D. Levermore, The zero dispersion limit for the KdV equation, Proc. Nat. Acad. Sci. USA 76 (1979),

3602-3606.

7. Lax, P. D. and C. D. Levermore, The small dispersion limit for the KdV equation I, Comm. Pure Appl. Math. 36 (1983), 253-290; II, Comm. Pure Appl. Math. 36 (1983), 571-594; III, Comm. Pure Appl. Math. 36 (1983), 809-829.

8. Venakides, S., The zero dispersion limit of the Korteweg de Vries equation for initial potentials with non-trivial reflection coefficient, preprint.

9. G. B. Whitham, ":Nonlinear dispersive waves," Proc. Royal Soc. London A 283, 238-261 (1965).

10. G. B. Whitham, Linear and Nonlinear Dispersive Waves, Wiley-Interscience, New York, 1974.

11. M. G. Forest and D. W. McLaughlin, "Modulations of sinh-Gordon and sine-Gordon wavetrains," Stud. Appl. Math. 68, 1983, 11-59.

12. N. Ercolani, M. G. Forest, and D. W. McLaughlin, "Modulational stability of two phase sine-Gordon wavetrains," Stud. Appl. Math.

13. M. G. Forest and D. W. McLaughlin, unpublished.

14. A. V. Gurevich and L. P. Pitaevskii, "Nonstationary structure of a collisionless shock wave," Soviet Phys. JETP 38(2) (1974), and B. Fornberg and G. B. Whitham, "A numerical and theoretical study of certain nonlinear wave phenomena," Phil. Trans. Roy. Soc. Lond. 289, 1978, pp. 373-404.

15. R. Haberman and R. Sun, "Nonlinear cusped caustics for dispersive waves", Stud. Appl. Math. pp. 1-37 (1984).

16. R. Haberman and D. Allgaier, "Slowly varying solitary wave tails: focusing cusped caustics, wave number shocks and birth of tails," to appear SIAM J. Appl. Math.

17. J. D. Cole, Q. Appl. Math. 9, 225-236 (1951).

18. E. Hopf, C.P.A.M. 3, 201-230 (1950).

19. R. Rosales, private communication.

20. S. Venakides, in preparation (1985).

21. Tartar, L., "Compensated compactness and applications to partial differential equations," in Research notes in mathematics, nonlinear analysis, and mechanics: Heriot-Watt Symposium, Vol. 4, Knops, R. J., (ed.) New York: Pitman Press, 1979, and "The compensated compactness method applied to systems of conservation laws," in Systems of nonlinear partial differential equations, Ball, J. M. (ed.) NATO ASI Series, C. Reidel Publishing Co. (1983).

22. R. Diperna, "Measure valued solutions to conservation laws," Duke Univ. Preprint (1984).

23. W. D. Hayes, "Group velocity and nonlinear dispersive wave propagation," Proc. Royal Soc. London A 332, 199-221 (1973).

24. M. G. Forest and D. W. McLaughlin, "Canonical variables for the periodic sine-Gordon equation and a method of averaging,"Report of the Los Alamos National Laboratory, LA-UR 78-3318, 1978.

25. N. Ercolani, M. G. Forest, D. W. McLaughlin, R. Montgomery, in preparation.

26. M. G. Forest and D. W. McLaughlin, "Modulations of perturbed KdV wavetrains," SIAM J. Appl. Math. 44, 287-300 (1984).

27. M. E. Schonbek, "Convergence of solutions to nonlinear dispersive equations," preprint, U. Rhode Island, 1981.

28. E. N. Pelinovsky and S. Kh. Shavratsky, "Breaking of Stationary Waves in Nonlinear Dispersive Media," Physica 1D, 317-328 (1980).

29. T. B. Benjamin, "The stability of solitary waves," Proc. Roy. Soc. Lond. A 328 (1972) 153-183; J. Bona, "On the stability of solitary waves," Proc. R. Soc. Lond. A 344 (1975) 363-374.

30. T. Cazenave, and P. L. Lions, "Orbital stability of standing waves for some nonlinear Schrodinger equations," Comm. Math. Phys. 85 (1982) 549-561.

31. M. I. Weinstein, "Lyapunov Stability of Ground States of Nonlinear Evolution Equations," preprint, Princeton University (1984).

32. J. Marsden, to appear (1985).

33. E. W. Laedke and K. H. Spatschek, "Stable three-dimensional envelope solitons," Phys. Rev. Lett. 52 (1984) 279-282.

34. J. Shatah, "Stable standing waves of nonlinear Klein-Gordon equations," to appear.

35. J. L. Bona, W. G. Pritchard, L. R. Scott, "A comparison of solutions of two model equations for long waves," Lectures in Appl. Math., 1983.

36. W. A. Strauss, "Nonlinear scattering theory at low energy," J. Func. Anal. 41 (1981), pp. 110-133; "Nonlinear scattering theory at low energy: sequel," J. Func. Anal. 43 (1981), pp. 281-293.

37. J. Ginibre and G. Velo, "On a class of nonlinear Schrodinger Equations II. Scattering Theory, general case," J. Func. Anal. 32 (1979), pp. 33-71.

38. H. Warschal, preprint, Univ. of Texas (1984).

39. V. N. Vlasov, I. A. Petrishchev, V. I. Talanov, Izv. Radiofizika 14 (1971), p. 1353.

40. R. T. Glassey, "On the blowing-up of solutions to the Cauchy problem for the nonlinear Schrodinger equation," J. Math. Phys. 18 (1977), pp. 1794-1797.

41. V. E. Zakharov and V. S. Synakh, "The nature of the self-focusing singularity," Sov. Phys. JETP 41 (1976), pp. 465-468.

42. F. John and S. Klainerman, C.P.A.M. 37 (4), 443-456 (1984).

43. K. Nozaki and N. Bekki, Phys. Rev. Lett. 50, 1226 (1983); Phys. Lett. 102(A), 383 (1984).

44. N. Bekki and K. Nozaki, "A soliton as an attractor of the driven, damped nonlinear Schrodinger equation," Proc. 7th Kyoto Summer Institute, Summer 1984, to appear in Springer Series in Synergetics (1985).

45. W. E. Ferguson, H. Flachka, and D. W. McLaughlin, Nonlinear normed modes for the Toda Chain," . Comp. Phys. 45, 157-209 (1981).

46. B. Holian, H. Flaschka, and D. W. McLaughlin, "Shocks in the Toda Lattice: Analysis," Phys. Rev. A. 24, 2595-2623 (1981).

47. D. W. McLaughlin and E. A. Overman II, "Breather annihilation by simple dissipation," Phys. Rev. A26, 3497-3507 (1982).

48. K. Fesser, D. W. McLaughlin, A. R. Bishop, B. L. Holian, "Chaos and Nonlinear Modes in a Perturbed Toda Chain," preprint (1984).

49. A. R. Osborne and P. L. Burch, "Internal Solitons in the Andaman Sea", Science, 208 451 (1980); plus recent preprints on the periodic spectral transform.

50. E. A. Overman II, D. W. McLaughlin, and A. R. Bishop, "Coherence and Chaos in the Driven, Damped Sine-Gordon Equation: Measurement of the Soliton Spectrum", to appear, Physica D (1985).

51. D. W. McLaughlin, J. V. Moloney, and A. C. Newell, "An Infinite Dimensional Map from Optical Bistability whose regular and Chaotic Attractors Contain Solitary Waves," in Chaos in Nonlinear Systems, edited by J. Chandra, SIAM (1984).

52. A. R. Bishop, K. Fesser, P. S. Lomdahl, W. C. Kerr, M. B. Williams, and S. E. Trullinger, "Coherent spatial structure versus time chaos in a perturbed sine-Gordon system," Phys. Rev. Lett. 50, 1095-1099 (1983).

53. A. R. Bishop, K. Fesser, P. S. Lomdahl, and S. E. Trullinger, "Influence of solitons in the initial state on chaos in the driven, damped sine-Gordon system." Physica 7D, 259-279 (1983).

54. A. R. Bishop, . C. Eilbeck, I. Satija and G. Wysin, "Pattern Selection and Low-Dimensional Chaos in Dissipative Many Degree-of-Freedom Systems," Los Alamos preprint LA-UR-84-1725.

55. D. J. Kaup and A. C. Newell, "Solitons as particles,...", Proc. Roy. Soc. London A 361, 413-446 (1978).

56. N. Ercolani, M. G. Forest, and D. W. McLaughlin, "Modulational Instabilities of Periodic Sine-Gordon Waves: A Geometric Analysis," this proceedings.

57. T. B. Benjamin and J. E. Feir, "The disintegration of wavetrains on deep water Part I theory," J. Fluid Mech. 27, 417 (1967); "Instability of Periodic Wavetrains in Nonlinear Dispersive Systems", Proc. Roy. Soc. A299, 59-75 (1967).

58. E. R. Tracy, "Topics in Nonlinear Wave Theory with Applications," Thesis, U. of Maryland (1984); E. R. Tracy, H. H. Chen, and Y. C. Lee, "Study of Quasiperiodic Solutions of the Nonlinear Schrodinger Equation and the Nonlinear Modulational Instability," Phys. Rev. Lett. 53, 218-221 (1984).

59. D. W. McLaughlin, J. V. Moloney and A. C. Newell, "Solitary Waves as Fixed Points of Infinite Dimensional Maps in an Optical Bistable Ring Cavity," Phys. Rev. Lett. 51, 75-78 (1983).

60. D. W. McLaughlin, J. V. Moloney and A. C. Newell, "A new class of instabilities in passive optical cavities", Phys. Rev. Lett. (1985).

61. S. Aubry, Physica D 7 (1983)

62. H. P. McKean and P. van Moerbeke, "The spectrum of Hill's Equation," Invent. Math. 30 (11), 217-274 (1975).

63. H. Flaschka and D. W. McLaughlin, "Canonically Conjugate Variables for the KdV...," Prog. Theor. Phys. 55, 438-456 (1976).

64. M. G. Forest and D. W. McLaughlin, "Spectral Theory for the Peirodic Sine-Gordon Equation: A Concrete Viewpoint," J. Math. Phys. 23, 1248-1277 (1982).

65. H. P. McKean, "The Sine-Gordon and Sinh-Gordon Equations on the circle," C.P.A.M. 34, 197 (1981).

66. N. M. Ercolani and M. G. Forest, "The Geometry of Real Sine-Gordon Wavetrains," to appear Comm. Math. Physics.

67. G. Springer, Introduction to Riemann Surfaces, Addison-Wesley, Chapter 10 (1957).

[1]Department of Mathematics
Ohio State University
Columbus, Ohio 43210

[2]Program in Applied Mathematics
University of Arizona
Tucson, Arizona 85721

Lectures in Applied Mathematics
Volume 23, 1986

Basic Form for Riemann Matrices

Allan Finkel

Harvey Segur

Abstract: A canonical form, called basic form is developed for 2×2 real Riemann matrices. The basic form provides a quick test for deciding if a 2×2 Riemann matrix is the period matrix of a genus 2 hyperelliptic curve.

1. Introduction Kadomtsev and Petviashvili [2] introduced the equation that now bears their names,

$$(u_t + 6uu_x + u_{xxx}) + 3\sigma u_{yy} = 0, \quad \sigma = \pm 1, \tag{1}$$

as a two-dimensional generalization of the Korteweg-de Vries (KdV) equation,

$$u_t + 6uu_x + u_{xxx} = 0. \tag{2}$$

Their original objective was to study the stability of KdV solitons to long transverse perturbations. It is now recognized that (1) is important in its own right, from several perspectives. Three separate motivations for studying (1) are the following:

(a) A physical model. Kadomtsev and Petviashvili recognized that wherever the KdV equation arises as an approximate model of a physical problem, the KP equation may also arise as a more general model. In water waves, for example, the KdV equation describes the evolution of long waves of moderate amplitude as they propagate in one direction in shallow water; the KP equation describes the evolution of the same waves when the requirement of strict one-dimensionality is relaxed.

(b) An integrable evolution equation. Not only is the KP equation a generalization of the KdV equation as a physical model, it also shares with (2) the rich mathematical structure that makes the KdV equation so interesting. One aspect of this structure is that both equations admit exact, quasi-periodic solutions of the form

$$u = 2\partial_x^2 \ln \theta\big(\phi_1, \dots, \phi_N; Z\big) \tag{3}$$

as shown by Krichever [3]. Here θ is a Riemann theta function of genus N, each ϕ_j is linear in (x, t) for (2), or in (x, y, t) for (1), and Z is a Riemann matrix (i.e., a symmetric, $N \times N$ matrix with negative definite real part). Krichever's construction begins with a compact Riemann surface of genus N (topologically, a sphere with N handles). Every such surface generates in a natural way a Riemann matrix (more precisely, an equivalence class of such matrices) which uniquely characterizes the surface. The Riemann theta function in (3) is constructed from such a Riemann matrix according to the N fold summation.

$$\theta = \sum_{-\infty}^{\infty} \cdots \sum_{-\infty}^{\infty} \exp\left\{ \frac{1}{2}\vec{m} \bullet Z \bullet \vec{m} + i\vec{m} \bullet \vec{\phi} \right\}, \tag{4}$$

where

$$\vec{m} = (m_1, \dots, m_N), \quad \vec{\phi} = (\phi_1, \dots, \phi_N)$$

$$\vec{m} \bullet Z \bullet \vec{m} = \sum_{i=1}^{N} \sum_{j=1}^{N} m_i Z_{ij} m_j, \quad \vec{m} \bullet \phi \bullet = \sum_{j=1}^{N} m_j \phi_j.$$

Krichever showed that every Riemann matrix that came from a Riemann surface generates solutions of the KP equation.

(c) The Schottky problem in algebraic geometry. It is known that many Riemann matrices are not related to any Riemann surface. The Schottky problem is to determine which Riemann matrices correspond to smooth compact Riemann surfaces. Novikov conjectured that an $N \times N$ Riemann matrix corresponds to a smooth compact Riemann surface of genus N if and only if it generates nontrivial solutions of the KP equation, *via* Krichever's construction. The validity of this conjecture was proved explicitly by Dubrovin [1] for low genera ($N = 1, 2, 3$), and for arbitrary genus by Shiota [6].

To summarize, the KP equation admits exact solutions in the form (3). These solutions provide explicit models of periodic (or quasi-periodic) waves in shallow water. At the same time, the same solutions also solve an outstanding problem in algebraic geometry This remarkable coincidence identifies the KP equation, and especially its quasi-periodic solutions, as objects worthy of further study.

In a companion paper, Segur and Finkel [5] noted that the real-valued, two-dimensional, periodic KP solutions of genus 2 are natural, nontrivial generalizations to two dimensions of the one-dimensional cnoidal waves (i.e.KdV solutions of genus 1) that have been used so successfully as models of "typical" one-dimensional periodic waves of moderate amplitude in shallow water. (See Sarpkaya and Isaacson, [4] for a discussion of cnoidal waves from an engineering viewpoint.) In that paper, we proposed that the real-valued, two-dimensional KP soutions of genus 2 might be useful as practical, engineering models of "typical" periodic waves of moderate amplitude in shallow water. The purpose of the present paper is to establish some of the facts about real 2×2 Riemann matrices that were used in the companion paper.

We begin with some background material,see [1] or [7] for details.Two $N \times N$ Riemann matrices $\dot{Z}$ and Z are said to be *equivalent* if they are related by a unimodular transformation of the form:

$$\dot{Z} = 2\pi i(AZ + 2\pi iB)(CZ + 2\pi iD)^{-1}, \tag{5}$$

where (A, B, C, D) are all $N \times N$ matrices of integers, with

$$\det\begin{pmatrix} A & B \\ C & D \end{pmatrix} = 1, \tag{6}$$

and

$$\begin{pmatrix} A & B \\ C & D \end{pmatrix} \begin{pmatrix} 0 & 1 \\ -1 & 0 \end{pmatrix} \begin{pmatrix} A^T & C^T \\ B^T & D^T \end{pmatrix} = \begin{pmatrix} 0 & 1 \\ -1 & 0 \end{pmatrix}, \tag{7}$$

and where $(\bullet)^T$ denote the transpose of $(\bullet)$. The matrices are then equivalent in the sense that their corresponding theta functions are related in a very simple way (See Dubrovin, [1] for more details). Moreover, the equivalence class of Riemann matrices that come from the same Riemann surface is given by (5). These transformations can be viewed as matrix generalizations of the linear fractional transformations of a complex variable.

It is desirable to have a single, canonical form to represent all of the Riemann matrices in a given equivalence group. It is especially helpful if the Riemann matrices are used to construct solutions of the KP equation, because all of the equivalent Riemann matrices correspond to the same Riemann surface, and generate the same set of KP solutions.

We now define such a canonical form for real 2×2 Riemann matrices. Let $\mathscr{R}$ denote the set of all real 2×2 Riemann matrices and $\tilde{\mathscr{R}}$ the set of all real 2×2 Riemann matrices with rational entries. If $Z \in \mathscr{R}$, then Z has the form

$$Z = \begin{pmatrix} z_{11} & z_{12} \\ z_{12} & z_{22} \end{pmatrix}$$

where z_{ij} is real, $z_{11} < 0, z_{22} < 0$, and $\det(Z) > 0$. Let

$$\lambda_1 = \frac{z_{12}}{z_{11}}, \quad \lambda_2 = \frac{z_{12}}{z_{22}}. \tag{8}$$

Definition

A real, 2×2 Riemann matrix is said to be *basic* if

$$-\frac{1}{2} < \lambda_i \leq \frac{1}{2}, \quad i = 1, 2. \tag{9}$$

Perhaps the most important questions about any Riemann matrix is whether it corresponds to a Riemann surface. For 2×2 matrices, it is known that either the matrix corresponds to a Riemann surface of genus 2, or it is equivalent to a diagonal matrix.

Theorem 1

Let $Z \in \mathscr{R}$ be in basic form. Then Z is equivalent to a diagonal matrix if and only if it is diagonal; i.e., only if

$$\lambda_i = 0, \quad i = 1, 2. \tag{10}$$

(The proof of this and the other theorems given in this Introduction are deferred to Sec. 2 and Sec. 3.)

An equivalent statement of Theorem 1 is

Theorem 1a

Suppose $Z \in \mathscr{R}$ is diagonalizable. Then there is a 2×2 unimodular matrix U such that $UZU^T = D$ where D is a diagonal matrix.

Corollary

Every basic Riemann matrix with $\lambda_i \neq 0$ is the period matrix of a compact Riemann surface of genus 2. Moreover, every such matrix generates real KP solutions of genus 2.

Theorem 1 shows the significance of the basic form. Next, we shown that every equivalence class of real, 2×2 Riemann matrices contains a basic one.

Theorem 2

Let $Z \in \mathscr{R}$. Then there is a 2×2 unimodular transformation U such that $UZU^T = B$, where B is in basic form.

Note that the general transformation of a 2×2 matrix, given by (5), involves 4×4 matrices. Theorem 2 asserts that it is always sufficient either to set $A = D = 0$ in (5), or to set $B = C = 0$, provided Z is real. Moreover, we give below a *reduction algorithm,* by which the transformation matrix (U) in Theorem 2 is constructed explicitly in a finite number of steps. This reduction algorithm may be found in the proof of Theorem 2.

The reduction algorithm, when combined with Theorem 1, provided an explicit method to determine whether any $Z \in \mathscr{R}$ can be diagonalized. Dubrovin [1] gave another method, in which one evaluates the determinant of a 4×4 matrix. The method given here was invented because we found Dubrovin's approach rather delicate in numerical computations.

Within an equivalence class of Riemann matrices, the basic form is effectively unique, as we show next. Let $J = \begin{pmatrix} 0 & 1 \\ -1 & 0 \end{pmatrix}$.

Theorem 3

(i) The basic form of any $Z \in \mathscr{R}$ unique under 2×2 unimodular transformations up to a transformation by $\pm J$. If the basic form is diagonal, these transformations amount to reordering the diagonal elements.

(ii) Let $Z \in \tilde{\mathscr{R}}$ be in basic form, and not diagonal $\left[\text{i.e., } \lambda_i \neq 0 \text{ in } (8)\right]$. Z is unique within the full group defined by (5), up to transformations in the subgroup generated by $\pm J$ and inversion, $4\pi^2()^{-1}$.

(iii) Let $Z \in \mathscr{R}$ be basic. The set (λ_1, λ_2) is invariant under transformations generated by $\pm J$ and inversion, $4\pi^2()^{-1}$.

In addition, in a remark at the end of the next section we give a characterization of the possible diagonal forms that a real diagonalizable Riemann matrix can take. The result asumes that the determinant of the Riemann matrix is not a rational multiple of π^2 .

To summarize, for real 2×2 Riemann matrices, every class of equivalent matrices contains an (effectively) unique element in basic form. Because any $Z \in \mathscr{R}$ can be reduced to its basic form, it is sufficient to work only with basic Riemann matrices. This simplification is especially helpful if the Riemann matrices are being used to generate KP solutions. We believe that the concept of basic form can be generalized at least to real, $N \times N$ Riemann matrices, but we have not proved it. The theorems stated in this Introduction are proved in the following sections. For an application of these results to a problem in water waves, see Segur and Finkel [5].

2. Diagonalizing Riemann Matrices

In this section we show that a real 2×2 Riemannn matrix which is diagonalizable by a 4×4 unimodular transformation can also be diagonalized by a 2×2 unimodular transformation. This result is proven in Theorem 1a. The results of the next section will show that Theorem 1a is equivalent to Theorem 1. We give two different proofs of Theorem 1a here, one slightly more general than the other. The following sequence of lemmas will help establish Theorem 1a and the Remark following it. That remark characterizes possible diagonal forms for a subclass of Riemann matrices whose determinants are not rational multiples of π^2. We let

$$M = \begin{bmatrix} A & B \\ C & D \end{bmatrix}$$

represent a 4×4 unimodular matrix and Z represent a real Riemann matrix. By $\dot{Z} = M[Z]$ we mean the transformation (5).

The following lemma may be found in [7] .

Lemma 1

Let

$$M = \begin{bmatrix} A & B \\ C & D \end{bmatrix}$$

be a unimodular matrix. Suppose $\det A \neq 0$. Then

$$M = \begin{bmatrix} A & AS_1 \\ S_2A & (A^T)^{-1} + S_2AS_1 \end{bmatrix} \tag{11a}$$

so that,

$$M = \begin{bmatrix} A & 0 \\ S_2A & (A^T)^{-1} \end{bmatrix} \begin{bmatrix} I & S_1 \\ 0 & I \end{bmatrix} \tag{11b}$$

where S_1 and S_2 are symmetric **rational** matrices.

Proof: The conditions (7) that M be a symplectic matrix imply

$$A^TC = C^TA$$

$$AB^T = BA^T$$

and,

$$A^TD - C^TB = I$$

determine the decomposition. For example, the first relationship implies that CA^{-1} is symmetric, so that, $CA^{-1} = S_2$ where S_2 is a symmetric and not necessarily integer matrix, and $C = S_2A$. We may solve for B and D similarly to complete the proof.

□

Remark: If we assume that det $D \neq 0$ we obtain a similar decomposition for M:

$$M = \begin{bmatrix} (D^T)^{-1} & S_2 D \\ 0 & D \end{bmatrix} \begin{bmatrix} I & 0 \\ S_1 & I \end{bmatrix}. \tag{11c}$$

Let $Z \in \mathscr{R}$. Successive applications of the two matrices in (11c) to Z imply,

$$M[Z] = \left(DZ^{-1}D^T + \frac{1}{2\pi i} DS_1D^T \right)^{-1} + 2\pi i S_2. \tag{12a}$$

Now suppose det $A \neq 0$. Successive applications of the two matrices in (11a) to Z imply,

$$M[Z] = \left((AZA^T + 2\pi i AS_1A^T)^{-1} + \frac{1}{2\pi i} S_2 \right)^{-1}. \tag{12b}$$

Lemma 2

Let

$$M = \begin{bmatrix} A & B \\ C & D \end{bmatrix}$$

be a unimodular matrix, and $Z \in \mathscr{R}$, with det Z not a rational multiple of π^2. Suppose $\dot{Z} \equiv M[Z]$ is diagonal. Then

(i) if det $A \neq 0$ then AZA^T is diagonal.

(ii) if det $D \neq 0$ then $(D^{-1})^T Z D^{-1}$ is diagonal.

Proof: Suppose $\det D \neq 0$, then $\dot{Z}$ may be written in the form (12a). Replace Z^{-1} and S_1 in (13) by $DZ^{-1}D^T$ and DS_1D^T. Lemma 3, part iii. implies that DS_1D^T, S_2 and $(D^{-1})^TZD^{-1}$ are diagonal.

If $\det A \neq 0$, then (12b), (14) and Lemma 3 part iii. imply that AS_1A^T, S_2 and AZA^T are diagonal.

□

Lemma 3

Let $Z \in \mathscr{R}$ and let S_1 and S_2 be real symmetric rational matrices. Set

$$\dot{Z} \equiv \left(Z^{-1} + \frac{1}{2\pi i}S_1\right)^{-1} + 2\pi i S_2 \tag{13}$$

Then

(i) If $Z \in \tilde{\mathscr{R}}$ then $\dot{Z}$ is real only if $S_1 = 0$ and $S_2 = 0$,

(ii) If $Z \in \tilde{\mathscr{R}}$ and $\dot{z}_{12}$ is purely imaginary then $z_{12} = 0$,

(iii) If $\det Z$ is not a rational multiple of π^2. then $\dot{Z}$ is diagonal only if S_1, S_2 and Z are diagonal.

Furthermore, when

$$\dot{Z} = \left((Z + 2\pi i S_1)^{-1} + \frac{1}{2\pi i}S_2\right)^{-1} \tag{14}$$

then the conclusions (i) (ii) (iii) still hold.

Proof: We consider $\dot{Z}$ given by (13) only. The calculations for $\dot{Z}$ given by (14) are similar.

A calculation shows that

$$\dot{Z} = \lambda\left(Z + \frac{\det Z}{2\pi i}\begin{bmatrix} (S_1)_{22} & -(S_1)_{12} \\ -(S_1)_{12} & (S_1)_{11} \end{bmatrix}\right) + 2\pi i S_2 \qquad (15)$$

where $\lambda = \dfrac{1}{1 - \dfrac{\det S \det Z}{4\pi^2} + \dfrac{tr(S_1 Z)}{2\pi i}}$

To establish (i) note that the right side of (15) may be regarded as a rational function of $\left(\frac{1}{2\pi i}\right)$ with rational coefficients, because Z, S_1 and S_2 all have rational entries.

Since $\frac{1}{2\pi i}$ is transcendental, we may balance powers of $\frac{1}{2\pi i}$ to see that Z is real only when $S_2 = 0$. By (13) this implies that $S_1 = 0$.

To establish (ii), suppose $\dot{z}_{12}$ is pure imaginary. Taking the real part of (15), we see

$$z_{12}\left(1 - \frac{\det Z \det S}{4\pi^2}\right) + \frac{(S_1)_{12} \det Z \, tr S_1 Z}{4\pi^2} = 0.$$

This equation implies that $z_{12} = 0$ by balancing powers of $(2\pi i)$

To establish (iii) note that $Im \frac{\dot{z}_{12}}{\lambda} = 0$ implies

$$\left[(2\pi)^2 (S_2)_{12} + \det Z((S_1)_{12} - (\det S_1)(S_2)_{12}) = 0 \right]$$

which can be satisfied if $\det Z$ is not a rational multiple of π^2 only if $(S_1)_{12} = 0$ and $(S_2)_{12} = 0$.

□

Lemma 4

Let $Z \in \mathscr{R}$ and M a unimodular transformation such that $M[Z]$ is diagonal Then there is a unimolular transformation $M' = \begin{bmatrix} A' & B' \\ C' & D' \end{bmatrix}$ with det $A' \neq 0$ such that $M'[Z]$ is diagonal. Furthermore from M we can build a 2×2 matrix u such that uZu^T is diagonal, provided det Z is not a rational multiple of π^2.

Proof: If det $A \neq 0$ then $M' = M$ and Lemma 2 implies AZA^T is diagonal. If det $A = 0$, let

$$M_i = V_i^{-1}M, \; M = V_iM_i \quad i = 1, 2, 3$$

Where $V_1 = \begin{bmatrix} 0 & I \\ -I & 0 \end{bmatrix}$

$$V_2 = \begin{bmatrix} 1 & 0 & 0 & 0 \\ 0 & 0 & 0 & 1 \\ 0 & 0 & 1 & 0 \\ 0 & -1 & 0 & 0 \end{bmatrix} \qquad V_3 = \begin{bmatrix} 0 & 0 & 1 & 0 \\ 0 & 1 & 0 & 0 \\ -1 & 0 & 0 & 0 \\ 0 & 0 & 0 & 1 \end{bmatrix}$$

The matrices V_i are all unimodular and we assert that for at least one i, $V_i^{-1}M$ has det A_i nonzero.

To see this fact, note that the 4×2 matrix,

$$H = \begin{bmatrix} A \\ C \end{bmatrix}$$

obtained by taking the first two columns of M has rank 2. The matrices V_i manipulate the rows of H so that the two independent rows appear on top.

A calculation shows that for any Riemann matrix $\tilde{Z}$, $V_i[\tilde{Z}]$ is diagonal if and only if $\tilde{Z}$ is diagonal. Therefore if M_i has det $A_i \neq 0$, $M_i[Z]$ is diagonal and by Lemma 2 $A_i Z A_i^T$ is diagonal.

□

Remark: The matrix u of Lemma 4 is not necessarily unimodular.

Lemma 5

Suppose M diagonalizes $Z \in \mathcal{R}$, det $A \neq 0$ and det Z is not a rational multiple of π^2 Then,

$$M = \begin{bmatrix} A & S_1(A^T)^{-1} \\ S_2 A & S_3(A^T)^{-1} \end{bmatrix} \tag{16}$$

where S_1, S_2, S_3 are diagonal matrices. Furthermore S_1, S_3, S_1S_2, S_2S_3 are integer matrices and, $S_3 - S_2S_1 = I$

Proof: By Lemma 1,

$$M = \begin{bmatrix} A & A\dot{S}_1 \\ S_2 A & (A^T)^{-1} + S_2 A\dot{S}_1 \end{bmatrix}$$

From Lemma 2 and the proof of Lemma 3 we know that S_2 and $A\dot{S}_1A^T$ are diagonal.

Therefore, $DA^T = I + S_2A\dot{S}_1A^T$, so that DA^T is diagonal and $D = S_3(A^T)^{-1}$ with S_3 diagonal. Next note that $A\dot{S}_1A^T = BA^T = S_1$ is diagonal, so that $B = S_1(A^T)^{-1}$

To see the second part of the Lemma note that

$$\begin{bmatrix} I & 0 \\ -S_2 & I \end{bmatrix} M = \begin{bmatrix} A & S_1(A^T)^{-1} \\ 0 & (S_3 - S_2S_1)(A^T)^{-1} \end{bmatrix}$$

The right hand side is a special case of (11a) with $C = 0$, therefore $S_3 - S_2S_1 = I$.

Since, $A(S_1(A^{-1})^T)^T$ is an integer matrix, we see that S_1 is an integer matrix. Similar manipulations establish the other statements in the Lemma.

□

We next present two proofs of Theorem 1a. The first proof is shorter, but requires Theorem 3 before we can establish the equivalence of Theorem 1a and Theorem 1. This equivalence is established in the next section. The first proof also assumes that the determinant of the Riemann matrix is not a rational multiple of π^2. The second proof of Theorem 1a makes use of the fact (shown independently in Theorem 2) that any $Z \in \mathscr{R}$ can be put in basic form by a 2×2 unimodular matrix. The second proof of Theorem 1a actually establishes Theorem 1 directly and under less restrictive assumptions. Either proof may be skipped by the reader, without loss.

Theorem 1a

Suppose $Z \in \mathscr{R}$ is diagonalizable. Then there is a 2×2 unimodular matrix U such that $UZU^T = D$ where D is a diagonal matrix.

Proof: Without loss, we suppose that $\det A \neq 0$ and that $\det Z$ is not a rational multiple of π^2. Then in the notation of Lemma 5, it follows from that Lemma that

$$S_2 = \begin{bmatrix} \dfrac{p_1}{q_1} & 0 \\ 0 & \dfrac{p_2}{q_2} \end{bmatrix} \tag{17a}$$

$$S_1 = \begin{bmatrix} q_1 s_1^1 & 0 \\ 0 & q_2 s_2^1 \end{bmatrix} \tag{17b}$$

and,

$$S_3 = \begin{bmatrix} q_1 s_1^3 & 0 \\ 0 & q_2 s_2^3 \end{bmatrix}. \tag{17c}$$

where (q_1, p_1) and (q_2, p_2) are chosen to be relatively prime. It then follows from the relationship $S_3 - S_2 S_1 = I$ of Lemma 5 that,

$$q_i s_i^3 - p_i s_i^1 = 1 \tag{8}$$

where $i = 1$ or 2.

Since $S_2 A$ is an integer matrix, we see,

$$\begin{bmatrix} \dfrac{1}{q_1} & 0 \\ 0 & \dfrac{1}{q_2} \end{bmatrix} A \tag{17d}$$

is an integer matrix which we call $\dot{A}$.

We see that $\det A = q_1 q_2 \det \dot{A}$ and

$$(A^T)^{-1} = \frac{1}{\det \dot{A}} \begin{bmatrix} \dfrac{\dot{a}_{22}}{q_1} & -\dfrac{\dot{a}_{21}}{q_1} \\ -\dfrac{\dot{a}_{12}}{q_2} & \dfrac{\dot{a}_{11}}{q_2} \end{bmatrix}$$

Furthermore, $S_1(A^T)^{-1}$ and $S_2(A^T)^{-1}$ are integer matrices, so that

$$\frac{1}{\det \dot{A}}\begin{bmatrix} s_1^i\dot{a}_{22} & -s_1^i\dot{a}_{21} \\ -s_2^i\dot{a}_{12} & s_2^i\dot{a}_{11} \end{bmatrix}$$

are integers, where $i = 1$ or 3.

Now from eq. 1, we see that

$$\frac{\dot{a}_{ij}}{\det \dot{A}}$$

$i,j = 1, 2$, are integers. This implise that $\dot{A}^{-1}$ is an integer matrix so that $\det \dot{A} = \pm 1$. From Lemma 3, AZA^T is diagonal. Therefore $\dot{A}Z\dot{A}^T$ is also diagonal.

Next we consider the case the det Z is a rational multiple of π^2. The argument we are about to give does not require this assumption and is in fact a full proof of Theorem 1. It does not, however, provide us with the information we get from Lemma 5. This argument constitutes the second proof of this theorem.

proof 2: Without any loss of generality, suppose Z is basic form. It may be put in basic form by the procedure outlined in the next section using 2×2 unimodular transformations. Therefore let

$$Z = \begin{pmatrix} b & b\lambda \\ b\lambda & g + b\lambda^2 \end{pmatrix}$$

with $\lambda^2 \leq 1/4$ and $g/b + \lambda^2 \geq 1$ so that $g/b \geq 3/4$.

Now suppose Z can be put in diagonal form by a unimodular transformation M with

$$M = \begin{bmatrix} A & B \\ C & D \end{bmatrix}.$$

By Lemma 4 (slightly adapted) we may assume that $\det D \neq 0$ so that

$$M = \begin{bmatrix} (D^T)^{-1} + S_2 D S_1 & S_2 D \\ D S_1 & D \end{bmatrix}$$

If $M[Z] = \begin{pmatrix} \alpha + 2\pi i\beta & 0 \\ 0 & \gamma + 2\pi i\delta \end{pmatrix}$ then taking the real and imaginary parts of

$$\begin{pmatrix} \alpha + 2\pi i\beta & 0 \\ 0 & \gamma + 2\pi i\delta \end{pmatrix} (CZ + 2\pi i D) = 2\pi i(AZ + 2\pi i B)$$

we obtain

$$\begin{aligned} B &= \begin{pmatrix} \beta & 0 \\ 0 & \delta \end{pmatrix} D - \frac{1}{4\pi^2} \begin{pmatrix} \alpha & 0 \\ 0 & \gamma \end{pmatrix} C Z^{-1} \\ S_2 &= \begin{pmatrix} \beta & 0 \\ 0 & \delta \end{pmatrix} - \frac{1}{4\pi^2} \begin{pmatrix} \alpha & 0 \\ 0 & \gamma \end{pmatrix} D S_1 Z^{-1} D^{-1} \end{aligned} \tag{19}$$

and

$$A = \begin{pmatrix} \beta & 0 \\ 0 & \delta \end{pmatrix} C + \begin{pmatrix} \alpha & 0 \\ 0 & \gamma \end{pmatrix} D Z^{-1} \tag{20}$$

Substituting the second part of (19) into (12a) gives the equation

$$\begin{pmatrix} \alpha & 0 \\ 0 & \gamma \end{pmatrix} = \left(D Z^{-1} D^T + \frac{1}{4\pi^2} C Z C^T \right)^{-1} \tag{21}$$

Equation (21) may be written as 3 equations in 8 integer unknowns D and C

$$\alpha\left(gd_{11}^2 + b(d_{12} - \lambda d_{11})^2 + \frac{b^2 g}{4\pi^2}(c_{11} + \lambda c_{12})^2 + \frac{bg^2}{4\pi^2}c_{12}^2\right) = bg \quad (22)$$

$$gd_{21}d_{11} + b(d_{12} - \lambda d_{22})(d_{22} - \lambda d_{21}) + \frac{b^2 g}{4\pi^2}(c_{11} + \lambda c_{12})(c_{21} + \lambda c_{22}) + \frac{gb^2}{4\pi^2}c_{12}c_{22} = 0 \quad (23)$$

$$\gamma\left(gd_{21}^2 + b(d_{22} - \lambda d_{21})^2 + \frac{b^2 g}{4\pi^2}(c_{21} + \lambda c_{22})^2 + \frac{gb^2}{4\pi^2}c_{22}^2\right) = bg \quad (24)$$

We now use a case by case analysis to show that (22), (23) and (24) constrained by (19) and (20) has no integer solutions if $\lambda \neq 0$. That is Z cannot be diagonalized if its basic form is not a diagonal matrix.

Case 1

$d_{11} = 0,\ d_{12} \neq 0,\ d_{21} \neq 0$

1. if $c_{11} + \lambda c_{12} = 0$

   This is impossible for irrational λ, so let $\lambda = \frac{p}{q}$

   Since, $\lambda^2 \leq \frac{1}{4}$ we have $q \geq 2$, and $c_{12} = mq$ (possibly $m = 0$)

   a. (19) implies $b_{11} = 0$

   b. (20) implies $a_{11} + \lambda a_{12} = 0$, so $a_{12} = nq$

   c. Because $b_{11} = 0$ and $A^T D - C^T B = I$, we see, $a_{12}d_{12} = b_{12}c_{12} + 1$ or $(nd_{12} - mb_{12})q = 1$

   But this equation has no integer solutions with $q \geq 2$

2. if $c_{12} = 0, c_{11} \neq 0$

   a. (22) implies $\alpha\left[d_{12}^2 + \frac{bg}{(2\pi)^2}c_{11}^2\right] = g$

   b. step a. and (19) imply $c_{11}b_{11} = \frac{-\frac{bg}{(2\pi)^2}c_{11}^2}{d_{12}^2 + \frac{bg}{(2\pi)^2}c_{11}^2}$

   so $0 < (-b_{11}c_{11}) < 1$, and there are no integer solutions.

3. if $c_{11} + \lambda c_{12} \neq 0, \ c_{12} \neq 0$

   set $c_{11} + \lambda c_{12} = \xi$ where $\xi \neq 0$

   a. (22) implies $\alpha\left[b_{12}^2 + \frac{bg}{(2\pi)^2}\xi^2 + \left(\frac{gc_{12}}{2\pi}\right)^2\right] = g$

   b. (19) and step a. imply

$$b_{11} = -\frac{\xi}{\xi^2 + (gc_{12}^2)/b + (2\pi d_{12})^2/bg}$$

Because $(gc_{12}^2)/b \geq 3/4$, b_{11} cannot attain any integer vaule except zero, which was excluded by hypothesis. Therefore we conclude there are no integer solutions with $d_{11} = 0$

<u>Case 2</u>

$d_{11} \neq 0$

1. if $c_{12} = 0$, and $d_{12} - \lambda d_{11} = 0$

a. (20) implies $a_{12} = 0$,

b. (19) implies $b_{12} - \lambda b_{11} = 0$

This is impossible for irrational λ , so let $\lambda = p/q$, then $b_{11} = mq$, $d_{11} = nq$

c. $A^T D - C^T B = I$ implies $a_{11}d_{11} = b_{11}c_{11} + 1$ and $(na_{11} - mc_{11})q = 1$

Since $q \geq 2$ there are no integer solutions.

2. if $c_{12} = 0$, $d_{12} - \lambda d_{11} \neq 0$

a. (22) implies $\alpha \left[g d_{11}^2 + b(d_{12} - \lambda d_{11})^2 + \frac{b^2 g}{(2\pi)^2} c_{11}^2 \right] = bg$

set $d_{12} - \lambda d_{11} = \xi \neq 0$

b. From (20), $a_{12} = \dfrac{\xi}{\xi^2 + \frac{g}{b} d_{11}^2 + \frac{bg}{(2\pi)^2} c_{11}^2}$

As in Case 1, a_{12} cannot attain nonzero integer values.

3. if $c_{12} \neq 0$

a. (19) and (20) imply

$$a_{12}d_{11} - b_{11}c_{12} = \frac{\alpha}{g} d_{11}(d_{12} - \lambda d_{11}) + \frac{\alpha b}{(2\pi)^2} c_{12}(c_{11} + \lambda c_{12})$$

$$= \frac{d_{11}(d_{12} - \lambda d_{11}) + \frac{bg}{(2\pi)^2} c_{12}(c_{11} + \lambda c_{12})}{\frac{g}{b} d_{11}^2 + (d_{12} - \lambda d_{11})^2 + \frac{bg}{(2\pi)^2}(c_{11} + \lambda c_{12})^2 + \left(\frac{g c_{12}}{2\pi}\right)^2}$$

b. Set $\sqrt{\frac{g}{b}}\, d_{11} = r \sin \theta$

$d_{12} - \lambda d_{11} = r \cos \theta$

$$\frac{\sqrt{bg}}{2\pi}(c_{11} + \lambda c_{12}) = p \sin \phi$$

$$\frac{g c_{12}}{2\pi} = p \cos \phi$$

c. Then $a_{12}d_{12} - b_{11}c_{11} = \dfrac{1}{\sqrt{\frac{g}{b}}} f$, where

$$2f = \frac{r^2 \sin 2\theta + p^2 \sin 2\phi}{r^2 + p^2}$$

and the only integer solution has

$$a_{12}d_{11} - b_{11}c_{12} = 0$$

d. Now (19) and (20) imply

$$(a_{11} + \lambda a_{12})d_{11} - b_{11}(c_{11} + \lambda c_{12})$$
$$= \frac{\alpha}{b} d_{11}^2 + \frac{\alpha b}{(2\pi)^2}(c_{11} + \lambda c_{12})^2.$$

Therefore,

$$a_{11}d_{11} - b_{11}c_{11} = \frac{gd_{11}^2 + \frac{gb^2}{(2\pi)^2}(c_{11} + \lambda c_{12})^2}{gd_{11}^2 + b(d_{12} - \lambda d_{11})^2 + \frac{gb^2}{(2\pi)^2}(c_{11} + \lambda c_{12})^2 + \frac{g^2 b}{(2\pi)^2}c_{12}^2}$$

Because $d_{11} \neq 0$ and $c_{12} \neq 0$ it follows that $1 > a_{11}d_{11} - b_{11}c_{11} > 0$ which has no integer solutions.

All cases are now covered and we conclude there is no unimodular transformation which diagonalizes a Riemann matrix in basic form when $\lambda \neq 0$. This statement establishes Theorem 1 and implies Theorem 1a because any $Z \in \mathcal{R}$ may be put in basic form by a 2×2 unimodular matrix, as Theorem 2 will show.

□

Theorem 4 and the remark following it concern the uniqueness of the diagonal form of a matrix $Z \in \mathcal{R}$.

Theorem 4

Let $Z \in \mathcal{R}$ be diagonalizable. Then up to a reordering of the diagonal elements, the diagonal form is unique, under 2×2 unimodular transformations.

Proof: Suppose Z can be transformed into diagonal matrices D and $\dot{D}$. Then there is a transformation A such that $ADA^T = \dot{D}$ where A is a 2×2 unimodular matrix. Let

$$D = \begin{bmatrix} d_1 & 0 \\ 0 & d_2 \end{bmatrix}$$

Upon working out the expression for the off diagonal element of $\dot{D}$ we get the equation,

$$d_1 a_{11} a_{21} + d_2 a_{12} a_{22} = 0$$

Lemma 6 and the fact that $d_1 < 0$ and $d_2 < 0$ show that this equation can only be satisfied if $d_1 a_{11} a_{21} = 0$ and $d_2 a_{12} a_{22} = 0$. So either $A = I$ or $A = \pm J$

$$J = \begin{bmatrix} 0 & 1 \\ -1 & 0 \end{bmatrix} \tag{25}$$

□

Remark: Let Z be a real diagonalizable Riemann matrix whose determinant is not a rational multiple of π^2. Lemma 5 and the proof of the first part of Theorem 1 enable us to construct all 4×4 unimodular matrices L such that $L[Z]$ is diagonal. We proceed as follows. First let U be a 2×2 unimodular transformation such that UZU^T is diagonal. By Theorem 4 the only transformations with this property are U and $\pm JU$ where J is given by (9).

Next choose integers, $q_1, p_1, s_1^1\, s_1^3, q_2, p_2, s_2^1\, s_2^2$ such that

$$q_i\, s_i^3 - p_i s_i^1 = 1 \tag{8}$$

where $i = 1, 2$ and q_1, q_2 are nonzero. In analogy with (17d) let

$$A = \begin{bmatrix} q_1 & 0 \\ 0 & q_2 \end{bmatrix} U \tag{17e}$$

Define S_1, S_2 and S_3 by (17a), (17b) and (17c). Let the unimodular transformation M be given by (16).

Let L equal either M or V_iM, $i = 1, 2, 3$, where V_i is defined in Lemma 4. For each choice of L, $L[Z]$ is diagonal. These are in fact the only unimodular transformations which diagonalize Z.

To see this fact we use Lemma 4 to assert that we need only consider transformations $M = \begin{bmatrix} A & B \\ C & D \end{bmatrix}$ which diagonalize Z and have det A nonzero. Lemma 5 asserts that M must have the form (16), which we have specified. The proof of Theorem 1 shows us that S_1 , S_2 and S_3 must be in the form of (17a), (17b) and (17c). Also relationship (8) must hold. Finally, since the proof of Theorem 1 shows that det $A = \pm 1$ and that $\dot{A}Z\dot{A}^T$ is diagonal (17d) justifies defining A by (17e).

□

3. Basic Form

In the previous section, we proved that any diagonalizable real Riemann matrix can be diagonalized by a 2 × 2 unimodular transformation, instead of the full 4 × 4 transformation. In this section, we turn our attention to basic form and prove Theorem 2. The reduction algorithm may be found in the proof of Theorem 2.

Theorem 2

Let $Z \in \mathscr{R}$. Then there is a 2×2 unimodular transformation U such that $UZU^T = B$, where B is in basic form.

Proof:

For integer k, let

$$U(k) = \begin{bmatrix} 1 & k \\ 0 & 1 \end{bmatrix}.$$

The matrices $U(k)$ and their transposes generate the two dimensional unimodular group, so that successive applications of these transformations to Z will generate the transformation U which puts Z in basic form.

We proceed as follows with the *reduction algorithm* .

Let,

$$Z_1 = U^T(k_0)ZU(k_0)$$

where we choose k_0 such that $| \lambda_1(Z_1) | \leq \frac{1}{2}$. Now let,

$$Z_2 = U(k_1)Z_1U^T(k_1)$$

where we choose k_1 so that $| \lambda_2(Z_2) | \leq \frac{1}{2}$.

In general, we let,

$$Z_{2n+1} = U^T(k_{2n})Z_{2n}U(k_{2n})$$

where k_{2n} is chosen so that $| \lambda_1(Z_{2n+1}) | \leq \frac{1}{2}$ and set

$$Z_{2n} = U(k_{2n-1})Z_{2n-1}U^T(k_{2n-1}),$$

where k_{2n-1} is chosen so that $| \lambda_2(Z_{2n}) | \leq \frac{1}{2}$.

In other words we want to pick k_{2n} so as to minimize $| \lambda_1(Z_{2n+1}) |$. We may do this by choosing k_{2n} to be the integer which is closest to $\lambda_1(Z_{2n})$. This algorithm then minimizes $| (z_{12})_n |$. We choose k_{2n-1} so as to minimize $| \lambda_2(Z_{2n}) |$,by letting k_{2n-1} be the integer closest to $\lambda_2(Z_{2n-1})$. We will show that for any $Z \in \mathcal{R}$ there is a number N such that Z_N is in basic form. All the succeeding numbers k_n with $n > N$ are actually 0.

We suppose the process of defining Z_n continues to infinity and we obtain an infinite sequence $\{ Z_n \}$. The sequence has the following properties.

$$| z_{12;2n} | \leq \frac{1}{2} | z_{22;2n} |$$

$$| z_{12;2n-1} | \leq \frac{1}{2} | z_{11;2n-1} |$$

$$| z_{ij;n+1} | \leq | z_{ij;n} |$$

Therefore Z_n converges to a matrix $\dot{Z}$ in basic form. By iterating backwards, we may write,

$$Z_n = A_n Z A_n^T$$

where $\det A_n = 1$ and

$$A_n = \begin{bmatrix} a_n & b_n \\ c_n & d_n \end{bmatrix}$$

Upon multiplying we see,

$$z_{11;n} = a_n^2 z_{11} + 2a_n b_n z_{12} + b_n^2 z_{22} \tag{26a}$$

$$z_{22;n} = c_n^2 z_{11} + 2c_n d_n z_{12} + d_n^2 z_{22} \tag{26b}$$

We now show that there is a number M such that

$$a_n^2 + b_n^2 + c_n^2 + d_n^2 < M \tag{27}$$

for each integer n. We show only that $\{ a_n \}$ is bounded. The other three cases are similar. Suppose then $\{ a_n \}$ is an unbounded sequence which we may in fact take to converge to $+ \infty$. Since $\{ Z_n \}$, converges (26a) shows that $\frac{b_n}{a_n}$ forms a bounded sequence which does not converge to zero. We may then assume without loss that

$$b_n = \kappa a_n + o(a_n)$$

where κ is a nonzero constant. Substituting this expression for b_n into (26a) gives to leading order

$$a_n^2 \left(z_{11} + 2\kappa z_{12} + \kappa^2 z_{22} \right)$$

This expression cannot be bounded in n though because Z is negative definite and $\{a_n\} \to \infty$. But this contradicts the convergence of $\{z_{11;n}\}$ and so our original assumption that $\{ a_n \}$ was unbounded must have been incorrect and we see that (27) is in fact bounded in n.

It then follows that the sequence $\{A_n\}$ is bounded. Since it is an integer sequence, it must repeat itself infinitely often and there is a subsequence $\{A_{j_n}\}$ which is constant. Let $A = A_{j_n}$ then

$$\dot{Z} = \lim_{n'\infty} Z_{j_n} = \lim_{n'\infty} A_{j_n} Z A_{j_n}^T = AZA^T .$$

Therefore Z can indeed be put in basic form by the transformation A.

□

Remark: Preliminary consideration suggests that the number of steps the algorithm takes can be estimated as follows. Let b be the diagonal entry of Z closest to zero and let $g = \det Z/b$. If $g/b > 1/n$ for some integer n, the algorithm takes at most n steps.

Next we show that the basic form of a Ricmann matrix is unique modulo some trivial transformations. This is accomplished by the next two results.

Lemma 6

Let A be a 2×2 unimodular matrix with

$$A = \begin{bmatrix} a & b \\ c & d \end{bmatrix}$$

and suppose that all of the entries of A are nonzero. Then *ac* and *bd* have the same sign and *ad* and *bd* have the same sign. Furthermore

$$2 \mid ac + bd \mid > \mid ad + bc \mid + 2 \tag{28}$$

Proof: Since $\det A = 1$ it follows that *ad* and *bc* must have the same sign. This fact implies that *ac* and *bd* have the same sign. To see the inequality (28) let l_1, l_2, l_3, l_4 be the elements of A in decreasing order of magnitude, $|l_1| \geq |l_2 \geq |l_3| \geq |l_4|$ and suppose without loss that $l_1 = a$. Then we must have,

$$A = \begin{bmatrix} l_1 & l_2 \\ l_3 & l_4 \end{bmatrix}$$

or,

$$A = \begin{bmatrix} l_1 & l_3 \\ l_2 & l_4 \end{bmatrix}$$

In the first case, since,

$$2 \mid l_1 l_3 \mid \geq \mid l_2 l_3 \mid + \mid l_1 l_4 \mid \geq \mid l_2 l_3 + l_1 l_4 \mid$$

we see that the inequality

$$2 \mid l_1 l_3 + l_2 l_4 \mid > \mid l_2 l_3 + l_1 l_4 \mid + 2$$

holds because of the parity of signs. But this is just the inequality (28). The second case is handled similarly.

□

Lemma 7

Suppose $Z \in \tilde{\mathscr{R}}$ is nondiagonalizable and M is a unimodular matrix. Then $\dot{Z} = M[Z]$ is real only if and only if

$$M = \begin{bmatrix} A & 0 \\ 0 & (A^T)^{-1} \end{bmatrix} \text{ or } M = \begin{bmatrix} 0 & B \\ -(B^T)^{-1} & 0 \end{bmatrix}$$

Proof: Suppose first that det $D \neq 0$. By Lemma 3, because

$$\dot{Z} = \left(DZ^{-1}D^T + \frac{1}{2\pi i}DS_1D^T\right)^{-1} + 2\pi i S_2$$

we see $S_1 = 0$ and $S_2 = 0$ so that

$$M = \begin{bmatrix} A & 0 \\ 0 & (A^T)^{-1} \end{bmatrix}.$$

If det $D = 0$, then we let $M = Vi(Vi^{-1}M)$ $[Z]$ where $V_1 = \begin{bmatrix} 0 & I \\ -I & 0 \end{bmatrix}$

$$V_2 = \begin{bmatrix} 1 & 0 & 0 & 0 \\ 0 & 0 & 0 & 1 \\ 0 & 0 & 1 & 0 \\ 0 & -1 & 0 & 0 \end{bmatrix} \qquad V_3 = \begin{bmatrix} 0 & 0 & 1 & 0 \\ 0 & 1 & 0 & 0 \\ -1 & 0 & 0 & 0 \\ 0 & 0 & 0 & 1 \end{bmatrix}$$

Lemma 4 implies that for some i $V_i^{-1}M$ will have det $D_i \neq 0$.

The case V_1 gives us

$$M = \begin{bmatrix} 0 & B \\ -(B^T)^{-1} & 0 \end{bmatrix}.$$

For the cases V_2 and V_3 we see that for $\dot{Z}$ to be real we need $\left[(V_i^{-1}M) [Z] \right]_{12}$ to be purely imaginary. By Lemma 3 this can happen only if Z is diagonalizable.

□

Theorem 3

(i) The basic form of any $Z \in \mathcal{R}$ unique under 2×2 unimodular transformations up to a transformation by $\pm J$. If the basic form is diagonal, these transformations amount to reordering the diagonal elements.

(ii) Let $Z \in \tilde{\mathscr{R}}$ be in basic form, and not diagonal [i.e., $\lambda_i \neq 0$ in (8)]. Z is unique within the full group defined by (5), up to transformations in the subgroup generated by $\pm J$ and inversion, $4\pi^2()^{-1}$.

(iii) Let $Z \in \mathscr{R}$ be basic. The set (λ_1, λ_2) is invariant under transformations generated by $\pm J$ and inversion, $4\pi^2()^{-1}$.

Proof: We establish (1) first. Suppose

$$| \dot{z}_{12} | \leq | z_{12} | . \tag{29}$$

Let, $AZA^T = \dot{Z}$ where

$$A = \begin{bmatrix} a & b \\ c & d \end{bmatrix}$$

Therefore,

$$\dot{z}_{12} = acz_{11} + (ad + bc)z_{12} + bdz_{22}.$$

Let $\lambda_1 = \dfrac{z_{12}}{z_{11}}$ and $\lambda_2 = \dfrac{z_{12}}{z_{22}}$. Since Theorem 4 handles the case of equivalent diagonal matrices, we may assume in light of (29) that z_{12} is not equal to zero. Then the following inequality holds,

$$\left| \frac{ac}{\lambda_1} + (ad + bc) + \frac{bd}{\lambda_2} \right| = \left| \frac{\dot{z}_{12}}{z_{12}} \right| \leq 1 \tag{30}$$

where $| \lambda_1 | \leq \frac{1}{2}$ and $| \lambda_2 | \leq \frac{1}{2}$.

Suppose first that all of the entries of A are nonzero so that Lemma 6 applies. It then follows that,

$$\left| \frac{ac}{\lambda_1} + (ad + bc) + \frac{bd}{\lambda_2} \right| \geq | 2(ac + bd) | - | (ad + bc) | > 2$$

The first inequality follows from the parity of the signs of *ac* and *bd*, and eq (28). The second inequality follows from equation (28). We see then that for inequality (30) to hold, the transformation A must have at least one zero entry. The unimodular transformations with zero entries are given by the transformations $U(k)$ of Theorem 2 and by transformations of the form $\pm U(k)J$. Close examination of the inequality (30) for matrices A of this form shows that when $| \lambda_1 | < \frac{1}{2}$ and $| \lambda_2 | < \frac{1}{2}$ the inequality (30) can hold only when $A = I$ or $A = \pm J$. When either λ_1 or λ_2 has modulus $\frac{1}{2}$ special cases hold. For example if $\lambda_2 = \frac{1}{2}$ (30) can be satisfied by the transformation

$$A = \begin{bmatrix} 1 & -1 \\ 0 & 1 \end{bmatrix}$$

To avoid these possibilities the definition of basic form is specialized when either λ_1 or λ_2 has modulus $\frac{1}{2}$.

Part (ii) follows from Part (i) and Lemma (7).

□

The proof of Theorem 3 also establishes part of Theorem 1, since diagonal matrices are a subclass of basic matrices and the basic form is unique.

Theorem 1

Let $Z \in \mathscr{R}$ be in basic form. Then Z is equivalent to a diagonal matrix if and only if it is diagonal; i.e., only if

$$\lambda_i = 0, \quad i = 1, 2. \tag{10}$$

Proof: The second part of the theorem follows from the discussion in the first section.

□

Finally we note that Theorem 3 is equivalent to Theorem 3 of the introduction.

The permanent address of H. Segur is ARAP, Box 2229, Princeton, NJ 08540. H. Segur's work was supported in part by a grant from the U.S. Army Research Office.

References

[1] Dubrovin, B.A., "*Theta Functions and Nonlinear Equations*", Russian Math Survey **36:2,** 1981, pp. 11-92.

[2] Kadomtsev, B.B. and V.I. Petvicshvili, Soviet Physics, Doklady, **15,** 1970, pp. 539-541.

[3] Krichever, I.M., "*Integration of Nonlinear Equations by the Methods of Algebraic Geometry*", Functional Analysis Appl., **11-1,** 1977, pp. 12-26.

[4] Sarpkaya, T. and M. Isaacson, "*Mechanics of Wave Forces on Offshore Structures*", Van Nostrand Reinhold, 1981.

[5] Segur, H. and A. Finkel, "*An Analytical Model of Periodic Waves in Shallow Water*", preprint.

[6] Shiota, T., "*Characterization of Jacobian Varieties in terms of Soliton Equations*", preprint 1984.

[7] Siegel, C.L., "*Topics in Complex Function Theory*", **Vol. 2.** Wiley Interscience, 1971.

Allan Finkel

Mathematical Sciences Dept.

IBM T.J. Watson Research Center

Yorktown Heights, New York

and

Harvey Segur

Institute for Theoretical Physics

University of California

Santa Barbara, California

Typed by: Barbara A. Newman

Lectures in Applied Mathematics
Volume 23, 1986

A Review of Superintegrable Systems

Boris A. Kupershmidt

The University of Tennessee Space Institute
Tullahoma, TN 37388

Center for Nonlinear Studies,
Los Alamos National Laboratory
Los Alamos, NM 87544

Summary. The current developments in superintegrable systems and associated supermathematics are reviewed, including: Classical, nonclassical, and nonstandard integrable and superintegrable systems, both continuous and discrete; Various superextensions of the Korteweg-de Vries equation and classification of all possible superextensions of the Lie algebra of vector fields on the line; Variational calculus with anticommutative variables; Super Hamiltonian formalism; Classification of affine super-Hamiltonian operators via generalized two-cocycles on differential-difference Lie superalgebras; Semidirect product Lie superalgebras and canonical quadratic maps associated with representations of Lie superalgebras.

§1. Introduction.

Superdynamical systems are those containing anticommuting variables = odd variables = odd elements of a Grassmann algebra = fermions. Superintegrable systems are, in the first approximation, those superdynamical systems which have an infinite set of conservation laws and/or supercommuting flows. We shall be a bit more precise later on.

Superdynamics made its first appearance in a little noticed at the time paper of Martin [1] who analyzed, in particular, the question of quasiclassical limits of fermionic quantum mechanical

0075-8485/86 $1.00 + $.25 per page

systems. His idea (revolutionary at the time), to allow Hamiltonians ("functions on a phase space") to take values in a Grassmann algebra instead of a number field, has become by now a naturally accepted fact of life (in physics and mathematics).

In fact, the movitation for developing supermathematics in general and integrable (and nonintegrable) field theories in particular, has come and is continuing to come from physics, where the idea of superunification of fundamental forces on the basis of supersymmetry is now being fervently tested and all commutative notions are being re-examined and supersymmetrized. It would be probably not an exaggeration to say that the current super-revolution in physics is comparable in its consequences to the one experienced by mathematics due to the introduction of complex numbers into real analysis. Naturally, the proof (if any) of the latter statement may only be arrived at gradually, accumulating particular results and ideas found during various superactivities, especially so since the serious mathematical development of the subject has just begun. And although the most of the effective uses of superthinking are concentrated in physics, there are already some of mathematical developments worth mentioning with an eye of future possibilities:

1) Classification by V. Kac of : simple complex finite-dimensional Lie superalgebras [2], and Kac superalgebras [3] (their infinite-dimensional Kac-Moody type analogs).

2) First applications, initiated by Manin, in algebraic geometry, Yang-Mills equations, and supergravity [4–6].

3) Construction of classical [7] and nonclassical [8] superintegrable systems (these will be discussed below).

As an example of the helpfulness of the superview on the usual (= commutative) mathematics, let me mention that nonclassical integrable systems have been overlooked in the theory of integrable systems, and were first noticed as supersystems.

Two immediate problems of a more practical nature include:

1) Fluids with fermionic degrees of freedom (for example, spin; see, e.g., [9]), and super-Clebsch representations associated with them (discussed in Appendix 3); Self-gravitating fluids whose spin interacts with torsion in generalized relataivity (U_4–theories).

2) Chaotic behavior arising from odd perturbations of integrable systems.

In this paper I review the present state of superintegrable systems and some of the associated mathematics. The plan of the paper is as follows. In §2 we recall the basic facts about continuous and discrete integrable systems of Lax type: classical, nonclassical, and nonstandard. In §3 we discuss what changes should be made in commutative theories to accomodate superintegrable systems. In §4 we look at various superextensions of the Korteweg-de Vries (KdV) equation and related objects. Basic supermathematical facts (distilled from [10]) are discussed in the appendices. Appendix 1 gives a digest of the calculus of variations in the presence of anticommutative variables. The super-Hamiltonian formalism is outlined in Appendix 2, together with characterization of affine super-Hamiltonian structures through generalized 2-cocycles on stable differential-difference Lie superalgebras. Appendix 3 contains a construction of supercanonical quadratic maps associated to representations of stable differential-difference Lie superalgebras; these maps are called "Clebsch representations" in physics (in a purely bosonic case). Appendices may be read independently of the main text in §2 – §4 (but not vice versa).

§2. Lax equations, commutative case

We start off with continuous systems.

Classical systems are of the form [11]:

$$L_t = [P_+, L] = [-P_-, L], \tag{2.11}$$

$$L = \sum_{i=0 \text{ or } -\infty}^{n} u_i \xi^i \, , \, n \in \mathbf{N}, \tag{2.2}$$

where: $\xi = \partial =$ "$\partial/\partial x$"; u_i's are ℓ x ℓ matrices: $u_i = (u_{i,\alpha\beta})\, ; P \in Z(L)$, the centralizer of L in the ring $Mat_\ell(\mathcal{O}_C)$ of matrix pseudo-differential operators with coefficients in the differential algebra $C = k[u_{i,\alpha\beta}^{(m)}]$, where k is a field of characteristic zero (say, $k = \mathbf{C}$) and the derivation ∂ acts on the generators of C as: $\partial(u_{i,\alpha\beta}^{(m+1)}) = u_{i,\alpha\beta}^{(m+1)}$, $m \in \mathbf{Z}_+$; P_+ stands for the "differential part" of P : if

$$P = \sum^{m} v_j \xi^j \, , \tag{2.3}$$

then $P_{\leq k} = \sum_{j \leq k} v_j \xi^j$, and analogously for $P_{\geq k}$, $P_{<k}$, $P_{>k}$, so that $P_+ = P_{\geq 0}$, $P_- = P_{<0}$; also, $Res\, P = v_{-1}$. The two highest coefficients in L satisfy the normalization conditions:

(2.4i) u_n is ∂-constant, invertible, diagonalizable matrix from $Mat_\ell(k)$,
(2.4ii) $u_{n-1} \in Im\, ad\, u_n$.
If u_n is already diagonal, which can be assumed without loss of generality, then the conditions (2.4) become:
(2.5i) $u_n = diag\,(c_1, \dots, c_\ell), 0 \neq c_\alpha \in k$,
(2.5ii) $u_{n-1,\alpha\beta} = 0$ whenever $c_\alpha = c_\beta$.

Let w be the grading on C, $\mathcal{O}_C$, $Mat_\ell(\mathcal{O}_C)$, defined as

$$(2.6) \qquad \mathrm{w}(u^{(m)}_{i,\alpha\beta}) = m + n - i\,; \mathrm{w}(\xi) = \mathrm{w}(\partial) = 1\,; \mathrm{w}(R) = = \mathrm{w}(R_{\alpha\beta}), R \in Mat_\ell(\mathcal{O}_C).$$

Then $\mathrm{w}(L) = n$. The description of $Z(L)$ is this [11]:
(2.7i) Each w-homogeneous element $P \in Z(L)$ is uniquely defined by its highest term $\mathrm{v}_m\xi^m$ in (2.3); and v_m is constant and belongs to the center of $Z(u_n)$ in $Mat_\ell(k)$; – equivalently, if u_n is diagonal, then

$$(2.7i') \qquad \mathrm{v}_{m,\alpha\alpha} = \mathrm{v}_{m,\beta\beta} \quad \text{whenever} \quad c_\alpha = c_\beta\,;$$

$Z(L)$ consists of (infinite) sums of its w-homogeneous elements; $Z(L)$ is abelian;
(2.7ii) If we denote by X_P the evolution derivation of C corresponding to (2.1), then $[X_P, X_Q] = 0$ for $P, Q \in Z(L)$; also, $X_P(Q) = [P_+, Q]$ for $\mathrm{w}(P) > 0$:
As a corollary, one gets [11]:
(2.8) All the equations (2.1) has a common infinite set of conservation laws (= $c.\ell.$'s) $\{str\, Res\, | P \in Z(L)\}$.
Also,
(2.9) All the equations (2.1) are Hamiltonian. (This is proved in [12] for the case when L is a differential operator, $L = L_+$, and in [10] for the general case.)

The properties (2.7)–(2.9) form the absolute minimum of information one must have in order to work with integrable systems. (By "integrable" I mean a system with an infinite number of $c.l.$'s and commuting fields.) Various important systems are specializations of the classical ones (including those associated with Kac-Moody algebras) and/or related to them.
Nonclassical systems have the Lax form (2.1) but with L given as

$$(2.10) \qquad L = \sum_{i=0}^{n} u_i\, \xi^i + \sum_s \varphi_s\, \xi^{-1} \psi_s^t\,.$$

Special cases of such systems have appeared previously in [13–16], the general form (2.10) was given in [8]. A new feature here, with respect to classical systems, is to make sure that the equations (2.1) make sense, that is, that the R.H.S. of (2.1) looks like a derivation of (2.10); more precisely, we have to worry only about the pseudo-differential tail of (2.10). From a different point of view, if we rewrite L in (2.10) as

$$(2.10') \qquad L = \sum_{i=0}^{n} u_i\,\xi^i + \sum_{\mathrm{k}=0}^{\infty} \Big(\sum_{s} \varphi_s \psi_s^{(\mathrm{k})t}\,(-1)^{\mathrm{k}}\Big)\xi^{-\mathrm{k}-1},$$

and treat it as a specialization of L in (2.2), we have to make sure that the R.H.S. of (2.1) looks like a derivation of (2.10′). From this interpretation it is clear that to establish in a purely algebraic fashion the properties (2.7*ii*), (2.8) for such an L would be a major undertaking (if at all possible); the only available route [17] consists in using the methods of the Hamiltonian formalism. (However, the second Hamiltonian structure [12] for the Lax equations with L given in (2.10), does <u>not</u> exist.)
<u>Nonstandard systems</u> have the form

$$(2.11) \qquad L_t = \left[((P^\dagger)_{\geq \mathrm{k}})^\dagger, L\right] = \left[-((P^\dagger)_{<\mathrm{k}})^\dagger, L\right], \mathrm{k} = 0,1,2,$$

$$(2.12) \qquad L = \sum_{i=-\mathrm{k}\ \mathrm{or}\ -\infty}^{n} u_i \xi^i\,, n \in \mathbf{N},$$

where "†" stands for "adjoint", and u_i's in (2.12) are <u>scalars.</u> The normalization conditions on L are

$$(2.13) \qquad \mathrm{k} = 0 : u_n = 1,\ u_{n-1} = 0; \qquad \mathrm{k} = 1 : u_n = 1.$$

For $\mathrm{k} = 0$, (2.11) becomes (2.1) (for scalar L) since only for $\mathrm{k} = 0$ taking the adjoint commutes with projecting on $\mathcal{G}_{\geq 0}$ (or $\mathcal{G}_{<0}$), where $\mathcal{G}$ stands for the Lie algebra of pseudo-differential operators. The nonstandard systems were introduced in [18]; their theory, in many respects more rich than the theory of classical systems, was developed in [19]. The basic sceleton (2.7)–(2.9), is available for the systems (2.11) with $\mathrm{k} > 0$.

We now turn to the discrete systems [20,21].

Classical systems are of the same form (2.1), with

$$L = \varsigma^{\beta}\left(1 + \sum_{i=1}^{N \text{ or } \infty} \varsigma^{-i} u_i\right), \beta \in \mathbf{N}, \tag{2.14}$$

where: u_i's are scalars; P runs over $\{L^n | n \in \mathbf{N}\}$; in the ring $C = k[u_i^{(m)}], m \in \mathbf{Z}$, Δ acts as an automorphism, $\Delta^s(u_i^{(m)}) = u_i^{(m+s)}, s \in \mathbf{Z}$, and in $C((\varsigma^{-1})), \varsigma^s \mathrm{v} = \Delta^s(\mathrm{v})\varsigma^s$, $\mathrm{v} \in C, s \in \mathbf{Z}$; $Res\,(\sum \mathrm{v}_j \varsigma^j) = \mathrm{v}_0$.
The systems (2.1), (2.14) include infinite generalized Toda lattices, and many other important systems. The basic properties of these systems are:
(2.15) All the flows (2.1) commute;
(2.16) All the flows (2.1) have a common infinite set of $c.l.$'s $\{Res\, P\, | P = L^n, n \in \mathbf{N}\}$;
(2.17) All the systems (2.1) are Hamiltonian.
2.18. Example. Let $H = \sum_{n \in \mathbf{Z}} (\frac{1}{2} p_n^2 + e^{q_{n-1} - q_n})$ be the (nongeneralized) infinite Toda lattice, with $\dot{q}_n = -\partial H/\partial p_n = -p_n, \dot{p}_n = \partial H/\partial q_n = e^{q_n - q_{n+1}} - e^{q_{n-1} - q_n}$.
Introducing $\mathrm{v}_n = e^{q_{n-1} - q_n}$, we get

$$\dot{\mathrm{v}}_n = \mathrm{v}_n(p_n - p_{n-1}), \qquad \dot{p}_n = \mathrm{v}_{n+1} - \mathrm{v}_n, \tag{2.19}$$

which results also from $L_t = [P_+, L]$ with

$$L = \varsigma + p + \mathrm{v}\varsigma^{-1}, P = L, \tag{2.20}$$

where p and v are considered as functions on $\mathbf{Z}$.
Nonclassical systems are discrete Lax equations with

$$L = \varsigma^{\beta}(1 + \sum_{i=1}^{N} \varsigma^{-1} u_i + \sum_{s} \varphi_s \varsigma^{-N-1} \psi_s). \tag{2.21}$$

As in the continuous case, the operator L in (2.21) can be considered as a specialization of (2.14):

$$L = \varsigma^{\beta}(1 + \sum_{i=1}^{N+1} \varsigma^{-i} u_i), \quad u_{N+1} = \sum_{s} \varphi_s^{(N+1)} \psi_s, \tag{2.21'}$$

but this representation won't help in establishing the basic properties (2.15)–(2.17): one has to appeal to the Hamiltonian formalism to do that.

Nonstandard discrete systems exist only for k = 1. They have the form

$$L_t = [P_{\geq 1}, L] = [-P_{\leq 0}, L], \tag{2.22}$$

$$L = \sum_{-\infty}^{\beta} \varsigma^j v_j, \quad v_j\text{'s are scalars.} \tag{2.23}$$

(If the lower limit in (2.23) is finite, this case is isomorphic to the classical case k = 0 when $\varsigma \mapsto \varsigma^{-1}$.) All the properties (2.15)–(2.17) remain in force.

§3. Superintegrable systems

We begin with continuous systems first.
Classical systems [7,10] are of the form (2.1), (2.2), where L is now an even operator. This means that we fix two nonnegative integers ℓ_0 and ℓ_1 such that $\ell = \ell_0 + \ell_1$, and introduce the $\mathbf{Z}_2$-grading p in the space of matrices: $p(R) = p(R_{\alpha\beta}) + p(\alpha) + p(\beta)$, where $p(\alpha) = \overline{0} \in \mathbf{Z}_2$ for $\alpha \leq \ell_0$ and $p(\alpha) = \overline{1} \in \mathbf{Z}_2$ for $\alpha > \ell_0$. The matrix elements $u_{i,\alpha\beta}$'s of L take values in a Grassmann algebra, and $p(u^{(m)}_{i,\alpha\beta}) = p(\alpha) + p(\beta)$ makes C into a commutative superalgebra (see [2,22] for basics of linear superalgebra). That L is even means that in each matrix u_i, the upper left- and lower right-block entries are even, and the rest are odd elements (of a Grassmann algebra). The normalization conditions on L are (2.4), with k being now a commutative superalgebra, and

$$ad\, u_n \text{ is invertible on } Im\, ad\, u_n \tag{3.1}$$

(in the commutative case this requirement is automatically satisfied; in the supercase $ad\, u_n$ could act on its root spaces by multiplication by milpotent elements, and we preclude this possibility); equivalently if u_n is diagonal as in (2.5*i*), then (3.1) reads

$$\text{If } c_\alpha \neq c_\beta \text{ then } (c_\alpha - c_\beta) \text{ is invertible in } k. \tag{3.2}$$

The description of $\mathbf{Z}_2$-homogeneous elements in $Z(L)$ by (2.7*i*) remains true; its "coordinate" description (2.7*i'*), however, changes into

$$(-1)^{p(\alpha)\eta} v_{m,\alpha\alpha} = (-1)^{p(\beta)\eta} v_{m,\beta\beta}, \text{whenever } c_\alpha = c_\beta;\ \eta := p(v_m). \tag{3.3}$$

$Z(L)$ is now a commutative superalgebra, and it is generated by its even part $Z(L)_0$ over k. Thus, one can consider the Lax equations (2.1) with P's being even only. The property (2.7*ii*) then remains true, as well as (2.9). The property (2.8) survives when str is substituted instead of tr, where

$$str(R) = \sum_{\alpha=1}^{\ell} (-1)^{p(\alpha)[1+p(R)]} R_{\alpha\alpha}\,. \tag{3.4}$$

If one considers a special case of the above set-up, the so-called general zero-curvature equations associated to simple complex Lie superalgebras (classified in [2]) instead of just simple Lie algebras as was done in [23]: $[\partial - U, \partial_t - V] = 0$, with $U = u + \lambda F$ and F being even regular semisimple, then one can show that all the results of Wilson remain in force provided: a) one substitutes the matrix B_1 (A3.14) (from the Appendix 3) with $p(\alpha) = 0$ or 1 according to whether E_α is even or odd root vector, instead of the canonical matrix ((2.11) in [23]); and b) one considers only the basic classical Lie superalgebras (from [2]), i.e., only those for which there exists a nondegenerate ad-invariant bilinear form.
Nonclassical systems [8, 17] have the same form (2.10) where, however, for each s, φ_s and ψ_s are both even or both odd. These systems are integrable and Hamiltonian. In particular, if $\ell_1 = 0$, one obtains superextensions of all the usual (commutative) integrable systems. The case of the KdV equation will be considered in the next section.
3.5 Example. $\ell = \ell_0 = 1$, $L = \xi^3 + u\xi + v + \varphi^t \xi^{-1}\psi$, where φ and ψ are vectors, $P = L^{3/2}$. The (super-Boussinesq) equations are:

$$\begin{cases} \dot{u} = -u_{xx} + 2v_x, \dot{v} = v_{xx} - 2u_{xxx}/3 - 2uu_x/3 + 2(\varphi^t\psi)_x, \\ \dot{\varphi} = \varphi_{xx} + 2u\varphi/3, \dot{\psi} = -\psi_{xx} - 2u\psi/3. \end{cases} \tag{3.6}$$

As in the commutative case, v can be eliminated resulting in

$$\ddot{u} + (u_{xx} + 2u^2)_{xx}/3 - 4(\varphi^t\psi)_{xx} = 0. \tag{3.7}$$

3.8 Example. Let $L = \xi + \varphi^t \xi^{-1}\psi$, where φ and ψ are vectors. For $P = L^2$ we obtain

$$\dot{\varphi_{xx}} + 2\varphi(\varphi^t\psi), \dot{\varphi} = -\psi_{xx} - 2\psi(\varphi^t\psi), \tag{3.9}$$

which results, for $\theta = \varphi + i\psi$, in the super NLS equation

$$\dot{\theta} = \overline{\theta}_{xx} + i\overline{\theta}(\theta^t\theta), \tag{3.10}$$

where $\overline{\theta}$ is the complex conjugate of θ.

No superanalog is known at the present time for the nonstandard systems, either continuous or discrete.

Not much is known about discrete superintegrable systems as well, except for the nonclassical ones which have the form (2.21) but where, for each s, φ_s and ψ_s are both even or both <u>odd</u>. These systems are integrable and Hamiltonian.

3.11 <u>Example</u>. Let $L = \varsigma + u_0 + \varsigma^{-1}u + \varphi^t\varsigma^{-2}\psi\,, P = L$. The corresponding super-Toda equations are

$$\begin{cases} \dot{u}_0 = (1-\Delta^{-1})u_1, \dot{u}_1 = u_1(\Delta - 1)u_0 + (1-\Delta^{-1})\varphi^{(2)t}\psi, \\ \dot{\varphi} = \varphi u_0, \dot{\psi} = -\psi u_0, \end{cases} \tag{3.12}$$

where $(\Delta^s\varphi)(n) = \varphi(n+s)$, $n, s \in \mathbf{Z}$.

There exists a discrete superintegrable system which does not fit so far in any of the theories. It has

$$L = \varsigma + \varphi^t(1 + \varsigma^{-1} + \varsigma^{-2} + \cdots)\psi\,, \tag{3.13}$$

where φ and ψ are vectors and, for each s, φ_s and ψ_s are both even or both odd. (When φ and ψ are both even, these systems have been studied by J. Gibbons.) All the flows with L in (3.13) are Hamiltonian. For $P = L^2$, the motion equations are

$$\dot{\varphi} = \mp\varphi^{(1)} + \varphi(\varphi^t\psi), \dot{\psi} = \mp[\psi^{(1)} + \psi(\varphi^t\psi)], \tag{3.14}$$

where the upper (resp., lower) sign is taken for even (resp., odd) components of φ and ψ.

Notice that (3.9) and (3.14) can be considered as purely fermionic integrable systems, and not as fermionic extensions of bosonic ones.

§4. <u>Super–Korteweg–de Vries systems</u>

A s–KdV system is a superintegrable system containing in itself, upon vanishing of some variables, the KdV equation. It is likely that a great many different superextensions exist for the KdV equation (and other commutative integrable systems), but it is not clear how to classify them. At the present time there are two extensions available [13, 8] (given below), in addition to the non-classical extension (2.10) of the KdV equation. They result from the classification of s–KdV systems with the following properties: a) Galilean invariance; b) An existense of a Miura map, from the corresponding s–mKdV system, which can be transformed into a deformation of the s–KdV system itself.

The first s–KdV system is of nonclassical type (2.10), with L specialized to be

$$L = -\xi^2 + u + \varphi^t \xi^{-1} \varphi + z^t \xi^{-1} y - y^t \xi^{-1} z, \tag{4.1}$$

$$P = L^{3/2}, \tag{4.2}$$

$$p(u) = p(z_j) = p(y_j) = 0, \; p(\varphi_i) = 1, \tag{4.3}$$

$$\begin{cases} u_t = \partial(3u^2 - u_{xx} + 12E_1), \\ \varphi_t = P_1(\varphi), z_t = P_1(z), y_t = P_1(y), \end{cases} \tag{4.4}$$

$$P_1 = P_1[u] = 3(u\partial + \partial u) - 4\partial^3, \tag{4.5}$$

$$E_1 = E_1[\varphi, z, y] = \varphi^t \varphi_x + 3(z^t y_x - y^t z_x). \tag{4.6}$$

The corresponding s–mKdV system is

$$\begin{cases} v_t = \partial(2v^3 - v_{xx} + 6vE_2 + 3E_{2,x}), \\ \alpha_t = P_2(\alpha), \; a_t = P_2(a), \; b_t = P_2(b), \end{cases} \tag{4.7}$$

$$p(v) = p(a_j) = p(b_j) = 0, \; p(\alpha_i) = 1, \tag{4.8}$$

$$P_2 = P_1[v^2 - v_x] + 6E_2\partial - 3E_{2,x}, \tag{4.9}$$

$$E_2 = E_1[\alpha, a, b]. \tag{4.10}$$

The Miura map is

$$u = v^2 + v_x + E_2, \; \varphi = (\partial + v)(\alpha), \; z = (\partial + v)(a), \; y = (\partial + v)(b). \tag{4.11}$$

Applying methods from [24], we arrive at the following deformation of the s–KdV system (4.4):

$$\begin{cases} U_t = \partial\left(3U^2 - U_{xx} + 2\epsilon^2 U^3 + 12(1 + 2\epsilon^2 U)E_3 + 12\epsilon E_{3,x}\right), \\ \Phi_t = P_3(\Phi), \; Z_t = P_3(Z), \; Y_t = P_3(Y), \end{cases} \tag{4.12}$$

(4.13) $p(U) = p(Z_j) = p(Y_j) = p(\epsilon) = 0,\ p(\Phi_i) = 1,$

(4.14) $E_3 = E_1[\Phi, Z, Y],$

(4.15) $P_3 = P_1[U - \epsilon U_x + \epsilon^2 U^2] + 24\epsilon^2 E_3 \partial - 12\epsilon^2 E_{3,x},$

together with its contraction onto (4.4):

(4.16)
$$\begin{cases} u = U + \epsilon U_x + \epsilon^2 U^2 + 4\epsilon^2 E_3, \\ \varphi = P_4(\Phi),\ z = P_4(Z),\ y = P_4(Y),\ P_4 = 1 + 2\epsilon\partial + 2\epsilon^2 U. \end{cases}$$

Since U is a c.l. in (4.12), inverting (4.16) in the form $U = u + \sum_{n\geq 1} \epsilon^n H_n[u, \varphi, z, y]$, provides an infinity of c.l.'s H_n for the s-KdV system (4.4).

4.17. Remark. The Lax operator L in (4.1) is self-adjoint, $L^\dagger = L$, and it is a specialization of the general (nonself-adjoint) Lax operator

$$L = -\xi^2 + u + \mu^t \xi^{-1} \nu, \tag{4.18}$$

which also leads to a superintegrable hierarchy (being of general nonclassical type). However, the Miura map for this hierarchy does not seem to allow a change into a deformation.

The second s-KdV system is

$$\begin{cases} u_t = \partial(3u^2 - u_{xx} + 3E_4), \\ w_t = P_5(w),\ \sigma_t = P_5(\sigma),\ f_t = P_5(f), \end{cases} \tag{4.19}$$

(4.20) $p(u) = p(f_i) = 0,\ p(w_j) = p(\sigma_j) = 1,$

(4.21) $E_4 = E_4[w, \sigma, f] = w^t \sigma + f^t f,$

(4.22) $P_5 = P_5[u] = 6\partial u - \partial^3,$

while the corresponding s-mKdV system is

$$\begin{cases} v_t = \partial(2v^3 - v_{xx} + 6vE_5), \\ c_t = P_6(c),\ \beta_t = P_6(\beta),\ \gamma_t = P_6(\gamma), \end{cases} \tag{4.23}$$

(4.24) $p(v) = p(c_i) = 0,\ p(\beta_j) = p(\gamma_j) = 1,$

(4.25) $E_5 = E_4\,[\beta, \gamma, c],$

(4.26) $P_6 = P_5\,[v^2] + 6E_5\partial,$

and the Miura map looks like

$$(4.27)\qquad \begin{cases} u = v^2 + v_x + E_5, \\ w = P_7(\beta),\ \sigma = P_7(\gamma),\ f = P_7(c),\ P_7 = \partial + 2v. \end{cases}$$

Again, using (4.23) and (4.27), we construct a deformation of (4.19)

$$(4.28)\qquad \begin{cases} U_t = \partial\left(3U^2 - U_{xx} + 2\epsilon^2 U^3 + 3(1 + 2\epsilon^2 U)\,E_6\right), \\ \Omega_t = P_8(\Omega),\ \Sigma_t = P_8(\Sigma),\ F_t = P_8(F), \end{cases}$$

$$(4.29)\qquad p(U) = p(F_i) = p(\epsilon) = 0,\ p(\Omega_j) = p(\Sigma_j) = 1,$$

$$(4.30)\qquad E_6 = E_4\left[\Omega,\ \Sigma,\ F\right],$$

$$(4.31)\qquad P_8 = P_5[U + \epsilon^2 U^2] + 6\epsilon^2 E_6 \partial,$$

and its contraction $C(\epsilon)$ into (4.19):

$$(4.32)\qquad \begin{cases} u = U + \epsilon U_x + \epsilon^2 U^2 + \epsilon^2 E_6, \\ w = P_9(\Omega),\ \sigma = P_9(\Sigma),\ f = P_9(F),\ P_9 = 1 + \epsilon\partial + 2\epsilon^2 U. \end{cases}$$

Since U is a c.l. for (4.28), inverting (4.32) we obtain $U = u + \sum_{n\geq 1} \epsilon^n H_n[u, w,\ \sigma, f]$, and thus an infinity of c.l.'s H_n for (4.19).

4.33. Remark. A linear problem (and the nature) of the systems (4.19) and (4.23), as well as an interpretation of the Miura maps (4.27) and (4.11), is not known. (Factorizations of (4.1) and (4.18) lead to different, from (4.11), Miura maps.)

4.34. Remark. The system (4.28) depends upon ϵ^2 while the contraction $C(\epsilon)$ in (4.32) depends upon ϵ. This allows us to obtain a Bäcklund transformation (i.e., an infinitesimal automorphism) of the s-KdV system (4.19) as simply $C(-\epsilon) \circ C(\epsilon)^{-1}$.

4.35. Remark. To prove the integrability of the s-KdV systems (4.4) and (4.19), we still have to demonstrate the commuting hierarchy. This can be done with the help of the (super-) Hamiltonian formalism (see Appendix 2).

The system (4.4) can be written in the form

$$(4.36)\qquad \begin{cases} u_t = \partial\left(\frac{\delta H}{\delta u}\right),\ \varphi_{i,t} = \frac{1}{4}\frac{\delta H}{\delta \varphi_i}, \\ z_{j,t} = -\frac{1}{12}\frac{\delta H}{\delta y_j},\ y_{j,t} = \frac{1}{12}\frac{\delta H}{\delta z_j}, \end{cases}$$

$$(4.37)\qquad H = u^3 + \frac{1}{2}ux^2 + 12uE_1 - 8\varphi^t\varphi_{xxx} - 48z^t y_{xxx}.$$

By Theorem A2.36, (4.36) is a Hamiltonian system (super-skew-symmetric constant-coefficient structure).

The system (4.19) can be cast into the form

$$(4.38)\qquad \begin{cases} u_t = \partial\left(\frac{\delta H}{\delta u}\right),\ f_{i,t} = \partial\left(\frac{\delta H}{\delta f_i}\right), \\ w_{j,t} = -2\partial\left(\frac{\delta H}{\delta \sigma_j}\right),\ \sigma_{j,t} = 2\partial\left(\frac{\delta H}{\delta w_j}\right), \end{cases}$$

$$(4.39)\qquad H = u^3 + \frac{1}{2}ux^2 + 3uE_4 + \frac{1}{2}(f^t f_{xx} + w_x^t\sigma_x).$$

Hence, (4.19) is also a Hamiltonian system.

In the commutative theory [12], the Miura maps are canonical between the Hamiltonian structure of the modified equations and the second Hamiltonian structure of the orginal (unmodified) equations. In the supercase, the situation is more complicated. We first notice that the s-KdV systems (4.4) and (4.19) may be also represented in the following forms:

$$\begin{aligned}
u_t &= \left[2(u\partial + \partial u) - \partial^3\right](\delta H/\delta u) + \\
&\quad + \Sigma(2\varphi_i\partial + \partial\varphi_i)(\delta H/\delta\varphi_i) + \\
&\quad + \Sigma(2y_j\partial + \partial y_j)(\delta H/\delta y_j) + \\
&\quad + \Sigma(2z_j\partial + \partial z_j)(\delta H/\delta z_j), \\
(4.40)\ \varphi_{i,t} &= (2\partial\varphi_i + \varphi_i\partial)(\delta H/\delta u) + (u - \partial^2)(\delta H/\delta\varphi_i), \\
y_{j,t} &= (2\partial y_j + y_j\partial)(\delta H/\delta u) + 3^{-1}(u - \partial^2)(\delta H/\delta z_j), \\
z_{j,t} &= (2\partial z_j + z_j\partial)(\delta H/\delta u) - 3^{-1}(u - \partial^2)(\delta H/\delta y_j), \\
(4.41)\quad H &= u^2/2 + 2E_1; \\
u_t &= \left[2(u\partial + \partial u) - \partial^3\right](\delta H/\delta u) + \\
&\quad + \Sigma\, 2(f_i\partial + \partial f_i)(\delta H/\delta f_i) + \\
&\quad + \Sigma\, 2(w_j\partial + \partial w_j)(\delta H/\delta w_j) + \\
&\quad + \Sigma\, 2(\sigma_j\partial + \partial\sigma_j)(\delta H/\delta\sigma_j), \\
(4.42)\ f_{i,t} &= 2(f_i\partial + \partial f_i)(\delta H/\delta u) + \\
&\quad + \left[2(u\partial + \partial u) - \partial^3\right](\delta H/\delta f_i), \\
w_{j,t} &= 2(w_j\partial + \partial w_j)(\delta H/\delta u) -
\end{aligned}$$

$$-2\left[2(u\partial+\partial u)-\partial^3\right](\delta H/\delta\sigma_j),$$

$$\sigma_{j,t}=2(\sigma_j\partial+\partial\sigma_j)(\delta H/\delta u)+$$

$$+2\left[2(u\partial+\partial u)-\partial^3\right](\delta H/\delta w_j),$$

(4.43) $H=(u^2+f^tf+w^t\sigma)/2.$

It is easy to check that both matrix operators corresponding to (4.40) and (4.42) are super-skew-symmetric (see Appendix 1). In addition, both these operators are affine. Hence, by Theorem A2.67, these operators are Hamiltonian if and only if the corresponding algebras are Lie superalgebras, and the constant parts of the operators represent respective 2-cocycles on these algebras. Checking the Jacobi identity, we quickly find that the second Hamiltonian structure exists in only the following cases:

(4.44) y, z are absent, φ is a scalar;

(4.45) w and σ are absent, f is a scalar.

The later case corresponds to a pair of noninteracting KdV fields $(u\pm f)$, and is not interesting. Before discussing the case (4.44) further, a few remarks are in order.

4.46. Remark. For the s-mKdV system (4.7) and (4.23) the Hamiltonian situation is analogous: they are not Hamiltonian, except when: 1) a, b are absent, α is a scalar; and 2) β,γ are absent, c is a scalar. The Miura maps (4.11) and (4.27), then, are not canonical except for these two cases; hence, the deformations (4.16) and (4.28) do not have a Hamiltonian character.

4.47. Remark. The s-KdV systems (4.4) and (4.19) can be called twone Hamiltonian systems, since they each have more than one but less than two Hamiltonian structures.

We now concentrate on the exceptional case of the system (4.40), when φ is a scalar and y and z are absent (the first reduced s-KdV system):

(4.48)
$$\begin{cases} u_t=\left[2(u\partial+\partial u)-\partial^3\right](\delta H/\delta u)+(2\varphi\partial+\partial\varphi)(\delta H/\delta\varphi),\\ \varphi_t=(2\partial\varphi+\varphi\partial)(\delta H/\delta u)+(u-\partial^2)(\delta H/\delta\varphi).\end{cases}$$

Let K be a commutative differential algebra with a derivation ∂ : K $\to$ K. Denote by D a free one-dimensional K-module with the Lie algebra structure $[X,Y]=X\partial(Y)-Y\partial(X)$ ("vector fields on the line"). Let $\lambda \in \mathrm{K}_c = Ker\, \partial|_{\mathrm{K}}$. Denote F_λ a free one-dimensional K- and D-module with the action of D given by the formula

$$(4.49) \qquad X : f \mapsto X\partial(f) - \lambda\partial(X)f$$

("the action of vector fields on $(dx)^{-\lambda}$ -densities"). Let L_λ denote the semidirect product $D \propto F_\lambda$ or $D \propto \oplus_i F_{\lambda(i)})$ when λ is a vector. If one wishes to make L_λ into a Lie superalgebra whose even part is D (see Appendix 2), one has first [2] to classify all possible D-homomorphisms $F_\lambda \otimes F_\mu \to D$. There are not many of them:

4.50. Theorem. Let $\Theta : F_\lambda \otimes F_\mu \to D$ be a homomorphism of D-modules, given by a bilinear differential operator with coefficients in $\mathrm{K}_c = Ker\, \partial|_{\mathrm{K}}$. Then Θ can be only of one of the following form (modulo multiplication by a constant from K_c):

$$(4.50.1) \qquad \lambda + \mu = 1\,, \Theta(f \otimes g) = fg;$$
$$(4.50.2) \qquad \lambda + \mu = 2\,, \Theta(f \otimes g) = \lambda f\partial(g) - \mu\partial(f)\cdot g;$$
$$(4.50.3a) \quad \lambda = 0, \mu = 3\,, \Theta(f \otimes g) = f^{(1)}g^{(1)} + 3f^{(2)}g;$$
$$(4.50.3b) \quad \lambda = 3, \mu = 0\,, \Theta(f \otimes g) = f^{(1)}g^{(1)} + 3fg^{(2)},$$

where $(\cdot)^{(\mathrm{k})} = \partial^{\mathrm{k}}(\cdot)$.

4.51. Corollary. There exists only one, up to isomorphism, Lie superalgebra extension of D, namely $L^s_{1/2} = D \propto F_{1/2}$, with the commutator

$$(4.52) \qquad [(X;\alpha),(Y;\beta)] = (XY^{(1)} - X^{(1)}Y + \alpha\beta;\, X\beta^{(1)} - \\ -\frac{1}{2}X^{(1)}\beta - \alpha^{(1)}Y + \frac{1}{2}\alpha Y^{(1)}).$$

It is easy to see that the following 2-form is a (unique nontrivial) 2-cocycle on $L^s_{1/2}$:

$$(4.53) \qquad \omega\left((X;\alpha),(Y;\beta)\right) = X\partial^3(Y) + \alpha\partial^2(\beta).$$

Therefore, by Theorem A2.67, one gets a super-Hamiltonian structure out of this ω on $L^s_{1/2}$: it is nothing but (4.48).

4.54. Remark. Since D is a subalgebra in $L^s_{1/2}$, $D \approx D \oplus \{0\}$, we can restrict ω (4.53) on D to obtain a (unique nontrivial) 2-cocycle on D:

$$(4.55) \qquad \bar{\omega}\,(X,Y) = X\partial^3(Y).$$

4.56. Remark. If we "localize" K, $L^s_{1/2}$, and ω, by considering $K_c = \mathbf{C}$, $K = \mathbf{C}[x, x^{-1}]$, $D = \{\Sigma X_n x^{1-n}\partial | X_n \in \mathbf{C}\}$, $\partial = \partial/\partial x$, $F_\lambda = \{\Sigma f_n x^{\gamma - n} | f_n \in \mathbf{C},\ \gamma = \gamma(\lambda)\}$, $\omega(\cdot,\cdot) \mapsto Res\,\omega(\cdot,\cdot)$, where Res singles out the x^{-1}-coefficient, we obtain the Neveu-Schwarz Lie superalgebra. Analogously, localizing K, D, and $\bar{\omega}$, one obtains the Virasoro algebra (which is also the even part of the Neveu-Schwarz algebra).

4.57. Remark. If one allows the even part of a Lie superalgebra be larger than D, then a variety of very interesting and nontrivial superextensions (together with nontrivial 2-cocycles on them) exists whose localizations are some dual string models [25].

We conclude by discussing relations between the Lie superalgebra $L^s_{1/2}$ and the first reduced s-KdV system. These relations belong to the following general scheme. Let $\mathcal{G}$ be a differential-difference Lie superalgebra (or Lie algeba), see Appendix 2, and $C = K[q_i^{q|\nu}]$ be the corresponding commutative superalgebra. Let $B = B(\mathcal{G})$ be the (super)-Hamiltonian matrix corresponding to $\mathcal{G}$. Let ω_1 and ω_2 be two (generalized) 2-cocycles on $\mathcal{G}$, and let $b(\omega_1)$ and $b(\omega_2)$ be the corresponding operators. Fix $\alpha, \beta \in K_c$ (K_c is the subring of constants in K). Let $B^1 = b(\omega_1) + \alpha B$, $B^2 = b(\omega_2) + \beta B$. Then B^1 and B^2 are Hamiltonian, by Theorem A2.67. Let X^i_H, $i = 1, 2$, denote the Hamiltonian vector field in C with the Hamiltonian $H \in C$, with respect to the Hamiltonian structure $B^i, i = 1, 2$. Suppose there exists a sequence $\{H_i \,|\, H_i \in C,\ i \in \mathbf{Z}_+\}$ such that $X^1_{H_{i+1}} = X^2_{H_i}$ and $X_{H_0} = 0$. Then the Hamiltonians $\{H_i\}$ are in involution with respect to both Hamiltonian structures B^1 and B^2. The question is, how to show that the sequence $\{H_i\}$ exists? In practice, if no other information is available, one considers only H_0, H_1, and H_2, and tries to construct a deformation of $X^1_{H_2} = X^2_{H_1}$ by finding first a canonical map $\Phi : C \to C_1$ with respect to B^2, and then bending Φ into a deformation. (In the case of twone Hamiltonian systems, with $\alpha = 0$, Φ is not canonical anymore, but still is a Miura map into $X^1_{H_2}$.) The choice of $\omega_1, \omega_2, \alpha, \beta$, and H_0, then makes the problem definite. In practice, ω_2 is a nontrivial 2-cocycle, and ω_1 is a trivial 2-cocycle.

4.58. Examples. 1) $\mathcal{G} = D$, $B = u\partial + \partial u$, $\omega_2 = \bar{\omega}$ in (4.54), ω_1 is "the first component" - cocycle, $\omega_1(X, Y) = X\partial(Y)$. Taking $\alpha = 0$, $\beta = -2$, $H_0 = u$, results in the KdV hierarchy. Taking

$B^1 = B$, $B^2 = \partial^3$, $H_0 = u^{1/2}$, results in the so-called Harry Dim hierarchy. (One has to extend C into $C\left[u^{1/2}, u^{-1/2}\right]$.) 2) $\mathcal{G} = L^{s}_{1/2}$, with B^2 given by (4.48), and B^1 is "the first component" - cocycle:

$$\omega_1\left((X;\alpha),(Y;\beta)\right) = 4X\partial(Y) + \alpha\beta. \tag{4.59}$$

Taking $H_0 = u$, $H_1 = u^2 + 4\varphi\varphi^{(1)}$, results in the first reduced s-KdV system. (In this way it was originally found in [13].)

Appendix 1. Grassmann-valued calculus

The material in this and the next appendix is taken from [10]. Most of the proofs are omitted but easily reconstructable using, as a model, commutative methods from [21].

Let k be a commutative ring with unity and K a commutative superalgebra over k, also with unity (see [2]). Let G be a discrete group acting by automorphisms on K/k and let $\partial_1, \ldots, \partial_m : K/k \to K/k$ be even left derivations of K over k which commute between themselves and with the action of G. Let $I_{\bar{0}}$ and $I_{\bar{1}}$ be two disjoint countable sets, $I = I_{\bar{0}} \cup I_{\bar{1}}, N = |I|$. Let $C = C_q = K[q_i^{(g|\nu)}], i \in I, g \in G, \nu \in \mathbf{Z}_+^m$, be a commutative superalgebra with the $\mathbf{Z}_2$-grading ([22]) p on C defined by: $p(q_i^{(g|\nu)}) = p(i)$, where $p(i) = \bar{0} \in Z_2$ for $i \in I_{\bar{0}}$ and $p(i) = 1 \in \mathbf{Z}_2$ for $i \in I_{\bar{1}}$. The actions of G (by automorphisms) and ∂'s (by derivations) are naturally extended on $C : \hat{h}(q_i^{(g|\nu)}) = q_i^{(hg|\nu)}, h \in G$, and $\partial^\sigma(q_i^{(g|\nu)}) = q_i^{(g|\sigma+\nu)}$, where $(\pm\partial)^\sigma := (\pm\partial_1)^{\sigma_1} \ldots (\pm\partial_m)^{\sigma_m}$ for $\sigma = (\sigma_1, \ldots, \sigma_m) \in \mathbf{Z}_+^m$. These actions still commute. (An informal model: $K = \oplus_G(C^\infty(M) \otimes \Lambda(r))$, where M is a manifold with coordinates $(x_1, \ldots, x_m)$, and $\Lambda(r)$ is a Grassmann algebra with r generators. G acts on K by moving summands according to the left action of G on itself; $\partial_i = \partial/\partial x_i$. Then $q_i^{(g|\nu)}$'s are 'coordinates' on the infinite tower of the jet bundles whose fiber coordinates are q_i's. (See [26,27].)) Let $Der(C)$ be the Lie superalgebra of all left derivations of C over K. A Lie superalgebra $D^{ev} = D^{ev}(C)$ consists of those derivations which commute with the actions of G and ∂'s: they are called <u>evolution</u> derivations (or vector fields). If $\partial/\partial_{q_i}^{(g|\nu)}$ denotes the natural left derivation of C/K, of the grading $p(i)$, then every evolution field X can be uniquely written as $X = \sum \hat{g}\partial^\nu(X_i) \cdot \partial/\partial_{q_i}^{(g|\nu)}$, where $X_i = X(q_i)$ and $q_i := q_i^{(e|0)}$ (e is the unit element in G).

Let $\Omega^1 = \Omega^1(C) = \{\sum dq_i^{g|\nu}\varphi_i^{(g|\nu)} | \varphi_i^{g|\nu} \in C$, finite sums$\}$ be the right $\mathbf{Z}_2$-graded C-module of 1 - forms on C and $\Omega_0^1 = \Omega_0^1(C) =$

$\{\sum dq_i\varphi_i | \varphi_i \in C$, finite sums$\}$ be the right $\mathbf{Z}_2$-graded C-module of reduced forms. The $\mathbf{Z}_2$-grading on Ω^1 is given by $p(dq_i^{(g|\nu)}) = p(i)+\bar{1}$. Let $d : C \to \Omega^1(C)$ be the universal left odd derivation of C over K defined by $d(q_i^{(g|\nu)}) = dq_i^{(g|\nu)} : dH = \sum dq_i^{(g|\nu)} \partial H / \partial q_i^{(g|\nu)}$. The actions of $D^{ev}(C), G$ and ∂'s are uniquely extended on $\Omega^1(C)$ such that they mutually commute between themselves and with d.

Denote $ImD = \sum_{g\in G} Im(\hat{g} - \hat{e}) + \sum_s Im\, \partial_s$. Elements of ImD are called trivial. We write $a \sim b$ if $(a - b) \in ImD$, and say that a is equivalent to b. We define the projection $\hat{\delta} : \Omega^1 \to \Omega^1_0$ by the rule

$$\hat{\delta}(dq_i^{(g|\nu)} f) = dq_i(-\partial)^\nu \hat{g}^{-1}(f). \tag{A1.1}$$

Obviously ('integrating by parts'),

$$\hat{\delta}\omega \sim \omega\,, \forall \omega \in \Omega^1. \tag{A1.2}$$

The Euler-Lagrange operator $\delta : C \to \Omega^1_0(C)$ is defined as

$$\delta = \hat{\delta} d : \delta(H) = \sum d\, q_i \frac{\delta H}{\delta q_i}, \tag{A1.3}$$

where, by (A1.1),

$$\frac{\delta H}{\delta q_i} = \sum (-\partial)^\nu \hat{g}^{-1} \left(\frac{\partial H}{\partial q_i^{(g|\nu)}} \right). \tag{A1.4}$$

Obviously,

$$\hat{\delta}(ImD) = 0,\ \delta(ImD) = 0,\ \frac{\delta}{\delta q_j}(ImD) = 0. \tag{A1.5}$$

Define the pairing $< Der\, C, \Omega^1(C) >$ by the formula

$$< \sum Y_i^{g|\nu} \frac{\partial}{\partial q_i^{(g|\nu)}}, \sum dq_j^{(h|\sigma)} \varphi_j^{h|\sigma} >= \sum Y_i^{g|\nu} \varphi_i^{g|\nu}.$$

It is easy to see that if $w \in \Omega^1_0(C)$ and $< Der\, C, w >\sim 0$ then $w = 0$. The first basic result in the calculus is

A1.6. Theorem. a) If $\omega \in \Omega^1_0$ and $\omega \sim 0$ then $\omega = 0$; b) If $\omega \in \Omega^1$ then $\omega \sim 0$ if and only if $< D^{ev}, \omega >\sim 0$; c) The projection $\hat{\delta} : \Omega^1 \to \Omega^1_0$ can be uniquely defined by the formula $< X, \hat{\delta}\omega > \sim < X, \omega >, \forall X \in D^{ev}$;

d) $Ker\,\widehat{\delta} = ImD$ in Ω^1; e) $Ker\delta = ImD + K$ in C.

For $X \in D^{ev}$, $H \in C$, we denote

$$(A1.7) \qquad \overline{X} = (X_i)\,,\, X_i := X(q_i),\ \frac{\delta H}{\delta \overline{q}} = \left(\frac{\delta H}{\delta q_i}\right),$$

the corresponding column vectors. Then, $\forall X \in D^{ev}$,

$$(A1.8) \quad X(H) = < X, dH > \sim < X, \delta H > = \sum X_i \frac{\delta H}{\delta q_i} = \overline{X}^t \frac{\delta H}{\delta \overline{q}},$$

and this relation uniquely defines the vector $\frac{\delta H}{\delta q}$ even when it is satisfied for only even X's.

Let $R = (R_r | R_r \in C)$ be a column vector. Its commutative Fréchet derivative $D(R)\ (= D_q(R))$ is a matrix operator with the matrix elements

$$(A1.9) \qquad [D(R)]_{ri} = D_i(R_r) = \sum \frac{\partial R_r}{\partial q_i^{(g|\nu)}} \widehat{g}\, \partial^\nu$$

Let $f \in C$ be a $\mathbf{Z}_2$–homogeneous element. Its even and odd Fréchet derivatives are the following row vectors of operators:

$$(A1.10) \quad [D^{\overline{0}}(f)]_i = D_i^{\overline{0}}(f) = \sum (-1)^{p(i)[p(f)+\overline{1}]} \frac{\partial f}{\partial q_i^{(g|\nu)}} \widehat{g}\, \partial^\nu,$$

$$(A1.11) \quad [D^{\overline{1}}(f)]_i = D_i^{\overline{1}}(f) = \sum (-1)^{p(f)[p(i)+\overline{1}]} \frac{\partial f}{\partial q_i^{(g|\nu)}} \widehat{g}\, \partial^\nu,$$

A1.12. Proposition. $X(f) = D^{p(X)}(f)(\overline{X}), X \in D^{ev}, f \in C$.

If $R = (R_r | R_r \in C)$ is a vector, then the even and odd Frécht derivatives of R are defined component-wise.

$$(A1.13) \qquad [D^\gamma(R)]_{ri} = [D^\gamma(R_r)]_i, \qquad \forall \gamma \in \mathbf{Z}_2.$$

Also, every $X \in D^{ev}$ acts on the vector R component-wise:

$$(A1.14) \qquad [X(R)]_r = X(R_r).$$

Let $R = (R_i | R_i \in C, i \in I)$ be a $\mathbf{Z}_2$-homogeneous vector. Its even and odd Fréchet derivatives are the following matrix operators:

(A1.15)

$$[D^0(R)]_{ij} = D_j^0(R_i) = \sum (-1)^{p(j)[p(R)+p(i)+1]} \frac{\partial R_i}{\partial q_j^{(g|\nu)}} \widehat{g}\, \partial^\nu,$$

(A1.16)
$$[D^1(R)]_{ij} = D^1_j(R_i) = \sum (-1)^{[p(j)+1][p(R)+p(i)]} \frac{\partial R_i}{\partial q_j^{(g|\nu)}} \widehat{g} \partial^\nu .$$

A1.17 Corollary. If $X \in D^{ev}$ and R are both $\mathbf{Z}_2$-homogeneous, then

(A1.18)
$$X(R) = D^{p(X)}(R)(\overline{X}).$$

A1.19. Definition. An operator $A : C^r \to C^s$ is a k-linear map of the form

(A1.20)
$$[A(v)]_a = \sum_b A_{ab}(v_b),$$

(A1.21)
$$A_{ab} = \sum A_{ab}^{g|\nu} \widehat{g}\, \partial^\nu, \text{ finite sum, } A_{ab}^{g|\nu} \in C,$$

where both C^r and C^s are considered consisting of column vectors (Although δ is not an operator in this sense, we shall continue to call it the Euler-Lagrange operator.)
A1.22. Definition. For a $\mathbf{Z}_2$-homogeneous operator $A : C \to C$, its adjoint is an operator $A^\dagger : C \to C$ satisfying

(A1.23)
$$[A^\dagger(u)]\, v \sim (-1)^{p(u)p(v)} [A(v)]\, u, \qquad \forall\, v, u \in C.$$

A1.24. Lemma. An adjoint operator exists and is unique.
A1.25. Definition. Let $A = (A_{ab})$ be an operator. Its commutative adjoint, $A^\dagger$, is defined as

(A1.26)
$$(A^\dagger)_{ba} = (A_{ab})^\dagger .$$

A1.27. Definition. For an operator $A : C^N \to C^N, N = |I|$, its super-adjoint, $A^{s\dagger}$, is defined as

(A1.28)
$$(A^{s\dagger})_{ji} = (-1)^{p(i)p(j)} (A_{ij})^\dagger .$$

A1.29. Proposition. Let $u, v \in C^N$ be even vectors. Then

(A1.30)
$$[A(u)]^t v \sim [A^{s\dagger}(v)]^t u .$$

Let $C_1 = K[u_j^{(g|\nu)}], j \in J = J_{\overline{0}} \cup J_{\overline{1}}, g \in G, \nu \in \mathbf{Z}_+^m$, be another commutative superalgebra. Let $\Phi : C \to C_1$ be an even homomorphism over K, commuting with the actions of G and ∂'s. Such a homomorphism is called a differential-difference homomorphism, or simply a homomorphism for brevity. We uniquely extend Φ to the map $\Phi : \Omega^1(C) \to \Omega^1(C_1)$ by requiring it to commute with d and the actions of G and ∂'s:

$$\Phi(dq_i^{(g|\nu)}) = d\,\Phi(q_i^{(g|\nu)}) = d\,\widehat{g}\,\partial^\nu\,\Phi(q_i) = \widehat{g}\,\partial^\nu\,d(\Phi_i) = \tag{A1.31}$$
$$= \widehat{g}\partial^\nu \sum du_j^{(h|\sigma)} \left(\frac{\partial \Phi_i}{\partial u_j^{(h|\sigma)}}\right),$$

where

$$\Phi_i = \Phi(q_i). \tag{A1.32}$$

In particular,
(A1.33)
$$\Phi(dq_i) = \sum du_j^{(h|\sigma)} \frac{\partial \Phi_i}{\partial u_j^{(h|\sigma)}} \sim \sum du_j\,\widehat{h}^{-1}(-\partial)^\sigma \left(\frac{\partial \Phi_i}{\partial u_j^{(h|\sigma)}}\right) =$$
$$= \sum du_j \frac{\delta \Phi_i}{\delta u_j} = \delta_1(\Phi_i)$$

where we use subscript '1' in δ to distinguish operations in C_1 from those in C.

Notice that since Φ commutes with the actions of G and ∂'s,

$$\Phi(ImD) \subset ImD, \quad \widehat{\delta}_1\Phi\widehat{\delta} = \widehat{\delta}_1\Phi. \tag{A1.34}$$

Denote by $\overline{\Phi}$ the following column vector:

$$(\overline{\Phi})_i = \Phi_i = \Phi(q_i). \tag{A1.35}$$

The natural properties of the calculus are given by
A1.36. Theorem. For any $H \in C$,

$$\frac{\delta\Phi(H)}{\delta\overline{u}} = D(\overline{\Phi})^{\dagger}\Phi\left(\frac{\delta H}{\delta\overline{q}}\right). \tag{A1.37}$$

Our last topic in this Appendix is to describe the image of the Euler-Lagrange operator $\delta : C \to \Omega_0^1(C)$. To do that we first construct a complex for δ.

Denote $\overline{C} = K[q_i^{(g|\overline{\nu})}]$, $i \in I$, $g \in G$, $\overline{\nu} \in \mathbf{Z}_+^{m+1}$, and let $\partial_{m+1} : \overline{C} \to \overline{C}$ be a new even left derivation which acts trivially on K, commutes with G and $\partial_1, ..., \partial_m$, and sends $q_i^{(g|\overline{\nu})}$ into $q_i^{(g|\overline{\nu}+1_{m+1})}$. Let us write $(g|\nu|s)$ instead of $(g|\overline{\nu})$ for $\overline{\nu} = \nu \oplus s, \nu \in \mathbf{Z}_+^m, s \in \mathbf{Z}_+$, where we identified $\mathbf{Z}_+^{m+1}$ with $\mathbf{Z}_+^m \oplus \mathbf{Z}_+$. The old notation $(g|\nu)$ will be used instead of $(g|\nu|0)$. This, in fact, means that we fix an injective homomorphism $C \hookrightarrow \overline{C}$ which we suppress from notation and hereafter consider C sitting inside $\overline{C}$.

Let $\overline{\tau} : \Omega^1(C) \to \overline{C}$ be an odd homomorphism of right C-modules given as

$$(A1.38) \qquad \overline{\tau}(dq_i^{(g|\nu)} f) = q_i^{(g|\nu|1)} f, \; f \in C.$$

Since $\overline{\tau}$ obviously commutes with the actions of G and $\partial_1, ..., \partial_m$, we see that

$$(A1.39) \qquad \overline{\tau}(ImD) \subset ImD.$$

A1.40. Lemma. $\partial_{m+1}(H) = \overline{\tau}d(H), \qquad \forall H \in C.$

Proof. We have

(A1.41)

$$\overline{\tau}d(H) = \overline{\tau}[\sum dq_i^{(g|\nu)} \frac{\partial H}{\partial q_i^{(g|\nu)}}] = \sum q_i^{(g|\nu|1)} \frac{\partial H}{\partial q_i^{(g|\nu)}} = \partial_{m+1}(H) \blacksquare$$

Let superscript '1' in the operators δ^1 and $\hat{\delta}^1$ refer to the algebra $\overline{C}$, so that $\delta^1 = \hat{\delta}^1 d$.

A1.42. Theorem. The sequence

$$(A1.43) \qquad C \xrightarrow{\delta} \Omega_0^1(C) \xrightarrow{\delta^1 \overline{\tau}} \Omega_0^1(\overline{C})$$

is a complex.

Proof. For any $H \in C, \delta(H) = \hat{\delta}(dH) \sim d(H)$ by (A1.2). Hence, $\overline{\tau}\delta(H) \sim \overline{\tau}(dH) = \partial_{m+1}(H)$ by (A1.39) and Lemma A1.40. Therefore,

$\delta^1 \overline{\tau} \delta(H) = \delta^1 \partial_{m+1}(H) = 0$ by (A1.5). $\blacksquare$

We now transform the equality $0 = \delta^1 \overline{\tau} \delta$ into a more convenient form.

A1.44. Definition. An operator $A : C^N \to C^N, N = |I|$, is called syper-symmetric (resp. super-skew-symmetric) if $A^{s\dagger} = A$ (resp. $A^{s\dagger} = -A$.)

A1.45. Lemma. Let us identify $\Omega_0^1(C)$ with $C^N, N = |I| : \sum dq_i R_i \leftrightarrow \overline{R} = (R_i)$. Then the equality $\delta^1 \overline{\tau}(\overline{R}) = 0$, for a $\mathbf{Z}_2$-homogeneous $\overline{R}$, is equivalent to $D(\overline{R})$ being super-symmetric:

$$(A1.46) \qquad D(\overline{R})^{s\dagger} = D(\overline{R}).$$

A1.47. Corollary For any $H \in C$, the commutative Fréchet derivative $D\left(\frac{\delta H}{\delta \overline{q}}\right)$ is super-symmetric:

$$D\left(\frac{\delta H}{\delta \overline{q}}\right)^{s\dagger} = D\left(\frac{\delta H}{\delta \overline{q}}\right). \tag{A1.48}$$

We can now describe a solution to "the inverse problem of the calculus of variations": to describe the Image of δ.

A1.49. Theorem. Let $\overline{R} \in C^N, N = |I|$, be a finite vector (i.e., with only a finite number of non-zero components) for which

$$D(\overline{R})^{s\dagger} = D(\overline{R}). \tag{A1.50}$$

Then there exists $H \in C$ such that $\overline{R} = \frac{\delta H}{\delta \overline{q}}$. In fact,

$$H = \sum_i q_i \int_0^1 A_t(R_i)dt, \tag{A1.51}$$

where $A_t : C \to C[t]$ is an additive map given by the formula $A_t(f) = t^{deg\, f} f$, where "deg" is the usual degree of a polynomial.

Appendix 2. Super-Hamiltonian formalism, and affine super-Hamiltonian structures.

Let k, K, and $C = K[q_i^{(g|\nu)}]$ be as in Appendix 1.

A2.1. Definition. An even k-linear map $\Gamma : C \to D^{ev}(C)$ is called super- Hamiltonian if the following conditions are satisfied:

$$\{H, F\} \sim -(-1)^{p(H)p(F)}\{F, H\}, \qquad \forall H, F \in C, \tag{A2.1i}$$

where $\{H, F\} = X_H(F)$ is called the Poisson bracket, and $X_H = \Gamma(H)$;

$$X_{\{H,F\}} = [X_H, X_F], \qquad \forall H, F \in C, \tag{A2.1ii}$$

where the commutator on the right is understood in the Lie superalgebra sense: $[a, b] = ab - (-1)^{p(a)p(b)}ba$;

(A2.1iii) There exist two operators $B^0, B^1 : C^N \to C^N, N = |I|$, such that

$$\overline{X}_H = B^{p(H)}\left(\frac{\delta H}{\delta \overline{q}}\right), \tag{A2.2}$$

where $\overline{X}_H = X_H(\overline{q})$ by (A1.7).

(A2.1i^4) The properties (A2.1i–iii) remain true for any (differential-difference) extension $K' \supset K$ over k, i.e., for any commutative superalgebra extension K' on which the action of G and ∂'s is compatible with their action on K.

A2.3. Remark. As usual in working with superobjects, one gives definitions and proves formulae for $\mathbf{Z}_2$-homogeneous elements only, and extends everything by additivity to all elements (see [2, 22]).

A2.4. Remark. The reader with roots in classical mechanics may wonder what has happened with two expected requirements on the Poisson bracket: a derivation property

(A2.5)
$$\{H, FR\} \sim \{H, F\}R + \{H, R\}F(-1)^{p(F)p(R)}, \ \forall H, F, R \in C;$$

and the graded Jacobi identity

(A2.6)
$$\{H, \{F, R\}\} \sim \{\{H, F\}R\} + (-1)^{p(H)p(F)}\{F, \{H, R\}\}, \forall F, R, H \in C.$$

The property (A2.5) is meaningless outside the area of classical mechanics (i.e., the case when G and ∂'s are absent) even (e.g., in models of classical field theory) when B^0 and B^1 are homomorphisms of right C-modules, since ImD is not a C-submodule in C. With respect to the graded Jacobi identity (A2.6), it is interchangeable with the basic Hamiltonian property (A2.1ii) but only when the stability condition (A2.1i^4) is invoked. At the moment let us notice that (A2.6) follows directly from (A2.1ii) by applying both parts of the equality (A2.1ii) to R.

A2.7. Remark. The only really new feature of the Hamiltonian formalism in the presence of Grassmann-type variables is the property (A2.1iii): one has now two defining matrices instead of just one in the purely commutative case.

A2.8. Lemma. Denote $B = B^0$. Then

$$(A2.9) \qquad B^{s\dagger} = -B\,, \ B^1_{ij} = (-1)^{p(i)} B_{ij}\,.$$

It follows that one can work with only one matrix, namely $B = B^0$, instead of both B^0 and B^1. If we write B in the block form as $B = \begin{pmatrix} \alpha & \beta \\ \gamma & \rho \end{pmatrix}$ where α is a N_0 x N_0-, β is a N_0 x N_1-, γ is a N_1xN_0-, and ρ is a N_1 x N_1- matrix, $N_0 = |I_0|, N_1 = |I_1|$, then (A2.9) means that $\alpha^\dagger = -\alpha$, $\rho^\dagger = \rho$, $\beta^\dagger = -\gamma$, with "$\dagger$" being the comutative adjoing defined by (A1.26).

A2.10. Corollary.

$$(A2.11)\qquad (\overline{X}_H)_i = (-1)^{p(i)p(H)} \sum_j B_{ij}\left(\frac{\delta H}{\delta q_j}\right),$$

$$(A2.12)\qquad \{H,F\} \sim \sum (-1)^{p(i)p(H)} B_{ij}\left(\frac{\delta H}{\delta q_j}\right)\cdot \frac{\delta F}{\delta q_i}.$$

From now on we assume that B is even super-skew-symmetric. Our main goal is to find necessary and sufficient conditions for a given super-skew-symmetric matrix B to be super-Hamiltonian, i.e., to define a super-Hamil tonian structure which means, in turn, that the formula (A2.1ii) is stably satisfied. We shall achieve this by subsequently transforming the nonoperator equality (A2.1ii) into an operator one. The reasoning goes as follows.

A2.13. Lemma. The equality (A2.1ii) is stably satisfied for all H, F if it is stably satisfied for all even H, F.

A2.14. Lemma. For even H and F, the relation (A2.1ii) is equivalent to

$$(A2.15)\quad B\frac{\delta}{\delta \overline{q}}\left\{\left[B\left(\frac{\delta H}{\delta \overline{q}}\right)\right]^t \frac{\delta F}{\delta \overline{q}}\right\} = D^0\left[B\left(\frac{\delta F}{\delta q}\right)\right] B\left(\frac{\delta H}{\delta \overline{q}}\right) - D^0\left[B\left(\frac{\delta H}{\delta \overline{q}}\right)\right] B\left(\frac{\delta F}{\delta \overline{q}}\right).$$

A2.16. Definition. For a column vector $R \in C^N$, we denote by R^{st} its super-transpose which is a row vector with components

$$(A2.17)\qquad (R^{st})_i = (-1)^{p(i)} R_i,$$

reserving the notation R^t for the usual transpose.

A2.18. Proposition. If R and S are even vectors in C^N then

$$(A2.19)\qquad R^t S = S^{st} R.$$

A2.20. Lemma. If $X \in D^{ev}, S \in C^N$, and $H \in C$ are all even then

$$(A2.21)\qquad \left[X\left(\frac{\delta H}{\delta \overline{q}}\right)\right]^{st} S \sim \overline{X}^t\left[D^0\left(\frac{\delta H}{\delta \overline{q}}\right)(S)\right].$$

A2.22. Definition. For an even evolution field $X \in D^{ev}$ and an operator $A : C \to C$, the action of X on A is defined as $X(\sum A^{g|\nu}\widehat{g}\,\partial^\nu) =$

$\sum X(A^{g|\nu})\,\hat{g}\,\partial^\nu$. If A is a matrix operator then X acts on A matrix elements-wise.

A2.23. Proposition. If $X \in D^{ev}$ is even, $B : C^N \to C^N$ is an even operator, and $R \in C^N$ is an even vector then

$$(A2.24)\qquad X(B)(R) = XB(R) - BX(R) =: [X,B](R) = \\ = ([D^0,B](R))(\overline{X}) := [D^0(BR) - BD^0(R)](\overline{X}).$$

A2.25. Corollary. With respect to each of the vectors R and $\overline{X}$, the expression $([D^0,B](R)(\overline{X})$ is an operator.

The main technical result of the super-Hamiltonian formalism (which is a generalization of Lemma 1.4 in [28]) is this:

A2.26. Theorem. For an even matrix $B : C^N \to C^N$ and even vectors $R, S \in C^N$, denote by $< B,R,S >$ a column vector defined as

$$(A2.27)\qquad < B,R,S >_i = (-1)^{p(i)} \left(([D^0,B](R))^{st}(S)\right)_i .$$

If B is even super-skew-symmetric then for any even H, $F \in C$,

$$(A2.28)\quad \frac{\delta}{\delta\overline{q}}\left[\left(B\left(\frac{\delta H}{\delta\overline{q}}\right)\right)^t \frac{\delta F}{\delta\overline{q}}\right] = D^0\left(\frac{\delta F}{\delta\overline{q}}\right) B\left(\frac{\delta H}{\delta\overline{q}}\right) - \\ - D^0\left(\frac{\delta H}{\delta\overline{q}}\right) B\left(\frac{\delta F}{\delta\overline{q}}\right) + < B, \frac{\delta H}{\delta\overline{q}}, \frac{\delta F}{\delta\overline{q}} > .$$

A2.29. Lemma. For even H and F, the relation (A2.1*ii*) is equivalent to

$$(A2.30)\qquad B < B, \frac{\delta H}{\delta\overline{q}}, \frac{\delta F}{\delta\overline{q}} > = \left([D^0,B]\left(\frac{\delta F}{\delta\overline{q}}\right)\right) B\left(\frac{\delta H}{\delta\overline{q}}\right) - \\ - \left([D^0,B]\left(\frac{\delta H}{\delta\overline{q}}\right)\right) B\left(\frac{\delta F}{\delta\overline{q}}\right) .$$

Now we can derive the main result of the super-Hamiltonian formalism:

(A2.31) Theorem. An even super-skew-symmetric matrix B is super-Hamiltonian if (A2.1*ii*) is stably satisfied for arbitrary even linear functions, i.e., for any even H and F of the form $H = \sum q_i X_i, F = \sum q_j Y_j$, with X_i's and Y_j's taken from arbitrary extension $K' \supset K$.

Proof. If $H = \sum q_i X_i$ and $F = \sum q_j Y_j$ are even and linear then the vectors $\frac{\delta H}{\delta\overline{q}} = X = (X_i)$ and $\frac{\delta F}{\delta\overline{q}} = Y = (Y_i)$ are even, and

(A2.30) becomes

(A2.32)

$$B < B,X,Y >= ([D^0,B](Y))\,B(X) - ([D^0,B](X))B(Y),$$

which we assume is satisfied for any even $X,Y \in K'^N$. Now, by (A2.27) and Corollary A2.25, each side of (A2.32) is a bilinear operator acting on components of X and Y. Fixing X (or Y) we obtain an equality involving two operators (on each side) acting on arbitrary $Y \in K'^N$ (or $X \in K'^N$). By Lemma A2.19, this implies that we have in fact an operator identity and, thus, (A2.32) is valid for arbitrary even $X,Y \in C'^N, C' = K'[q_i^{(g|\nu)}]$. In particular, (A2.32) is valid for $X = \frac{\delta H}{\delta \overline{q}}, Y = \frac{\delta F}{\delta \overline{q}}$ with arbitrary even $H,F \in C'$. Hence, (A2.30) is satisfied. ∎

A2.33. Corollary. For a given even super-skew-symmetric matrix B, to check the super-Hamiltonian property of B it is necessary and sufficient to check the following identity

$$(A2.34) \qquad B\frac{\delta}{\delta \overline{q}}[B(X)^t Y] = D^0(BY)B(X) - D^0(BX)B(Y)$$

in the superalgebra $K'[q_i^{(g|\nu)}]$, where $K' = K[X_i^{(g|\nu)}, Y_i^{(g|\nu)}], i \in I, p(X_i) == p(Y_i) = p(i)$.

We can now describe a large class of super-Hamiltonian structures.

A2.35. Definition. We say that an operator A is with coefficients in a (commutative superalgebra) $K' \supset K$ if all matrix elements of A are of the form $\{\sum \varphi \widehat{g}\,\partial^\nu | \varphi \in K'\}$.

A2.36. Theorem. If B is an even super-skew-symmetric matrix with coefficients in K then B is super-Hamiltonian.

A2.37. Remark. If $\mathcal{G}$ is a finite-dimensional Lie algebra over a field then the ring of polynomial functions on the dual space $\mathcal{G}^*$ to $\mathcal{G}$ possesses a natural Hamiltonian structure whose associated Poisson bracket can be defined very simply by: a) being a derivation with respect to each argument; and b) coinciding with the commutator in $\mathcal{G}$ for linear functions on $\mathcal{G}^*$ considered as elements of $\mathcal{G}$. Theorem A2.31 may be thought of as a nonlinear infinite-dimensional generalization of this definition.

We now consider the problem of canonical maps. Let $C_1 = K[u_j^{(g|\nu)}], j \in J = J_0 \cup J_1$, be another commutative superalgebra. Suppose matrices B and B_1 define super-Hamiltonian structures in C and C_1 respectively. Let $\Phi : C \to C_1$ be a homomorphism (see A1).

A2.38. Definition. Φ is (stably) canonical if evolution fields X_H (in C') and $X_{\Phi(H)}$ (in C'_1) are compatible for any $H \in C' = K'[q_i^{(g|\nu)}]$, for any $K' \supset K$:

$$(A2.39) \qquad \Phi X_H = X_{\Phi(H)}\Phi .$$

A2.40. Theorem. A homomorphism Φ is canonical if and only if

$$(A2.41) \qquad \Phi(B) = D^0(\overline{\Phi})B_1 D(\overline{\Phi})^{\dagger},$$

where $\overline{\Phi} = (\Phi_i = \Phi(q_i))$ and Φ acts on operators matrix elements-wise by the rule : $\Phi(\sum f\hat{g}\partial^{\nu}) = \sum \Phi(f)\hat{g}\partial^{\nu}$.
A2.42. Remark. Applying (A2.39) to arbitrary $F \in C$, we can rewrite (A2.39) in an equivalent form

$$(A2.43) \qquad \Phi(\{H,F\}) = \{\Phi(H),\Phi(F)\}, \qquad \forall H, F \in C.$$

We now turn to the last topic of this appendix: affine super-Hamiltonian operators and associated Lie superalgebras.
A2.44. Definition. A stable Lie superalgebra is a free $\mathbf{Z}_2$-graded K-module $\mathcal{G} = K^N = K^{N_0} \oplus K^{N_q}, N = N_0 + N_1$, together with an even multiplication $[\ ,\ ]$ and the grading $p : \mathcal{G} \to \mathbf{Z}_2$ defined by $p(X) = p(X_i) + p(i)$ where $p(i) = \overline{0}$ for $i \leq N_0$ and $p(i) = \overline{1}$ for $i > N_0$, satisfying the following properties:

$$(A2.44i) \qquad [X,Y] = -(-1)^{p(X)p(Y)}[Y,X], \qquad \forall X,Y \in \mathcal{G};$$

(A2.44ii)
$$[[X,Y],Z] = [X,[Y,Z]] - (-1)^{p(X)p(Y)}[Y,[X,Z]], \forall X,Y,Z \in \mathcal{G};$$

(A2.44iii) Let $K_c = \{\rho \in K|\hat{g}(\rho) = \rho, \forall g \in G; \partial_s(\rho) = 0, s = 1,\ldots,m\}$ be the subring of constants in K. Then $[X,Y\rho] = [X,Y]\rho$, $\forall X,Y \in \mathcal{G}$, $\forall \rho \in K_c$;
(A2.44i^4) Multiplication in $\mathcal{G}$ is an operator with respect to each argument, of the following form:
(A2.45)
$$[X,Y]_{\mathrm{k}} = \sum(-1)^{p(i)p(X)} c^{\mathrm{k}}_{i,h|\nu\,;j,g|\sigma}\, \hat{g}\,\partial^{\sigma}(X_j)\cdot \hat{h}\,\partial^{\nu}(Y_i), c^{\mathrm{k}}_{\ldots} \in K,$$
$$\forall X,Y \in \mathcal{G}.$$
In particular, the sum in (A2.45) is finite for each k, $1 \leq \mathrm{k} \leq N$, even if N is infinite. (In the case $N = \infty$, elements of K'^N

are finite vectors, i.e., vectors with only finite number on nonzero coordinates.);

($A2.44i^5$) The properties ($A2.44i - i^4$) remain true under arbitrary extension $K' \supset K$ which makes $\mathcal{G}$ into $\mathcal{G}' = K'^N$:

A2.46. Proposition. If formula (A2.45) is satisfied for all even $X, Y \in \mathcal{G}$ and the properties ($A2.44i, i^3$) hold stably, then (A2.45) is satisfied for all $X, Y \in \mathcal{G}$.

Thus, the multiplication in a stable Lie superalgebra can be reconstructed from the multiplication of only even elements in this superalgebra. (In 'coordinates', this is evident from (A2.45): the structure constants $c^{\mathrm{k}}_{\dots}$ are defined by the products of even elements only.) In a sense, then, the study of the stable Lie superalgebras is equivalent to the study of Lie algebras over commutative superalgebras instead of over commutative algebras.

Our plan now is this: to each stable superalgebra we associate a linear (in q's) super-Hamiltonian matrix, and vice versa; and then we show that affine Hamiltonian matrices are in one-to-one correspondence with generalized 2-cocycles on stable Lie superalgebras.

A2.47. Definition. An operator $A : C^a \to C^b$ is linear if each of its matrix elements is linear which means having the form $\sum \varphi \widehat{g}\, \partial^\nu$ with $\varphi = \sum c_i^{h|\sigma} q_i^{(h|\sigma)}, c_i^{h|\sigma} \in K$.

Thus, being linear is stable property.

A2.48. Definition. For each stable Lie superalgebra $\mathcal{G}$, let $B^\alpha = B^\alpha(\mathcal{G}), \alpha \in\in \mathbf{Z}_2$, be linear operators defined by the formula

(A2.49)

$$[B^{p(X)}(X)]^t Y \sim \vec{q}^t [X,Y] := \sum q_{\mathrm{k}} [X,Y]_{\mathrm{k}}, \quad \forall X, Y \in \mathcal{G}'.$$

A2.50. Theorem. The matrix $B = B^0$ is correctly defined and is super-Hamiltonian.

The converse to Theorem A2.50 is also true.

A2.51. Theorem. Suppose a linear matrix B defines a super–Hamiltonian structure in C. Define a multiplication [,] in K'^N by the formula (A2.49). Then this multiplication is correctly defined and it makes $\mathcal{G} = K'^N$ into a stable Lie superalgebra.

A2.52. Remark. Since the map between linear super-Hamiltonian structures and stable Lie superalgebras is given, in each direction, by the same formula (A2.49), we have, in fact, a one-to-one correspondence.

Now we turn to affine super-Hamiltonian operators.

A2.53. Definition. A bilinear form on K^N is a stable map K'^N x $K'^N \to K'$ which is an operator with coefficients in K with respect to each argument, and is a right K'_c-homomorphism with respect to the second argument, $\forall K' \supset K$.

A2.54. Definition. Two bilinear forms ω_1 and ω_2 are equivalent if $\omega_1(X,Y) \sim \omega_2(X,Y), \forall X,Y \in K'^N$

A2.55. Corollary. Every bilinear form is equivalent to a form of the type

$$(A2.56) \qquad \omega(X,Y) \sim \sum b^{p(X)}_{i;j,g|\sigma}\, \hat{g}\, \partial^\sigma (X_j) \cdot Y_i,\ b^{p(X)}_{\dots} \in K,$$

and such representation is unique.

A2.57. Definition. A bilinear form ω is called super-skew-symmetric if it satisfies the relation

$$(A2.58) \qquad \omega(X,Y) \sim -(-1)^{p(X)p(Y)}\omega(Y,X),\ \forall X,\ Y \in K'^N.$$

A2.59. Definition. An operator b with matrix elements $b_{ij} = \sum b^0_{i;j,g|\sigma}\hat{g}\partial^\sigma$ defined by (A2.56) is called associated with (or corresponding to) the bilinear form ω.

A2.60. Lemma. If ω is a bilinear super-skew-symmetric form and b is the associated operator then $p(b) = p(\omega)$,

$$(A2.61) \qquad \omega(X,Y) \sim \sum (-1)^{p(i)p(X)} b_{ij}(X_j) \cdot Y_i\,,$$

and b is super-skew-symmetric.

A2.62. Lemma. Let b be super-skew-symmetric. Then the bilinear form ω defined by (2.61) is super-skew-symmetric and $p(\omega) = p(b)$.

A2.63. Definition. A (generalized) two-cocycle on a stable Lie superalgebra $\mathcal{G} = K'^N$ is an even super-skew-symmetric form ω satisfying

$$(A2.64) \quad \omega([X,Y],Z) \sim \omega(X,[Y,Z]) - (-1)^{p(X)p(Y)}\omega(Y,[X,Z]),$$

$$\forall X,Y,Z \in \mathcal{G}.$$

A2.65. Theorem. Let ω be a two-cocycle on a stable Lie superalgebra $\mathcal{G}$, and let $b = b(\omega)$ be the operator corresponding to ω. Then the matrix

$$(A2.66) \qquad \tilde{B} = B(\mathcal{G}) + b(\omega)$$

is super-Hamiltonian.

A2.67. Theorem. Suppose $\tilde{B} = B + b$ is a super-Hamiltonian matrix, where B is an even linear super-skew-symmetric matrix and b

is an even super-skew-symetric matrix with coefficients in K. Define a multiplication $[\ ,\ ]$ in K'^N by the formula (A2.49). Then this multiplication is correctly defined, makes $\mathcal{G} = K'^N$ into a stable Lie superalgebra, and the bilinear form ω on $\mathcal{G}$ defined by the formula (A2.61) is a 2–cocycle on $\mathcal{G}$.

A2.68. Remark. Since the map between affine super-Hamiltonian operators and two-cocycles on stable Lie superalgebras is given, in each direction, by the same formulae (A2.49) and (A2.61), this map is, in fact, a one-to-one correspondence.

Appendix 3. Canonical quadratic maps associated with representations of stable differential-difference Lie superalgebras

We continue to use the notation of the previous two appendices.

Let $\mathcal{O}_K^+$ denote the set of scalar operators with coefficients in K (see A1.19), and $Mat_M(\mathcal{O}_K^+)$ denote the set of MxM matrices whose matrix elements belong to $\mathcal{O}_K^+$.

Let $V = K^M, M = M_{\bar{0}} + M_{\bar{1}}$, be a $\mathbf{Z}_2$-graded K-module, and $\mathcal{G} \approx K^N$ be a stable Lie superalgebra (see A2.44).

A3.1. Definition. An even k-linear map $\rho : \mathcal{G} \to Mat_M(\mathcal{O}_K^+)$ is called a representation of $\mathcal{G}$ on V if

$$(A3.1i) \qquad \rho([X,Y]) = [\rho(X), \rho(Y)], \qquad \forall X, Y \in \mathcal{G};$$

$$(A3.1ii) \qquad \rho(X\theta) = \rho(X)\hat{\theta}, \qquad \forall X \in \mathcal{G}, \qquad \forall \theta \in K_c,$$

where $\hat{\theta}$ is a diagonal matrix $:(\hat{\theta})_{\alpha\beta} = \delta_\beta^\alpha (-1)^{p(\alpha)p(\theta)}\theta;$

(A3.1i^3) The properties (A3.1$i - ii$) remain stable under any extension $K' \supset K$.

Recall that $X\theta$ is a column-vector with components $(X_i\theta)$.

A3.2. Proposition. For any $v \in V$,

$$(A3.3) \qquad \hat{\theta}v = (-1)^{p(v)p(\theta)}v\theta.$$

A3.4. Propositon. "Adjoint representation" is a representation.

Proof. Let $\rho : \mathcal{G} \mapsto Mat_N(\mathcal{O}_K^+)$ be the "adjoiont representation" map, $\rho(X)(Y) = [X,Y]$. We have,

$\rho(X\theta)(Y) = [X\theta, Y]$ [by (A2.44i)] $= -(-1)^{p(X\theta)p(Y)}[Y, X\theta]$ [by (A2.44i^3)] $=$

$$= -(-1)^{p(X)p(Y)+p(\theta)p(Y)}[Y,X]\theta \quad [\text{by (A2.44}i)] =$$

$$= (-1)^{p(\theta)p(Y)}[X,Y]\theta$$

[by (A2.44i^3)] $=$ $(-1)^{p(\theta)p(Y)}[X,Y\theta]$ [by (A3.3)] $=$
$= [X,\hat{\theta}Y] = \rho(X)\hat{\theta}(Y)$ ∎

A3.5. Definition. If V is a stable Lie superalgebra, then $\rho : \mathcal{G} \to Mat_M(\mathrm{O}_K^+)$ is called a representation of $\mathcal{G}$ on V if, in addition to (A3.1$i - i^3$), also

$$(A3.6) \qquad \rho(X)([v,w]) = [\rho(X)(v),w] +$$

$$+(-1)^{p(v)p(X)}[v,\rho(X)(w)], \forall X \in \mathcal{G}, \forall v,w \in V,$$

is satisfied, and stably at that.

A3.7. Definition. A semidirect product of $\mathcal{G} \propto V$ (with respect to a representation ρ of $\mathcal{G}$ on a stable Lie superalgebra V) is a $\mathbf{Z}_2$-graded K-module
K^{N+M}, $(N+M)_{\bar{0}} = N_{\bar{0}} + M_{\bar{0}}$, $(N+M)_{\bar{1}} = N_{\bar{1}} + M_{\bar{1}}$, with the multiplication

$$(A3.8) \qquad [\ell_1 \propto v_1, \ell_2 \propto v_2] = [\ell_1,\ell_2] \propto (\rho(\ell_1)(v_2) -$$

$$-(-1)^{p(\ell_2)p(v_1)}\rho(\ell_2)(v_1) + [v_1,v_2])$$

A3.9. Lemma $\mathcal{G} \propto V$ is a stable Lie superalgebra.

A3.10. Remark. The property (A3.1ii) guarentees that (A2.44i^3) is satisfied for $\mathcal{G} \propto V$.

In what follows we assume that V is abelian.

Writing ρ in coordinates, we have

$$(A3.11) \quad \rho(X)_{\alpha\beta} = \Sigma(-1)^{p(\beta)p(\mathrm{x})}\rho_{\alpha\beta}^{\mathbf{k};g,\nu|h,\sigma}X_{\mathbf{k}}^{h|\sigma}\hat{g}\partial^{\nu}, \rho_{\alpha\beta}^{\mathbf{k},\dots} \in K,$$

$$(A3.12) \qquad [\rho(X)(v)]_\alpha = \sum(-1)^{p(\beta)p(X)}\rho_{\alpha\beta}^{\mathbf{k};g,\nu|h,\sigma}X_{\mathbf{k}}^{h|\sigma}v_\beta^{g|\nu}.$$

Let $C = \mathrm{K}[q_i^{(g|\nu)}, c_\alpha^{(g|\nu)}], 1 \le i \le N, 1 \le \alpha \le M$, be the "ring of functions on L^*", $L := \mathcal{G} \propto V$, and let $B = B^0$ be the super-Hamiltonian matrix associated to L [see (A2.49)]:

$$(A3.13) \quad \left[B\begin{pmatrix}X\\u\end{pmatrix}\right]^t \begin{pmatrix}Y\\v\end{pmatrix} \sim \sum q_i[X,Y]_i + \sum c_\alpha(X\cdot v - Y\cdot u)_\alpha,$$

where: $X, Y \in K^N, u, v \in K^M; X \cdot v$ stands for $\rho(X)(v); p(X_i) = (Y_i) = p(i), p(u_\alpha) = p(v_\alpha) = p(\alpha); p(i) = \overline{0}$ for $i \le N_0, p(\alpha) = \overline{0}$ for $\alpha \le M_0; p(q_i) = p(i), p(c_\alpha) = p(\alpha)$.

Let $C_1 = K[a_\alpha^{(g|\nu)}, b_\alpha^{(g|\nu)}], 1 \le \alpha \le M, p(a_\alpha) = p(b_\alpha) = p(\alpha)$. Let $B_1 = B_1^0$ be a super-Hamiltonian matrix in C_1,

$$(A3.14) \qquad B_1 = \begin{matrix} & a_\alpha & b_\alpha \\ \begin{matrix} a_\beta \\ b_\beta \end{matrix} & \left(\begin{matrix} 0 \\ -\delta_\alpha^\beta \end{matrix}\right. & \left.\begin{matrix} (-1)^{p(\alpha)}\delta_\beta^\alpha \\ 0 \end{matrix}\right) \end{matrix}$$

so that $B = \begin{pmatrix} 0 & 1 \\ -1 & 0 \end{pmatrix}$ on even variables, and $B = \begin{pmatrix} 0 & -1 \\ -1 & 0 \end{pmatrix}$ on odd variables.

A3.15. Definition. For a column-vector $v \in K^M$, denote by $\tilde{v}$ another column vector: $(\tilde{v})_\alpha = v_\alpha(-1)^{p(\alpha)}$

A3.15. Definition. The elements $\varphi_i \in C_1, 1 \le i \le N$, are defined as

$$\varphi_i = -\sum \hat{h}^{-1}(-\partial)^\sigma \left(a_\gamma^{(g|\nu)} b_\alpha \rho_{\alpha\gamma}^{i;g,\nu|h,\sigma}\right).$$

A.3.18. Corollary.

$$(A3.19) \qquad \sum \varphi_i X_i \sim -\sum b_\alpha (X \cdot \tilde{a})_\alpha, \quad p(X) = 0,$$

where a is a column-vector with components $a_\alpha \in C_1$.

Proof. We have, by (A3.12),

$$-\sum b_\alpha (X \cdot \tilde{a})_\alpha = -\sum b_\alpha \rho_{\alpha\beta}^{i;g,\nu|h,\sigma} X_i^{h|\sigma)} a_\beta^{(g|\nu)} (-1)^{p(\beta)} =$$

$$= -\sum a_\beta^{(g|\nu)} b_\alpha \rho_\alpha^{i;g,\nu|h,\sigma} X_i^{h|\sigma} \sim -$$

$$\sim \sum \hat{h}^{-1}(-\partial)^\sigma (a_\beta^{(g|\nu)} b_\alpha \rho_\alpha^{i;g,\nu|h,\sigma}) \cdot X_i =$$

$= \sum \varphi_i X_i$. ∎

Denote $\dfrac{D\varphi_i}{Da_\gamma} = \sum \dfrac{\partial \varphi_i}{\partial a_\gamma^{(g|\nu)}} \hat{g} \partial^\nu$, etc., the commutative Fréchet derivatives.

A3.20. Lemma.

$$(A3.21) \qquad \frac{D\varphi_i}{Da_\gamma} = -\sum \hat{h}^{-1}(-\partial)^\sigma b_\alpha \rho_{\alpha\gamma}^{i;g,\nu|h,\sigma} \hat{g} \partial^\nu;$$

$$(A3.22) \quad (Y\cdot\tilde{a})_\alpha = -\sum\left(\frac{D\varphi_i}{Db_\alpha}\right)^\dagger (Y_i), \qquad \forall Y\in\mathcal{G}, p(Y)=0;$$

$(A3.23)$

$$\sum b_\alpha(X\cdot v)_\alpha \sim -\sum\left(\frac{D\varphi_i}{Da_\gamma}\right)^\dagger (X_i)\cdot v_\gamma, \forall X\in\mathcal{G}, \forall v\in V, p(X) =$$

$$= p(v) = 0.$$

Proof. The first equality (A3.21) follows directly from the definition (A3.17). Now,

$$(Y\cdot\tilde{a})_\alpha [\text{by (A3.12) and A3.15}] = \sum \rho_{\alpha\gamma}^{i;g,\nu|h,\sigma} Y_i^{(h|\sigma)} a_\gamma^{(g|\nu)}(-1)^{p(\gamma)} =$$

$$= \sum (-1)^{p(\gamma)[p(\alpha)+p(\gamma)]} a_\gamma^{(g|\nu)} \rho_{\alpha\gamma}^{i;g,\nu|h,\sigma}\hat{h}\partial^\sigma(Y_i)(-1)^{p(\gamma)} =$$

$$= \sum\left[\hat{h}^{-1}(-\partial^\sigma(-1)^{p(\alpha)p(\gamma)} a_\gamma^{(g|\nu)}\rho_{\alpha\gamma}^{i;g,\nu|h,\sigma}\right]^\dagger (Y_i) \ [\text{from (A3.17)}] =$$

$$= -\sum\left(\frac{D\varphi_i}{Db_\alpha}\right)^\dagger (Y_i),$$

which is (A3.22). Finally,

$$\sum b_\alpha (X\cdot v)_\alpha = \sum b_\alpha \rho_{\alpha\gamma}^{i;g,\nu|h,\sigma} X_i^{(h|\sigma)} v_\gamma^{(g|\nu)} \sim$$

$$\sim \sum\left[\hat{g}^{-1}(-\partial)^\nu b_\alpha \rho_{\alpha\gamma}^{i;g,\nu|h,\sigma}\hat{h}\partial^\sigma(X_i)\right]\cdot v_\gamma \ [\text{by (A3.21)}] =$$

$$= -\sum\left(\frac{D\varphi_i}{Da_\gamma}\right)^\dagger (X_i)\cdot v_\gamma,$$

which is (A3.23). ∎

Now we can state the main result of this Appendix.

A3.23′. Theorem. Let $\Phi : C \to C_1$ be a differential-difference homomorphism defined on generators as

$$(A3.24) \qquad \Phi(q_i) = -\varphi_i, \Phi(c_\alpha) = b_\alpha.$$

Then Φ is canonical between the super-Hamiltonian structures B (in C) and B_1 (in C_1).

Proof. Using Theorem A2.40, we have

$$D^0(\overline{\Phi}) = \begin{matrix} \\ \varphi_{\mathbf{k}} \\ \Phi(c_\gamma)\end{matrix} \begin{pmatrix} a_\beta & b_\beta \\ (-1)^{p(\beta)[p(\mathbf{k})+1]}\frac{D\varphi_{\mathbf{k}}}{Da_\beta} & (-1)^{p(\beta)[p(\mathbf{k})+1]}\frac{D\varphi_{\mathbf{k}}}{Db_\beta} \\ 0 & \delta_\gamma^\beta \end{pmatrix},$$

$$D(\overline{\Phi})^\dagger = \begin{array}{c} \\ a_\alpha \\ b_\alpha \end{array} \begin{array}{c|c} \varphi_i & \Phi(c_\beta) \\ \hline \left(\frac{D\varphi_i}{Da_\alpha}\right)^\dagger & 0 \\ \left(\frac{D\varphi_i}{Db_\alpha}\right)^\dagger & \delta^\alpha_\beta \end{array},$$

$$D^0(\overline{\Phi})B_1 D(\overline{\Phi})^\dagger = \begin{array}{c} \\ \varphi_{\mathbf{k}} \\ \overline{\Phi}(c_\gamma) \end{array} \begin{array}{c|c} \varphi_i & \overline{\Phi}(c_\beta) \\ \hline \sum_\alpha\left[(-1)^{p(\alpha)[p(\mathbf{k})+1]+1}\frac{D\varphi_{\mathbf{k}}}{Db_\alpha}\left(\frac{D\varphi_i}{Da_\alpha}\right)^\dagger + (-1)^{p(\alpha)p(\mathbf{k})}\frac{D\varphi_{\mathbf{k}}}{Da_\alpha}\left(\frac{D\varphi_i}{Db_\alpha}\right)^\dagger\right] & (-1)^{p(\beta)p(\mathbf{k})}\frac{D\varphi_{\mathbf{k}}}{Da_\beta} \\ -\left(\frac{D\varphi_i}{Da_\gamma}\right)^\dagger & 0 \end{array} \qquad (A3.25)$$

To show that $\Phi(B) = D^0(\overline{\Phi})B_1 D(\overline{\Phi})^\dagger$, it is enough to show that the relation $[\Phi(B)\binom{X}{u}]^t\binom{Y}{v} \sim [D^0(\overline{\Phi})B_1 D(\overline{\Phi})^\dagger\binom{X}{u}]^t\binom{Y}{v}$ is satisfied for all even X, Y, u, v. (This follows from the general properties of the calculus of variations and Theorem A1.6a).) Obviously, the above relation is satisfied for the terms bilinear in (X, v) and (Y, u), as (A3.23) shows. It, thus, remains to work out only (X, Y)-terms in this relation. For the R.H.S. we obtain

$$\sum(-1)^{p(\alpha)[p(\mathbf{k})+1]+1}\frac{D\varphi_{\mathbf{k}}}{Db_\alpha}\left(\frac{D\varphi_i}{Da_\alpha}\right)^\dagger(X_i)\cdot Y_{\mathbf{k}} + \qquad (A3.26a)$$

$$+\sum(-1)^{p(\alpha)p(\mathbf{k})}\frac{D\varphi_{\mathbf{k}}}{Da_\alpha}\left(\frac{D\varphi_i}{Db_\alpha}\right)^\dagger(X_i)\cdot Y_{\mathbf{k}}, \qquad (A3.26b)$$

while for the L.H.S. we get
$\sum\varphi_{\mathbf{k}}[X,Y]_{\mathbf{k}}$ [by (A3.19)] $\sim -\sum b_\alpha([X,Y]\cdot\tilde{a})_\alpha$ [since ρ is a representation of $\mathcal{G}$ on V] $= \sum b_\alpha[X\cdot(Y\cdot\tilde{a}) - Y\cdot(X\cdot\tilde{a})]_\alpha$ [by (A3.23)] $\sim$

$\sim \sum\left(\frac{D\varphi_i}{Da_\gamma}\right)^\dagger(X_i)\cdot(Y\cdot\tilde{a})_\gamma - \sum\left(\frac{D\varphi_i}{Da_\gamma}\right)^\dagger(Y_i)\cdot(X\cdot\tilde{a})_\gamma$ [by (A3.22)] $=$

$= -\sum\left(\frac{D\varphi_i}{Da_\gamma}\right)^\dagger(X_i)\cdot\left(\frac{D\varphi_{\mathbf{k}}}{Db_\gamma}\right)^\dagger(Y_{\mathbf{k}}) + \sum\left(\frac{D\varphi_{\mathbf{k}}}{Da_\gamma}\right)^\dagger(Y_{\mathbf{k}})\cdot\left(\frac{D\varphi_i}{Db_\gamma}\right)^\dagger(X_i) \sim$

$\sim -\sum\frac{D\varphi_{\mathbf{k}}}{Db_\gamma}\left(\frac{D\varphi_i}{Da_\gamma}\right)^\dagger(X_i)\cdot Y_{\mathbf{k}}(-1)^{p(\gamma)[p(\mathbf{k})+p(\gamma)]} +$

$+\sum(-1)^{p(\gamma)p(\gamma)}(-1)^{p(\gamma)[p(\mathbf{k})+p(\gamma)]}\frac{D\varphi_{\mathbf{k}}}{Da_\gamma}\left(\frac{D\varphi_i}{Db_\gamma}\right)^\dagger(X_i)\cdot Y_{\mathbf{k}},$

and these are exactly the expressions (A3.26a) and (A3.26b), respectively. ∎

A3.27. Remark. In the case of classical mechanics, i.e., when no G and no ∂'s are present, one could exchange the map $\Phi : C \to C_1$ into an equivalent map $R : V \oplus V^* \to (\mathcal{G} \propto V)^*$, such that $\overline{\Phi} = R^*$. The map R would be a canonical representation for the dual spaces of the semidirect product Lie superalgebras.

A3.28. Remark. The usual form of the Clebsch representation is obtained when $\mathcal{G}$ itself (or its even part) is the Lie algebra of vector fields on a manifold, and the representation is furnished by the Lie derivative on various tensor fields.

A3.29. Remark. The proof of Theorem A3.23 given above, is shorter than the proof of the purely commutaive version (Theorem VIII 4.33 in [21]).

A3.30 Remark. Theorem A3.23 describes canonical quadraic maps involving "dual spaces of Lie superalgebras" in the most general situations. However, for special Lie algebras one finds special canonical maps which are not quadratic but polynomial (and even rational). Such situations occur, for example, in fluid dynamical field theories.

A3.31. Remark. There exist quadratic canonical maps, for special Lie algebras and superalgebras, whose origin is not due to Theorem A3.23. Here are two examples.

A3.32. Let $C = K[u^{(n)}]$ correspond (see Appendix 2) to the Lie algebra D (see (4.49)), so that $B = B(D) = u\partial + \partial u$ is the corresponding Hamiltonian operator. Let $C_1 = K[v^{(n)}], p(v) = 1$, be the ring of functions on the differential Grassmann algebra $\Lambda(1)$ over K. Let $B_1 = 1/2$ be the Hamiltonian structure on C_1. Let $\Phi : C \to C_1$ be given as $\Phi(u) = vv^{(1)}$. Then Φ is canonical.

A3.33. Let $C = K[u^{(n)}, \varphi^{(n)}], p(u) = 0, p(\varphi) = 1$, correspond to the Lie superalgebra $L^s_{1/2}$ (4.52), with the super-Hamiltonian matrix B given by (4.48) (so that a 2-cocycle now enters the picture!). Let $C_1 = K[v^{(n)}, \alpha^{(n)}]$,

$p(v) = 0,\ p(\alpha) = 1$, and let $B_1 = \begin{pmatrix} \partial & 0 \\ 0 & 1 \end{pmatrix}$ be the super-Hamiltonian structure of the reduced $s - mKdV$ system (4.7) with a, b absent and α being scalar. Let $\Phi : C \to C_1$ be the reduced Miura map (4.11): $\Phi(u) = v^2 + v^{(1)} + \alpha\alpha^{(1)}, \Phi(\varphi) = \alpha^{(1)} + v\alpha$.Then [13] the map Φ is canonical.

A3.34. Remark. The quadratic character of the latter map Φ is a bit misleading, as can be seen from the following observation. The map Φ can be factored to induce a canonical map $\Phi' : C' = K[u^{(n)}] \to C'_1 = K[v^{(n)}], \Phi'(u) = v^2 + v^{(1)}$, which is nothing but the original Miura map for the KdV system. This map, however, is a member of the following family of polynomial canonica maps [12]: $\Phi'(\xi^{n+2} - \sum_0^n u_i \xi^i) = (\xi - v_0)(\xi - v_1)...(\xi - v_{n+2})$, with $\sum v_j = 0$. (For $n = 0, v = v_1$, one recovers the above map Φ').

Acknowledgement. Thanks are due to Darryl Holm, Mack Hyman, and Basil Nicols for the hospitality during the conference. This work was partly supported by the NSF and DOE.

References.

[1] J. L. Martin, "Generalized Classical Dynamics, and the 'Classical Analogue' of a Fermi oscillator", Proc. Roy. Soc. London A251 (1950) 536–542.

[2] V. G. Kac, "Lie Superalgebras", Adv. Math. 26 (1976) 8–96.

[3] V. G. Kac, "Contravariant Form for Infinite-Dimensional Lie Algebras and Superalgebras", Lect. Notes in Phys., v. 94 (1979) 441–445.

[4] Yu. I. Manin, "Flag Superspaces and Supersymmetric Yang-Mills Equations", in "Arithmetic and Geometry" vol. II (dedicated to I. R. Shafarevich) 175–198, "Progress in Mathematics", vol. 37, Birkhäuser, Boston (1983).

[5] Yu. I. Manin, "Holomorphic Supergeometry and Yang-Mills Superfields", "in Itogi Nauki i tekhniki. Sovremennye problemy mathematiki.", v. 27 (1984), VINITI, Moscow (in Russian).

[6] Yu. I. Manin, "Gauge Fields and Complex Geometry", Nauka (1984), Moscow (in Russian).

[7] B. A. Kupershmidt, "Superintegrable Systems," Proc. Natl. Acad. Sci. USA, 81 (1984).

[8] B. A. Kupershmidt, "Bosons and Fermions Interacting Integrably with the Korteweg-de Vries Field", J. Phys. A: Math & Gen. (to appear).

[9] B. A. Kupershmidt, "Supersymmetric Fluid in a Free One-Dimensional Motion", Lett. Nuovo Cim. (to appear).

[10] B. A. Kupershmidt, "Classical Superintegrable Systems", Preprint (1984).

[11] G. Wilson "Commuting Flows and Conservation Laws for Lax Equations", Math. Proc. Cambr. Philos. Soc. 86 (1979) 131–143.

[12] B. A. Kupershmidt and G. Wilson, "Modifying Lax Equations and the Second Hamiltonian Structure", Invent. Math. 62 (1981) 403–436.

[13] B. A. Kupershmidt, "A Super Korteweg-de Vries Equation: an Integrable System", Phys. Lett. 102A (1984) 213–215.

[14] J. Gibbons, "Related Integrable Hierarchies. I - Two Nonlinear Schrödinger Equations", Preprint (1981), DIAS–STP–81–06, Dublin.

[15] V. G. Drinfel'd and V. V. Sokolov, "Equations of KdV Type and Simple Lie Algebras", Dokl. Akad. Nauk SSSR 258 (1981) 11–16 (Russian); Sov. Math. Dokl. 23 (1981) 457–462 (English).

[16] V. G. Drinfel'd and V. V. Sokolov, "Lie Algebras and the Korteweg-de Vries Type Equations", in "Itogi Nauki i Tekhniki. Sovremennye Problemi Mathematiki." VINITI, Moscow (to appear).

[17] B. A. Kupershmidt, "Nonclassical Superintegrable Systems", Preprint (1984).

[18] B. A. Kupershmidt, "Normal and Universal Forms in Integrable Hydrodynamical Systems", in "Proc. of the Berkeley-Ames Conf. on Nonlin. Problems in Control and Fluid Dynamics", L. R. Hunt and C. F. Martin, ed-s, (1984) 357–378, Math Sci Press, Boston.

[19] B. A. Kupershmidt, "Mathematics of Dispersive Water Waves", Preprint (1984).

[20] B. A. Kupershmidt, "On Algebraic Models of Dynamical Systems", Lett. Math. Phys. 6 (1982) 85–89.

[21] B. A. Kupershmidt, "Discrete Lax Equations and Differential - Difference Calculus", E.N.S. Lecture Notes, Paris (1982); "Asterisque", Paris (1985).

[22] D. A. Leites, "Theory of Supermanifolds", Petrozavodsk (1983) (in Russian).

[23] G. Wilson, "The Modified Lax and Two–Dimensional Toda Lattice Equations Associated with Simple Lie Algebras", Ergod. Th & Dynam. Syst. 1 (1981) 361–380.

[24] B. A. Kupershmidt, "Korteweg-de Vries Surfaces and Băcklund Curves", J. Math. Phys. 23 (1982) 1427–1432.

[25] M. Ademollo and 12 coauthors, "Dual String with U(1) Colour Symmetry", Nucl. Phys. B111 (1976) 77–110.

[26] B. A. Kupershmidt, "Geometry of Jet Bundles and the Structure of Lagrangian and Hamiltonian Formalisms", Lect. Notes in Math., p.p. 162–218, Springer-Verlag (1980).

[27] Manin Yu, I. "Algebraic Aspects of Non-Linear Differential Equations", "Itogi Nauki i Tekhniki, ser. Sovremennye Problemi Mathematiki" 11 (1978) 5–152 (Russian); J. Sov. Math. 11 (1979) 1–122 (English).

[28] B. A. Kupershmidt and Yu. I. Manin, "Equations of Long Waves with a Free Surface. II. Hamiltonian Structure and Higher Equations", Funct. Anal. Appl. 12: (1978) 25–37 (Russian); 20–29 (English).

Lectures in Applied Mathematics
Volume 23, 1986

Integrable Systems in General Relativity

B. Kent Harrison[1]

ABSTRACT. In recent years there has been some success in finding systems of linear equations whose integrability conditions have been equations from general relativity. The best example of this is the discovery of such systems for the Ernst equation, which arises from the vacuum, stationary, axially symmetric Einstein equations. However, such systems have not been discovered for the full vacuum Einstein equations, with no symmetries. Both of these cases will be discussed. The results are most elegantly expressed in the formalism of differential forms. The theory can be used to find Bäcklund transformations, which generate new solutions from old ones. Examples of such solutions will be given.

INTRODUCTION. In this paper, we will first explain what is meant by "integrable systems;" then we will discuss the Wahlquist-Estabrook method [18] for finding integrable systems and Bäcklund transformations. Preliminary to this we will see how differential forms may be used in writing differential equations [9]. Then we will present the general relativity Ernst equation, which describes vacuum, axially symmetric, stationary (i.e. time-independent but rotating) gravitating systems, and we will review how the Wahlquist-Estabrook method was used to find Bäcklund transformations for the Ernst equation [8, 11]. Following that, we will see how recent work,

[1]1980 Mathematics Subject Classification. 83C15, 35G20, 35C05.
Supported by National Science Foundation grant number PHY-8308055.

particularly by F. J. Chinea and M. Gürses, has led to an elegant formulation of the full vacuum Einstein equations in terms of matrix 3-forms, and how this formulation can be used to write integrable systems and even Bäcklund transformations for those equations [1, 7].

2. INTEGRABLE SYSTEMS; DIFFERENTIAL EQUATIONS IN TERMS OF DIFFERENTIAL FORMS. "Integrable systems" here will be taken to mean auxiliary systems of differential equations, whose integrability conditions yield the field equations studied. Such situations are often tied closely to the existence of Bäcklund transformations (BT's). By an (auto-) Bäcklund transformation we mean a method of obtaining a new solution of a set of equations from an old one, generally by means of integrating a set of first-order differential equations (the "transformation" equations) [4].

In this research, it is convenient to write differential equations in terms of a set of differential forms [9]. As a very simple example, we so write the one-dimensional heat equation, $u_t = u_{xx}$, where subscripts indicate derivatives. We define $v = u_x$; then $v_x = u_t$. We now define two 2-forms in the space of variables u, v, x, t:

$$\alpha = du \wedge dt - v\, dx \wedge dt \tag{1}$$
$$\beta = dv \wedge dt + du \wedge dx.$$

If we restrict α and β to a two-dimensional subspace $v = v(x, t)$, $u = u(x, t)$ ("sectioning" α and β), we have $\alpha = (u_x - v) dx \wedge dt$ and $\beta = (v_x - u_t)\, dx \wedge dt$. The original differential equation(s) are now equivalent to $\alpha = \beta = 0$. We say that α and β are annulled.

We can write any system of differential equations in this manner, simply by inspection. However, the differential forms representing the system are usually not unique. In the above case, for example, we could have defined $w = u_t$ and written our set of forms as $\gamma = du - vdx - wdt$, $\delta = (-d\gamma) = dv_\wedge dx + dw_\wedge dt$, and β (as given above). α is then just $\gamma_\wedge dt$.

We treat our chosen set of forms as an ideal I. We require I to include enough forms (such as δ above) so that it is closed: $dI \subset I$ (equivalently, $dI = 0 (\text{mod } I)$).

3. WAHLQUIST-ESTABROOK METHOD. If we are searching for an integrable system or a Bäcklund transformation for a set of differential equations and wish to use the Wahlquist-Estabrook (WE) method, it is convenient to write the equations as an ideal I of differential forms as shown above (although the method can be used without doing so.)

In mathematical physics, equations often admit potentials. For example, the equation $\nabla \times E - 0$ for electric field may be formally satisfied by introduction of an electrostatic potential ϕ: $E = -\nabla\phi$. In terms of forms, we simply write: $dE = 0$ (where E is a 1-form in this case) implies $E = -d\phi$.

The ideal I given in (1) above admits a potential in the following manner. We note that the 1-form $\sigma = udx + vdt$ obeys $d\sigma = \beta$. Thus, when I is annulled, we have $\beta = 0$, so that $d\sigma = 0$, and thus we may write $\sigma = dz$ for a potential z. (This is just $u - z_x$, $v = z_t$, which satisfies our original equation $v_x = u_t$.) It is convenient to define a 1-form $\omega = \sigma - dz = -dz + udx + vdt$; then $\omega = 0$ defines the potential equation.

For the general ideal I (in which the independent variables are still x and t), one may search for potentials by writing

$$\omega = -dz + Fdt + Gdx$$

where F and G are functions of the field variables (and x and t explicitly, too, if needed) and requiring $d\omega = 0 \ (\text{mod } I)$. If we do this for (1), writing F and G as functions of u and v, we get

$$d\omega = (F_u du + F_v dv)\wedge dt + (G_u du + G_v dv)\wedge dx$$
$$= F_u(\alpha + v\, dx\wedge dt) + F_v(\beta - du\wedge dx) + G_u du\wedge dx$$
$$+ G_v dv\wedge dx = 0 \bmod (\alpha, \beta),$$

giving $F_u = G_v = 0$, $F_v = G_u$. These equations have the solution $F = av + b$, $G = au + c$, where a, b, c are constants. Choosing $b = 1$ and $a = c = 0$ gives the trivial case $\omega = -dz + dt$; but the choice $a = 1$, $b = c = 0$ yields the nontrivial case $\omega = -dz + vdt + udx$ discussed above.

In the WE procedure, we generalize in the following. We write

$$\omega = -dy + Fdt + Gdx \qquad (2)$$

(called a prolongation form) where F and G are functions of the field variables and also of y, and require

$$d\omega = 0 \ (\text{mod } I, \omega) \qquad (3)$$

We call y a pseudopotential. Following the same procedure as above, and substituting for dy from (2), we get

$$d\omega = F_u\ (\alpha + vdx\wedge dt) + F_v dv\wedge dt + F_y(-\omega + Gdx)\wedge dt$$
$$+ G_u(\beta - dv\wedge dt) + G_v dv\wedge dx + G_y(-\omega + Fdt)\wedge dx. \qquad (4)$$

(3) now gives (since $I = \{\alpha, \beta\}$)

$$0 = (vF_u + GF_y - FG_y)dx\wedge dt + (F_v - G_u)dv\wedge dt + G_v dv\wedge dx,$$

yielding the three differential equations for F and G:

$$G_v = 0, \quad F_v - G_u = 0, \quad vF_u + GF_y - FG_y = 0. \qquad (5)$$

A few calculations show that $G = uA + B$ and $F = vA + H$, where $H = u(AB' - BA') + C$; A, B, and C are functions of y; the prime means d/dy; and $GH_y - HG_y = 0$ must still be satisfied. If $A = 0$, we get a trivial case: F and G functions of y alone. So assume $A \neq 0$. Then we can divide ω by A and define a new pseudopotential y' by $dy' = dy/A$. Equivalently, we merely put $A = 1$. Then $G = u + B$, $H = uB' + C$, and the last equation gives $B'' = 0$, $C' = B'^2$, and $BC' = CB'$. If $B' = 0$, we get the potential case already worked out. So we assume $B' = a \neq 0$ and obtain finally $G = u + ay + b$, $F = v + au + a^2y + ab$. b may be set = 0 without loss of generality, and we obtain a prolongation form

$$\omega = -dy + (v + au + a^2y)\, dt + (u + ay)dx, \tag{6}$$

featuring an undetermined parameter a. (This is a typical and important feature.)

The above example illustrates the simplest version of the WE method, in which there is a single pseudopotential. In a second version, we take y to be a vector of undetermined dimension; then F and G are vector functions and ω is a vector-valued 1-form, in other words a vector made up of 1-forms. It is now convenient , as a third version, to let F and G be linear in the y; F = My, G = Ny, where M and N are matrix functions of the field variables.

For the second and third versions, we explore how the equations (5) are modified. We denote the components of y, etc., by y^α, etc. $G_v = 0$ now becomes $G_v^\alpha = 0$ for all α; we treat G as a vector and write this again as $G_v = 0$. In place of (5), we now have

$$G_v = 0, \quad F_v - G_u = 0, \quad vF_u + [G, F] = 0, \tag{7}$$

where F and G are now vectors and we define a Lie bracket,

$$[G, F]^\alpha = G^\beta\, \partial F^\alpha/\partial y^\beta - F^\beta \partial G^\alpha/\partial y^\beta \tag{8}$$

which is a vector. Finally, if F and G are linear in the y's, F = My and G = Ny, where M and N are matrices, then (8) becomes

$$[G, F]^\alpha = [M, N]^\alpha{}_\gamma\, y^\gamma \tag{9}$$

and (7) becomes

$$N_v = 0, \quad M_v - N_u = 0, \quad vM_u + [M, N] = 0, \tag{10}$$

where [M, N] is the ordinary matrix commutator.

To use these results to find a BT, one now assumes the existence of a new set of field variables, denoted here with primes (u', v' in the above example), and assumes these to be functions of the old field variables (u, v) and the pseudopotentials (y). One then assumes the new variables to satisfy the original equations, given that the old variables also satisfy them. In symbols,

$$I' = 0 \pmod{I,\ \omega}. \tag{11}$$

Thus in our example we put $u' = g(u, v, y)$ and $v' = h(u, v, y)$. Assume a single y. We have, mod I and ω, by putting u' and v' in place of u and v in α and β, and expanding,

$$\alpha' = (g_u du + g_v dv + g_y dy)\wedge dt - h dx\wedge dt = 0, \tag{12}$$

$$\beta' = (h_u du + h_v dv + h_y dy)\wedge dt + (g_u du + g_v dv + g_y dy)\wedge dx = 0.$$

Using equations (1) and (2) we obtain

$$g_v = 0 \quad g_u - h_v = 0 \quad h = vg_u + Gg_y \tag{13}$$

$$vh_u + Gh_y - Fg_y = 0.$$

If we solve these equations, using the F and G of (6), we find the BT

$$u' = cu + ky + m \tag{14}$$

$$v' = cv + ku + aky$$

where c, k, and m are new constants. In other words, we have the following. If u is a solution of the heat equation, then we can find a pseudopotential y by integrating $y_t = u_x + au + a^2y$ and $y_x = u + ay$ (Eq. (6)). Integrability is guaranteed by the heat equation itself. A new solution u' is then given by the first of Eqs. (14), and the second equation is consistent with the first.

Let us recapitulate. If we are given a set of differential equations, we may often be able to define a set of variables y^α, which we here call pseudopotentials, defined by certain first-order differential equations. The original set of differential equations may be written as an ideal I of differential forms. In that case, the equations for the y^α are simply 1-forms. These may usually be written so as to be linear in the y^α. Their integrability conditions are to be satisfied when $I = 0$ (or may even imply $I = 0$). The 1-forms ω may also be used to search for BT's. In this case, the y^α provide additional freedom in the assumed functional form, usually just enough to find the BT.

Another useful example, less trivial than the last, is the sine-Gordon equation [4]:

$$\phi_{uv} = \sin\phi \tag{15}$$

If $r = \phi_u$, then the 2-forms we use are:

$$\alpha = d\phi \wedge dv - r\,du \wedge dv \tag{16}$$

$$\beta = dr \wedge du - \sin\phi\, dv \wedge du$$

The 1-form for a single y is (after simplification similar to that above)

$$\omega = -dy + (k \sin y - r)\, du + k^{-1} \sin(y + \phi)dv. \tag{17}$$

We see that

$$d\omega = [k \cos y\, du + k^{-1} \cos(y + \phi)dv] \wedge \omega - \beta + k^{-1} \cos (y + \phi)\alpha.$$

We search for a BT, in which a new solution ϕ' may be found from an old one ϕ, by writing

$$\alpha' = d\phi' \wedge dv - r' du \wedge dv$$

$$\beta' = dr' \wedge du - \sin\phi'\, dv \wedge du$$

where we take $\phi' = f(r, \phi, y)$ and $r' = g(r, \phi, y)$. The search yields $\phi' = -\phi - 2y$ and the BT equations

$$\phi'_u = \phi_u + 2\lambda \sin \tfrac{1}{2} (\phi' + \phi) \tag{18}$$

$$\phi'_v = -\phi_v + 2\lambda^{-1} \sin \tfrac{1}{2} (\phi' - \phi)$$

A somewhat more elegant (fourth) version is possible in some cases [10, 11]. It may be possible to write I in terms of a set of 1-forms ξ_i, with some equations in I (annulled) being expressions of the form

$$d\xi_i = a_i{}^{k\ell}\, \xi_k \wedge \xi_\ell \tag{19a}$$

and the remaining ones having the form

$$0 = b_i{}^{k\ell}\, \xi_k \wedge \xi_\ell \tag{19b}$$

where the $a_i{}^{k\ell}$ and $b_i{}^{k\ell}$ are constants. We may then try writing ω (vector) as

$$\omega = -dy + (F^i \xi_i)\, y \tag{20}$$

where y is a vector and the F^i are matrices, constant or perhaps functions of some specially defined variable. $d\omega = 0$ (mod ω, I) typically gives an incomplete Lie algebra for the F^i (prolongation structure)-- which may be a Kac-Moody algebra. Then a BT may be sought by using $I' = 0$ (mod ω, I). (Note: the results obtained in this version should be the same as those obtained in, say, the third version--including the incomplete Lie algebra.)

For the sine-Gordon equation, we may choose $\xi_1 = du$, $\xi_2 = rdu$, $\xi_3 = \sin\phi dv$, $\xi_4 = \cos\phi dv$. Then the equations of type (19b) are $\xi_1 \wedge \xi_2 = \xi_3 \wedge \xi_4 = 0$, and those of type (19a) are: $d\xi_1 = 0$, $d\xi_2 = \xi_3 \wedge \xi_1$, $d\xi_3 = \xi_2 \wedge \xi_4$, and $d\xi_4 = \xi_3 \wedge \xi_2$. The prolongation structure is then $[F^2, F^3] = -F^4$, $[F^2, F^4] = F^3$, $[F^1, F^3] = -F^2$, and $[F^1, F^4] = 0$, for constant matrices F^i. If we choose F^4 proportional to F^1, we get the $s\ell(2,R)$ Lie algebra.

4. THE ERNST EQUATION OF GENERAL RELATIVITY The equations of general relativity are a set of second order nonlinear partial differential equations for the metric tensor components g_{ik}. In vacuum, the equations are simply $R_{ik} = 0$, where R_{ik} is the Ricci tensor. Simplification is often achieved by assuming particular symmetries. If we assume axial symmetry and time independence--but still rotation--the equations reduce to a central equation, called the Ernst equation [5], plus others. Let ϕ be the azimuthal angle and t be the time. Call the other two independent coordinates r and z, although these are not, in general, the usual cylindrical coordinates. Define a linear Hodge star operator $*$([6], p. 15) by $*dr = dz$, $*dz = -dr$ ($** = -1$.) Now write the metric as [8, 11]

$$ds^2 = -f(dt + Qd\phi)^2 + S^2 f^{-1} d\phi^2 + e^{2\gamma} f^{-1}(dr^2 + dz^2) \tag{21}$$

where f, Q, γ and S are functions of r and z only. (We use a -+++ signature.)

The field equation for Q may be formally solved by defining

a potential ϕ as follows:

$$*d\phi = S^{-1}f^2 dQ; \tag{22}$$

(i.e., $\phi_r = S^{-1}f^2Q_z$, $\phi_z = -S^{-1}f^2Q_r$). The equation for S is:

$$S_{rr} + S_{zz} = 0; \tag{23}$$

$S = r$ is usually taken. Now put $E = f + i\phi$; the equations for f and ϕ may now be written

$$(\mathrm{Re}E)\ \nabla^2 E = (\nabla E)^2, \tag{24}$$

the Ernst equation. Here

$$\nabla^2 E = E_{rr} + E_{zz} + S^{-1}(S_r E_r + S_z E_z)$$

and $(\nabla E)^2 = E_r{}^2 + E_z{}^2$.

The above treatment can easily be modified in the case that the independent variables are r and t; then (24) is a hyperbolic instead of an elliptic equation.

It is useful to write (23) and (24) as 2-forms:

$$\begin{aligned} &d(*dS) = 0 \\ &d(Sf^{-1} * df) + Sf^{-2}\, d\phi \wedge * d\phi = 0 \\ &d(dQ) = d(Sf^{-2} * d\phi) = 0. \end{aligned} \tag{25}$$

We formally satisfy the first two of these by writing $*dS = dR$ and $*df = S^{-1}f(d\eta + Qd\phi)$, where R and η are new potentials. We then define six 1-forms:

$$\begin{aligned} &\xi_1 = f^{-1}d\phi && \xi_4 = f^{-1}df \\ &\xi_2 = S^{-1}fdQ\ (= *\ \xi_1) && \xi_5 = S^{-1}dS \\ &\xi_3 = S^{-1}(d\eta + Qd\phi)(= *\xi_4) && \xi_6 = S^{-1}dR\ (= *\xi_5). \end{aligned} \tag{26}$$

Then the ideal (I = 0) of the field equations becomes:

$$\begin{aligned} &d\xi_1 = \xi_1 \wedge \xi_4 && d\xi_4 = 0 \\ &d\xi_2 = \xi_2 \wedge (\xi_5 - \xi_4) && d\xi_5 = 0 \\ &d\xi_3 = \xi_3 \wedge \xi_5 - \xi_1 \wedge \xi_2 && d\xi_6 = -\xi_5 \wedge \xi_6, \end{aligned} \tag{27}$$

which are identities and are of the type (19a), and

$$\begin{aligned} &\xi_3 \wedge \xi_1 - \xi_2 \wedge \xi_4 = 0 && \xi_5 \wedge \xi_1 + \xi_2 \wedge \xi_6 = 0 \\ &\xi_3 \wedge \xi_2 + \xi_1 \wedge \xi_4 = 0 && \xi_5 \wedge \xi_3 - \xi_4 \wedge \xi_6 = 0 \\ &\xi_5 \wedge \xi_2 - \xi_1 \wedge \xi_6 = 0 && \xi_4 \wedge \xi_5 + \xi_6 \wedge \xi_3 = 0, \end{aligned} \tag{28}$$

of type (19b). We also define a variable ζ (invariant under the symmetry group of the Ernst equation):

$$\zeta = \{[i\,(R + \ell) - S][i\,(R + \ell) + S]^{-1}\}^{1/2} \tag{29}$$

where ℓ is a parameter. (This parameter plays the same role as the free parameters in known Bäcklund transformations; for example, it determines speed and amplitude of generated solitons.)

We now assume an ω of the form in (20):

$$\omega = -dq + B^i \xi_i q \tag{30}$$

where the matrices B^i are functions of ζ. The implied sum on i in (30) would normally run from 1 to 6; but it can be shown that we may take $B^5 = B^6 = 0$ without loss of generality, so the sum effectively goes to 4 only. $d\omega = 0$ (mod I, ω) now yields:

$$\begin{aligned} &[B^1, B^2] = -B^3 && [B^1, B^4] + [B^2, B^3] = B^1 \\ &-[B^2, B^4] + [B^1, B^3] = B^2 && [B^3, B^4] = 0 \end{aligned} \tag{31}$$

and

$$\begin{aligned} &-aB^{1\prime} = bB^{2\prime} && bB^{1\prime} - aB^{2\prime} = -B^2 \\ &-aB^{4\prime} = bB^{3\prime} && bB^{4\prime} - aB^{3\prime} = -B^3 \end{aligned} \tag{32}$$

where $' = d/d\zeta$, $a = \frac{1}{4}\zeta^{-1}(\zeta^4-1)$, and $b = \frac{i}{4}\zeta^{-1}(\zeta^2-1)^2$. Solution of (32) yields

$$\begin{aligned} &B^2 = \frac{i}{2}\zeta^{-1}(\zeta^2 -1)\nu && B^3 = \frac{i}{2}\zeta^{-1}(\zeta^2-1)\tau \\ &B^1 = \frac{1}{2}\zeta^{-1}(\zeta^2 + 1)\nu + \theta && B^4 = \frac{1}{2}\zeta^{-1}(\zeta^2+ 1)\tau + \phi \end{aligned} \tag{33}$$

where ν, θ, τ and ϕ are constant matrices. Equations (31) now give

$$\begin{aligned} &[\nu, \phi] = [\tau, \phi] = 0 \\ &[\theta, \tau] = \nu \quad [\theta, \nu] = -\tau \\ &[\nu, \tau] + [\theta, \phi] = \theta \end{aligned} \tag{34}$$

an incomplete Lie algebra. If we take ϕ proportional to the unit matrix, (34) simplifies to

$$[\theta, \tau] = \nu \quad [\theta, \nu] = -\tau \quad [\nu, \tau] = \theta \tag{35}$$

the $s\ell(2, R)$ Lie algebra. The simplest representation is 2×2; thus (30) becomes a set of two equations for pseudopotentials q^1 and q^2, whose integrability conditions reduce to the Ernst equation. If we choose, as in [11], $\tau = -\sigma_z$, $\nu = \sigma_x$, $\theta = i\sigma_y$, and $\phi = 0$, then expansion of Eq. (30) yields

$$4dq_1 = - Aq_1 + (B + 2\xi_1)q_2$$

and

$$4dq_2 = (B - 2\xi_1)q_1 + Aq_2$$

where $A = \zeta^{-1}[(\zeta^2 + 1)\xi_4 + i(\zeta^2 - 1)\xi_3]$ and $B = \zeta^{-1}[(\zeta^2 + 1)\xi_1 + i(\zeta^2 - 1)\xi_2]$. If we put $q_2 = uq_1$, then we find the Riccati equation $4du = -(B + 2\xi_1)u^2 + 2Au + B - 2\xi_1$. As noted in [11], u may be written as a function of q and ζ, where q is the pseudopotential defined in [8]. The Bäcklund transformations for the Ernst equation then can be worked out by the method discussed in section 3; they are given in [8], [11], and [16].

These BT's may be related to other methods of solving the Ernst equation. These include introduction of an infinite family of potentials for the equation; inverse scattering methods; or reduction of the problem to a homogeneous Hilbert problem [2]. Various interesting solutions have been found by these methods, such as solutions represent two or more singularities on an axis [3].

5. THE GENERAL VACUUM EINSTEIN EQUATIONS AS FORMS. The previous section reviewed some earlier work. This section and the next, however, present work done primarily in the past year--although it harks back to earlier work by W. Israel [12]--by F. Chinea [1], M. Gürses [7], and the present author, working independently. The presentation here gives a little different

approach to the known results of Chinea and Gürses. Some related material has been given by N. Sanchez [17].

The basic question is this: Can one find, for the general Einstein equations, without symmetries, integrable equations and Bäcklund transformations as was done with the special case of the Ernst equation? (We specialize to the vacuum equations for now.) The answer is now known to be _yes_. How much can be done is not yet clear; but some results are known.

We essentially use the Wahlquist-Estabrook approach, although the results can be presented independently. In this section we will write the vacuum Einstein equations in terms of forms; in the next we will look at integrability conditions and Bäcklund transformations.

It is convenient to write the Riemannian metric for spacetime in terms of basic 1-forms ω^i (for a discussion of this approach, see [14], [13]).

$$ds^2 = g_{ik}\omega^i \otimes \omega^k. \tag{36}$$

We take the g_{ik} to be constant. Then the first Cartan structural equation

$$d\omega^k = \omega^i \wedge \Omega_i{}^k \tag{37}$$

defines the connection 1-forms $\Omega_i{}^k$. We raise and lower indices with g_{ik} and its inverse. The connection forms satisfy

$$\Omega_{ik} = -\Omega_{ki}, \tag{38}$$

and these two equations uniquely define the Ω_{ik}. Finally, we use Cartan's second structural equation

$$\theta_k{}^i = d\Omega_k{}^i - \Omega_k{}^\ell \wedge \Omega_\ell{}^i \tag{39}$$

to define the curvature 2-forms $\theta_k{}^i$. $\theta_k{}^i$ obeys the equations

$$\theta_{ki} + \theta_{ik} = 0 \tag{40}$$

and $\omega^\ell \wedge \theta_\ell{}^k = 0.$ (41)

The components $R_{ikm\ell}$ of the curvature tensor are given by

$$\theta_{ik} = \tfrac{1}{2} R_{ikm\ell}\, \omega^m \wedge \omega^\ell, \tag{42}$$

and the Bianchi identities (integrability conditions for (39)) are

$$d\theta_k{}^i = \Omega_k{}^\ell \wedge \theta_\ell{}^i - \theta_k{}^\ell \wedge \Omega_\ell{}^i . \tag{43}$$

We now note that the Ricci tensor is given by

$$R_{\ell m} = R^k{}_{\ell k m}, \tag{44}$$

and the vacuum equations are $R^k{}_n = 0$ as noted before. One asks: How can these equations be expressed in terms of forms How can one manipulate the 2-form θ_{ik} so as to produce an $R^i{}_k$? It does not seem possible, since θ_{ik} is antisymmetric but R_{ik} is symmetric.

Nevertheless, it can be done. W. Israel first wrote the vacuum equations in terms of a set of 4-forms [12]. The procedure for deriving these is rather involved, and the final equations do not lend themselves easily to the study of integrability conditions and so forth. But one can see a simple procedure by noting the fact that if one multiplies θ_{ik} by any two different ω^p, ω^q, then the right-hand side of (42) will become a single component of the curvature tensor, multiplied by the volume 4-form $\sigma = \omega^0 \wedge \omega^1 \wedge \omega^2 \wedge \omega^3$. Once we have the single components, then R_{ik} is easy to construct.

Formally, we write:

$$\theta_{ik} \wedge \omega^p \wedge \omega^q = \tfrac{1}{2} R_{ikm\ell}\, \omega^m \wedge \omega^\ell \wedge \omega^p \wedge \omega^q$$

$$= \tfrac{1}{2} \varepsilon^{m\ell pq} R_{ikm\ell}\, \sigma$$

where $\varepsilon^{m\ell pq}$ is the alternating symbol, with $\varepsilon^{0123} = \varepsilon_{0123} = 1$, which can be inverted to give

$$R_{ikrs}\, \sigma = \tfrac{1}{2} \varepsilon_{rspq}\, \theta_{ik} \wedge \omega^p \wedge \omega^q . \tag{45}$$

Einstein's vacuum equations $R_{ik}\,(\sigma) = 0$ now become the 4-forms [10]

$$g^{ir}\, \varepsilon_{rspq}\, \theta_{ik} \wedge \omega^p \wedge \omega^q . \tag{46}$$

An equation in 4-forms is rather what we expect. It seems that equations in n independent variables should produce n-forms. But this need not always hold. Maxwell's equations, in general, can be expressed as 4-forms [9]; but they can also be expressed as 3-forms: $dF = 0$, $d{*}F = 4\pi J$. It was shown by

Chinea [1] that the vacuum Einstein equations can be expressed in terms of 3-forms. In the present formalism, this is easily shown as follows. Define the 3-forms λ_q:

$$\lambda_q = \varepsilon_{npmq}\, \theta^{np} \wedge \omega^m \tag{47}$$
$$= \tfrac{1}{2}\, \varepsilon_{npmq}\, R^{np}{}_{rs}\, \omega^r \wedge \omega^s \wedge \omega^m .$$

Now write

$$\lambda_q \wedge \omega^\ell = \tfrac{1}{2}\, \varepsilon_{npmq}\, R^{np}{}_{rs}\, \omega^\ell \wedge \omega^r \wedge \omega^s \wedge \omega^m$$
$$= \tfrac{1}{2}\, \varepsilon_{npmq}\, \varepsilon^{\ell rsm}\, R^{np}{}_{rs}\, \sigma$$
$$= -2 g^{\ell m} (R_{mq} - \tfrac{1}{2} g_{mq} R)\, \sigma,$$

after expansion of the double permutation symbol and combination of terms. The full Einstein equations are $R_{mq} - \tfrac{1}{2} g_{mq} R = 8\pi G T_{mq}$, so that $\lambda_q \wedge \omega^\ell = -16\pi G T^\ell{}_q \sigma$. In vacuum, $T^\ell{}_q = 0$ so that $\lambda_q \wedge \omega^\ell = 0$ for all q, ℓ. This implies $\lambda_q = 0$. So, in vacuum,

$$\varepsilon_{npmq}\, \theta^{np} \wedge \omega^m = 0. \tag{48}$$

The most common choices for the metric g_{ik} are the Minkowski form

$$g = \pm \begin{bmatrix} -1 & & & \\ & 1 & & \\ & & 1 & \\ & & & 1 \end{bmatrix} , \tag{49}$$

yielding

$$ds^2 = \pm (-dt^2 + dx^2 + dy^2 + dz^2) \quad (dt^2 = dt \otimes dt, \text{ etc.}),$$

and the null tetrad form

$$g = \begin{bmatrix} & 1 & & \\ 1 & & & \\ & & & \pm 1 \\ & & \pm 1 & \end{bmatrix} \tag{50}$$

for which

$$ds^2 = \ell \otimes n + n \otimes \ell \pm m \otimes \bar{m} \pm \bar{m} \otimes m$$

where $\ell = \pm 2^{-1/2}(dx - dt)$, $n = 2^{-1/2}(dx + dt)$, $m = 2^{-1/2}(dy + i dz)$ and $\bar{m} = 2^{-1/2}(dy - i dz)$ are the basis 1-forms. (Bar denotes complex conjugate.) Chinea and Gürses use (50) in their papers. We shall remain general here, adopting a formalism

which allows either, or any other, choice. We note that $\det(g) = -1$ for (49) and $+1$ for (50).
Define $\lambda = \sqrt{\det(g)}$ (= 1 or i). The ± indicates different sign conventions for the metric.

We now define

$$P_{ik} = \tfrac{1}{2}(\theta_{ik} + \tfrac{1}{2}\lambda\varepsilon_{npik}\theta^{np}). \tag{51}$$

Then, by (41) and (48),

$$P_{ik}\wedge\omega^k = 0 \tag{52}$$

It can also be shown that

$$P_{ik} = d\Sigma_{ik} - \Sigma_i{}^{\ell}\wedge\Sigma_{\ell k}, \tag{53}$$

where $\Sigma_{\ell k} = \tfrac{1}{2}(\Omega_{\ell k} + \tfrac{1}{2}\lambda\varepsilon_{np\ell k}\Omega^{np})$, (54)

$$P_{ik} + P_{ki} = 0, \tag{55}$$

and $g_{mi}g_{n\ell}\,\varepsilon^{i\ell qk}\,P_{qk} = 2\lambda P_{mn}$, (56)

a self dual property. Equation (43) becomes

$$dP_i{}^k = \Sigma_i{}^{\ell}\wedge P_\ell{}^k - P_i{}^{\ell}\wedge\Sigma_\ell{}^k. \tag{57}$$

At this point it is convenient to introduce matrix notation. We introduce matrices $\Sigma = [\Sigma_{ik}]$, $g = [g_{ik}]$, $P = [P_{ik}]$, $\Omega = [\Omega_{ik}]$, $S = Pg^{-1}$, $\gamma = \Sigma g^{-1}$, and $\tau = \Omega g^{-1}$, and vectors $\omega = [\omega^i]$ and $\sigma = g\omega$. (We note that Σ, P and Ω are antisymmetric.) Then equations (37), (53), (52) and (57), respectively, become

$$d\sigma = \tau\wedge\sigma \tag{58}$$

$$S = d\gamma - \gamma\wedge\gamma \tag{59}$$

$$S\wedge\sigma = 0 \tag{60}$$

$$dS = \gamma\wedge S - S\wedge\gamma \tag{61}$$

(Note: Products of matrix differential forms are evaluated simply by treating matrix multiplication and form multiplication separately, or sequentially. We note that $\gamma\wedge\gamma \neq 0$, since γ is a matrix.)

Now τ and γ are related. Can we express one in terms of the other? To express τ in terms of γ, we consider a change of basis, from ω to ω', represented by $\omega' = C\omega$, where C is a matrix. Then it is easily shown that $\sigma' = C_t{}^{-1}\sigma$, $g' = C_t{}^{-1}gC^{-1}$, $\tau' = C_t{}^{-1}\tau C$, $\gamma' = C_t{}^{-1}\gamma C_t$, and $\lambda = \lambda' \det C$. We see also from (54) that if $\lambda = i$--which is true for the Minkowski metric--that $\Omega_{ik} = \Sigma_{ik} + \overline{\Sigma_{ik}}$, yielding $\Omega = \Sigma + \overline{\Sigma}$ and $\tau = \gamma + \overline{\gamma}$. Now choose ω to be the Minkowski metric basis. For general C, ω' will be a general basis. We see from the above transformations that $\tau' = \gamma' + B^{-1}\overline{\gamma}'B$, where $B = \overline{C}_t{}^{-1}C_t$. We drop primes and write simply

$$\tau = \gamma + \overline{B}\,\overline{\gamma}\,B \tag{62}$$

so that
$$d\sigma = (\gamma + \overline{B}\,\overline{\gamma}\,B)\wedge\sigma \tag{63}$$

where we have used these easy results for B:

$$\overline{B} = B^{-1}, \quad g = BgB_t \tag{64}$$

If C is real, then clearly B = 1. For the null tetrad metric (50),

$$B = \begin{bmatrix} 1 & & & \\ & 1 & & \\ & & & 1 \\ & & 1 & \end{bmatrix}$$

It is not needed to find the transformation matrix C. It suffices merely to write $\overline{\omega}$ in terms of ω: $\overline{\omega} = A\omega$. Then it is easily proved that $B = A_t{}^{-1}$ and $\overline{\sigma} = B\sigma$. We can use the expression for $\overline{\omega}$ (or $\overline{\sigma}$) to determine B when it is needed.

Can we express γ in terms of τ? Yes, of course; Eq. (54) gives the $\Sigma_{\ell k}$ in terms of the $\Omega_{\ell k}$, which yields the $\gamma_{\ell k}$ in terms of the $\tau_{\ell k}$. But for some purposes it would be useful to have Eq. (54) as a matrix equation. This amounts to writing the antisymmetric 4×4 matrix Γ with components $\varepsilon_{np\ell k}\,\Omega^{np}$ in terms of the matrix $\Phi = g^{-1}\,\Omega g^{-1}$, which has components Ω^{np}. It is a simple matter to show that, for general antisymmetric Ω (and Φ),

$$\Gamma = \frac{1}{2}\sum_{i=1}^{3}(\lambda_i\Phi\nu_i + \nu_i\Phi\lambda_i)$$

where the λ_i and ν_i are the antisymmetric matrices such that λ_i has a +1 in the top row, i^{th} column (columns run from 1 to 3), a -1 in the first column, i^{th} row, and zero elsewhere, and ν_i has a +1 in the (jk) position (ijk cyclic), a -1 in the (kj) position, and zero elsewhere. (As such, the λ_i and ν_i form a representation of the O(4) Lie algebra.)

Thus, from (54), $\Sigma = \frac{1}{2}(\Omega + \frac{1}{2}\lambda\Gamma)$. We note that $\Phi = g^{-1}\tau$, so that

$$\gamma = \Sigma g^{-1} = \frac{1}{2}(\tau + \frac{1}{2}\lambda\kappa) \tag{65}$$

$$\text{where}\quad \kappa = \Gamma g^{-1} = \frac{1}{2}\sum_{i=1}^{3}(\lambda_i g^{-1}\tau\nu_i + \nu_i g^{-1}\tau\lambda_i)g^{-1}$$

$$= -\frac{1}{2}\sum_{i=1}^{3}(b_i\tau\, a_i + a_i\tau\, b_i),\quad b_i = \lambda_i g^{-1},\ \text{and}\ a_i = -\nu_i g^{-1}.$$

If g is the Minkowski metric (49) (+ sign), then the a_i and b_i are generators of the Lorentz group algebra.

The above formulation may be useful if we have a mechanical method (e.g., computer program) for matrix algebra. However, the extraction of the connection forms $\Omega_i{}^k$ from (37) and (38) will almost certainly need to be done by hand.

The number of independent complex 1-forms in γ is three. (This means six real 1-forms, with a total of 24 components. This is exactly the right number of rotation coefficients that one finds in defining the connection γ.) S will have just three 2-forms. If we denote the three 2-forms in S as α_μ and the 1-forms in γ as β_μ, then we may write

$$S = A^\mu\alpha_\mu, \qquad \gamma = A^\mu\beta_\mu \tag{66}$$

where the A^μ are constant matrices, three in number (the _same_ matrices for both S and γ.) The A^μ satisfy

$$[A^\alpha, A^\beta] = \pm\, 2i\varepsilon^{\alpha\beta\gamma}A^\gamma \tag{67}$$

$$[A^\alpha, \overline{BA}\,{}^\beta B] = 0.$$

The integrability condition for (63) is satisfied if

$$\gamma \wedge \overline{B\gamma B} + \overline{B\gamma B} \wedge \gamma = 0 \tag{68}$$

This is used to prove the second equation of (67), given that 1-forms anticommute. Since 1-forms and 2-forms commute, we have also, for example,

$$\gamma \wedge \overline{BSB} - \overline{BSB} \wedge \gamma = 0. \tag{69}$$

Chinea and Gürses specialized, at the outset, to the null tetrad and wrote their equations in terms of 2 × 2 matrices. We illustrate with Gürses' notation. The original 1-form basis is written as a 2 × 2 matrix σ, with $\sigma^+ = \sigma$. The connection forms are written as a 2 × 2 matrix γ, with $\mathrm{Tr}\gamma = 0$. R is written instead of S. Then the analogs to equations (63), (59), (60), and (61) are

$$d\sigma = \gamma \wedge \sigma - \sigma \wedge \gamma^+ \tag{70}$$

$$R = d\gamma - \gamma \wedge \gamma \tag{71}$$

$$R \wedge \sigma = 0 \tag{72}$$

$$dR = \gamma \wedge R - R \wedge \gamma. \tag{73}$$

Chinea has essentially the same equations. The entries in γ are then just 1-forms whose coefficients are the Newman-Penrose spin coefficients (see [13], pp. 43-47, 83).

6. INTEGRABILITY CONDITIONS AND BACKLUND TRANSFORMATIONS. Since the equations with which we now work are 2-forms ((63), (59), and (68)) and 3-forms ((60), (61), and (69)) (these equations form our ideal I), we cannot use the simple prescription used before ((3) and (4)) to find pseudopotentials and their auxiliary equations. H. Morris [15] suggested a possible approach to such situations. Following his lead, we generalize it as follows [10].

We write, specializing the orders of the forms to the present problem,

$$\omega = \phi \wedge d\lambda + \tau \wedge \lambda, \tag{74}$$

where λ is a vector 1-form, ω is a vector 3-form, ϕ is a matrix 1-form, and τ is a matrix 2-form. Their dimension is not yet specified. Require as before

$$d\omega = 0 \pmod{I,\ \omega}. \tag{75}$$

We satisfy (75) by writing

$$d\omega - \rho_\wedge\omega = 0 \pmod{I} \tag{76}$$

where ρ is a matrix 1-form, to be determined. We substitute (74) into (76) and equate the coefficients of λ and $d\lambda$ to zero, obtaining (mod I)

$$\tau = \rho_\wedge\phi - d\phi \tag{77}$$

$$d\tau - \rho_\wedge\tau = 0.$$

Substituting the first of these into the second, we find

$$(d\rho - \rho_\wedge\rho)_\wedge\phi = 0 \pmod{I}. \tag{78}$$

For the present problem, from Eqs. (59) and (60), it appears that ρ is like γ and ϕ is like σ. We guess $\rho = F\gamma G + H\bar{\gamma}K$ and $\phi = L\sigma$ (there is no point in including another term $M\bar{\sigma}$, since $\bar{\sigma} = B\sigma$.) Investigation shows that there are two cases, essentially equivalent:

$$\omega = \sigma_\wedge d\mu - \overline{B\gamma B}_\wedge\sigma_\wedge\mu \tag{79a}$$

$$\omega = \sigma_\wedge d\mu - \gamma_\wedge\sigma_\wedge\mu \tag{79b}$$

where μ is a <u>scalar</u> (1×1) 1-form. μ can be generalized to a matrix 1-form Φ, in which case (79b) is just Chinea's (11b) (his η is our σ):

$$d\Phi_\wedge\eta - \Phi_\wedge\gamma_\wedge\eta = 0. \tag{80}$$

Chinea includes a second equation

$$d\xi = \Phi_\wedge\eta_\wedge\Phi^+, \tag{81}$$

in his notation. The two equations (80) and (81) form an integrable system. The relativity equations (70)-(73) are a sufficient condition for their integrability. Chinea postulates that one may be able to find a sufficient number of independent Φ such that the field equations are <u>implied</u> by the integrability conditions for (80) and (81). As an example, he chooses a particular form for Φ and reconstructs the pp-wave metric ([13], p. 233)

$$ds^2 = 2d\zeta d\bar{\zeta} - 2dudv - 2Hdu^2 \tag{82}$$

where $H = f(\zeta, u) + \bar{f}(\bar{\zeta}, u)$ and f is an arbitrary function of ζ and u. Chinea is pursuing further work in this direction.

The introduction of ω as written above may be used to find Gürses' formulation of the problem, but it is more straightforward to break it into two parts. We introduce four-dimensional vector 1-forms p, u, v, y, matrix 1-forms q, C, D, E, F, G, H, and matrix 2-forms α, β. Then we write

$$\begin{aligned} \omega_1 &= dp + C\wedge p + D\wedge \bar{p} + q\wedge u + \bar{q}\wedge v \\ \omega_2 &= (dq + E\wedge q + F\wedge \bar{q} + q\wedge G + \bar{q}\wedge H)\wedge y + \alpha\wedge p + \beta\wedge \bar{p} \end{aligned} \tag{83}$$

and require $d\omega_i = 0 \pmod{\omega_i, I}$. p and q are the "pseudopotentials" here, and represent Gürses' notation (although for him they are 2×2 matrices with $p^+ = p$ and $\mathrm{Tr}\, q = 0$.)

Various cases are possible. We present just one, which after simplification is:

$$\begin{aligned} \omega_1 &= dp + C\wedge p + (q + \bar{q}B)\wedge \sigma \\ \omega_2 &= (dq + E\wedge q + F\wedge \bar{q} + q\wedge G + \bar{q}\wedge H)\wedge \sigma + \alpha\wedge p. \end{aligned} \tag{84}$$

$d\omega_1 = 0 \pmod{\omega_i, I}$ yields

$$dC + C\wedge C - \alpha - \bar{\alpha}T = 0$$

where $\bar{p} = Tp$ defines T and the following pairs anticommute: q and C-E, q and $\bar{F}$, $\bar{q}$ and C-$\bar{E}$, and $\bar{q}$ and F. Further work shows that C is real, F = 0, and so forth. One finds eventually that

$$\begin{aligned} \omega_1 &= dp - (\gamma + \overline{B\gamma B})\wedge p - (q + \overline{BqB})\wedge \sigma \\ \omega_2 &= (dq - \gamma\wedge q - q\wedge\gamma)\wedge \sigma + S\wedge p \end{aligned} \tag{85}$$

where $\bar{p} = Bp$ and q is a sum of the A^μ defined in (66) above. B is the matrix used in (63). This is equivalent to Gürses' system, which he writes as

$$\omega_1 = dp - \gamma_\wedge p + p_\wedge \gamma^+ - q_\wedge \sigma + \sigma_\wedge q^+ \tag{86}$$
$$\omega_2 = (dq - \gamma_\wedge q - q_\wedge \gamma)_\wedge \sigma + R_\wedge p.$$

One sees that $-\overline{B\gamma}B_\wedge p \to p_\wedge \gamma^+$, etc.

A BT can now be found easily, as Gürses shows. Write new forms

$$\sigma' = \sigma + p, \; \gamma' = \gamma + q \tag{87}$$

(p, q defined by $\omega_1 = \omega_2 = 0$), and require (as a sufficient condition)

$$q_\wedge p = 0 \tag{88}$$
$$\overline{Bq}B_\wedge q_\wedge \sigma = 0.$$

Then σ' and γ' are solutions of the field equations. In Gürses' notation, equations (88) are

$$q_\wedge p = 0 \tag{89}$$
$$q_\wedge \sigma_\wedge q^+ = 0$$

Gürses uses such a transformation to derive the generalized Kerr-Schild metric ([13], p. 304), a decidedly nontrivial metric. So the power of this particular transformation is clear.

Several comments are in order. First: The requirement(s) that $d\omega = 0 \bmod (I, \omega)$ are necessary and not sufficient for the existence of pseudopotentials, or 1-forms such as μ in (79). Actual calculations for certain metrics seem to indicate that such quantities do not always exist. Furthermore, the integrations needed to find such quantities may be difficult. Eq. (85) (or (86)) is a set of coupled partial differential equations--linear, but with variable coefficients--for the components of p and q. However, equation (88) (or (89)) will limit the number of independent components and will make solution easier. (This advantage was used by Gürses in his derivation of generalized Kerr-Schild.)

Second: It was noted above that one could use the above ω to find Gürses' formulation. This is easily pursued by using the null tetrad basis (ℓ, m, $\bar{m}$, n) and expanding all matrix equations. If one uses Gürses' 2 × 2 matrices, written as $\sigma = \begin{bmatrix} \ell & m \\ \bar{m} & n \end{bmatrix}$, $\gamma = \begin{bmatrix} A & B \\ C & -A \end{bmatrix}$, and $R = \begin{bmatrix} \alpha & \beta \\ \delta & -\alpha \end{bmatrix}$, then Eq. (20) gives ($\ell$ and n are real, all other forms are complex)

$$d\ell = (A + \bar{A})\wedge\ell + B\wedge\bar{m} + \bar{B}\wedge m,$$
$$dn = -(A + \bar{A})\wedge n + C\wedge m + \bar{C}\wedge\bar{m},$$
$$dm = (A - \bar{A})\wedge m + B\wedge n + \bar{C}\wedge\ell,$$

and its complex conjugate. Eq. (71) gives $\alpha = dA - B\wedge C$, $\beta = dB - 2A\wedge B$, and $\delta = dC - 2C\wedge A$. Eq. (72) yields $\alpha\wedge\ell + \beta\wedge\bar{m} = 0$, $\alpha\wedge m + \beta\wedge n = 0$, $\delta\wedge\ell - \alpha\wedge\bar{m} = 0$, and $\delta\wedge m - \alpha\wedge n = 0$. The last four 3-form equations, their complex conjugates, and the 2-form equations for $d\ell$, dn, dm, and $d\bar{m}$ comprise the ideal I. One now uses the approach of Eqs. (74)-(78). We assume a rather natural expansion for ϕ and p:

$$\phi = a\ell + bn + fm + g\bar{m} \tag{90}$$
$$\rho = T\alpha + U\beta + V\gamma + W\bar{\alpha} + X\bar{\beta} + Y\bar{\gamma} \; ;$$

then equation (77) yields an algebraic structure for the matrices a, b, . . . T, U, . . . Y. Columns of the matrices a, b, f, g act as eigenvectors for certain combinations of the matrices T, . . . Y. One can, if desired, extract a Lie algebra from the latter set of matrices. If one does this, the other conditions require that one of the sets (T, U, V) and (W, X, Y) vanishes and the other satisfies the $s\ell(2, R)$ Lie algebra. In this case one simply gets Chinea's equations (see (79)-(81) above). Further investigation is contemplated.

Third: The integrable systems obtained for equations with two independent variables, and their associated Bäcklund transformations, have often been linked to solitons or such structures. There has usually been a free parameter in the transformation which determines the speed/amplitude of the soliton. However, there is no such parameter in the methods of

Gürses and Chinea, or in any transformation that the author has studied, and possible interpretation of new solutions, in particular as anything like a soliton, is unclear.

Fourth: Possible connection of this work to other work, such as that of Sanchez [17] on general BT's for general relativity, has yet to be explored.

REFERENCES

1. F. J. Chinea, "Einstein Equations in Vacuum as Integrability Conditions," Phys. Rev. Lett. 52, (1984), 322-324.

2. Cosgrove, C. M., "Relationships Between the Group-Theoretic and the Soliton-Theoretic Techniques for Generating Stationary Axisymmetric Gravitational Solutions," J. Math. Phys. 21 (1980), 2417-2447; "Bäcklund Transformations in the Hauser-Ernst Formalism for Stationary Axisymmetric Spacetimes," J. Math. Phys. 22 (1981), 2624-2639; "Relationship Between the Inverse Scattering Techniques of Belinskii-Zakharov and Hauser-Ernst in General Relativity," J. Math. Phys. 23 (1982), 615-633.

3. W. Dietz and C. Hoenselaers, "Stationary System of Two Masses Kept Apart by Their Gravitational Spin-Spin Interaction," Phys. Rev. Lett. 48 (1982), 778-780.

4. A. Dold and B. Eckmann, eds., Bäcklund Transformations, Springer-Verlag, Berlin-Heidelberg, 1976, for example. Several references on the s-G equation exist, but not too many discuss the application of the WE method to it. But see H. C. Morris, "Bäcklund Transformations and the sine-Gordon Equation," R. Hermann, ed., in The 1976 Ames Research Center (NASA) Conference on the Geometric Theory of Non-Linear Waves, Math Sci. Press, Brookline, Mass., 1977, p. 105.

5. F. J. Ernst, "New Formulation of the Axially Symmetric Gravitational Field Problem," Phys. Rev. 167, (1968), 1175-1178.

6. H. Flanders, Differential Forms, Academic Press, New York, 1963.

7. M. Gürses, "Prolongation Structure and a Bäcklund Transformation for Vacuum Einstein's Field Equations," Phys. Lett. A 101A (1984), 388-390. See also M. Gürses, "Integrability of the Vacuum Einstein Equations," to be published.

8. B. K. Harrison, "Bäcklund Transformation for the Ernst Equation of General Relativity," Phys. Rev. Lett. 41, (1978), 1197-1200.

9. B. K. Harrison and F. B. Estabrook, "Geometric Approach to Invariance Groups and Solution of Partial Differential Systems," J. Math. Phys. 12 (1971), 653-666.

10. B. K. Harrison, "Prolongation Structures and Differential Forms," invited paper presented at the Workshop on Exact Solutions of Einstein's Equations, Retzbach, West Germany, November 1983. (to be published)

11. B. K. Harrison, "Unification of Ernst-Equation Bäcklund Transformations Using a Modified Wahlquist-Estabrook Technique," J. Math. Phys. 24, (1983), 2178-2187.

12. W. Israel, Differential Forms in General Relativity, 2nd ed., Dublin Institute for Advanced studies, Dublin, 1979.

13. D. Kramer, H. Stephani, E. Herlt, and M. MacCallum, Exact Solutions of Einstein's Field Equations, Cambridge Univ. Press, Cambridge, 1980.

14. C. W. Misner, K. S. Thorne, and J. A. Wheeler, Gravitation, W. H. Freeman, San Francisco, (1973).

15. See for example H. C. Morris, "Prolongation Structure and Nonlinear Evolution Equations in Two Spatial Dimensions," J. Math. Phys. 17, (1976) 1870-1872, and "Inverse Scattering Problems in Higher Dimensions: Yang-Mills Fields and the Supersymmetric Sine-Gordon Equation," J. Math. Phys. 21, (1980), 327-333.

16. G. Neugebauer, "Bäcklund Transformations of Axially Symmetric Stationary Gravitational Fields," J. Phys. A: Math. Gen. 12, (1979), L67-L70.

17. N. Sanchez, "Einstein Equations, Self-Dual Yang-Mills Fields and Non-Linear Sigma Models," preprint, 1983.

18. H. D. Wahlquist and F. B. Estabrook, "Bäcklund Transformation for Solutions of the Korteweg-de Vries Equation," Phys. Rev. Lett. 31, (1973), 1386-1390, and "Prolongation Structures of Nonlinear Evolution Equations," J. Math. Phys. 16, (1975), 1-7.

DEPARTMENT OF PHYSICS AND ASTRONOMY
BRIGHAM YOUNG UNIVERSITY
PROVO, UTAH 84604

Lectures in Applied Mathematics
Volume 23, 1986

MODULATIONAL INSTABILITIES OF PERIODIC SINE GORDON WAVES: A GEOMETRIC ANALYSIS

N. M. Ercolani[1]
M. G. Forest[1]
D. W. McLaughlin[2]

Abstract
We study the stability of the periodic sine-Gordon equation. Linearizing about an arbitrary N-phase wavetrain, we characterize all unstable solutions of this linearized equation and derive explicit formulae for the growth rates. Moreover, using the geometric structure of the sine-Gordon phase space, we describe the nonlinear saturation of these instabilities.

1980 Mathematics Subject Classification: 35B10,35B20,35Q20,58G25
[1]Supported in part by NSF MCS-8202288, DMS 8411002
[2]Supported in part by AFOSR 830227, NSF DMS 8403187

I. Introduction

We study the stability of solutions to the sine-Gordon (SG) equation,

$$u_{tt} - u_{xx} + \sin(u) = 0 \tag{I.1a}$$

where $u(x,t)$ is a C^∞ function satisfying periodic boundary conditions,

$$u(x+L,t) = u(x,t) \pmod{2\pi} . \tag{I.1b}$$

Our results may be viewed as a generalization of the classical "modulation instability" [7,8,9]. Let us review the classical argument as it applies to the sine-Gordon equation: One begins with a very special solution of (I.1a,b), which is independent of x and periodic in t,

$$u_0(t;k) = 2 \arcsin(k\, \mathrm{sn}(t;k)). \tag{I.2a}$$

Here $\mathrm{sn}(t;k)$ is the Jacobi elliptic function of modulus k. Since u_0 is independent of x, the linearized (SG) equation about u_0,

$$\tilde{u}_{tt} - \tilde{u}_{xx} + (\cos u_0)\tilde{u} = 0, \tag{I.2b}$$

can be solved by Fourier analysis. Inserting the Fourier expansion

$$\tilde{u} = \sum_{n=-\infty}^{\infty} \hat{u}^{(n)}(t) e^{iK_n x}, \quad K_n \equiv \frac{2\pi n}{L}, \tag{I.3a}$$

into (I.2b) yields a Hill equation for each Fourier mode,

$$(\partial_{tt} + \cos u_0)\hat{u}^{(n)}(t) = -K_n^2 \hat{u}^{(n)}. \tag{I.3b}$$

Using the explicit formula (I.2a) for u_0, this can be written in the form of the classical Lamé equation,

$$(-\partial_{tt} + 2k^2 \mathrm{sn}^2(t;k))\hat{u}^{(n)}(t) = (K_n^2 + 1)\hat{u}^{(n)}. \qquad \text{(I.4)}$$

The Floquet spectral theory of this Hill equation, with Lamé potential $2k^2 \mathrm{sn}^2(t;k)$, has precisely one instability gap. As a consequence, the mode $\hat{u}^{(n)}(t;K_n)$ is exponentially growing in t if and only if $K_n^2 \in (0,k^2)$. That is, the x-independent solution u_0, (I.2a), is linearly unstable to long wavelength perturbations.

Here we show that this classical instability is a special case of a very general instability - one which arises for rather general periodic solutions of the sine-Gordon equation. Furthermore, it arises in a number of other physically important integrable wave equations [8,9,10]. In our analysis of the stability of general sine-Gordon solutions, we use new methods from the mathematical theory of integrable soliton wave equations. Thus, in our work, methods for special integrable equations replace the classical methods for special (x-independent) solutions.

A key mathematical feature of these soliton equations is that the periodic phase space is stratified by tori, which are invariant manifolds for the integrable flows. This geometrical structure is the basis of our analysis. For sine-Gordon, McKean [2] has shown that in the space of L-periodic, C^∞ functions of x, an x-independent solution (like u_0 above) has a neighborhood which is stratified by invariant tori, most of which are infinite-dimensional. A similar result should hold in the neighborhood of any of the special solutions known as N-phase waves. These solutions are expressible in terms of hyperelliptic functions [5]. An N-phase wave is a point on an N-dimensional torus. Again, a generic torus in the neighborhood of an N-phase wave will be infinite-dimensional and have all its degrees of freedom excited. (In the classical example, the x-independent solution has one degree of freedom and corresponds to a circle or 1-torus. The fact that all Fourier modes can be excited ($\hat{u}(n,t) \neq 0 \ \forall n$) indicates that, in its neighborhood, the generic torus is infinite dimensional.) There are also infinite-dimensional tori in which some degrees of freedom are not excited. We will call all non-generic tori degenerate. Because of technical difficulties which arise in establishing the analytic structure of this stratification when $N > 1$, this work must be regarded as formal.

How can linear instabilities emerge from the regular structure of invariant tori? Certainly they will not arise from perturbations of a generic torus. In principle one can construct action-angle variables near a generic torus by Liouville's method. In these variables the time flows are linear on the torus. Hence linearization will just reproduce these flows which are manifestly bounded in time.

It seems, then, that the instability has to do with degenerate tori. But this can't be the whole story. The KdV equation has degenerate solutions, but these are all stable [7]. What is crucial is the geometry of the stratification near a degenerate torus which, for S.G., differs considerably from that of KdV. For a thorough mathematical discussion of this geometry we refer the reader to our forthcoming paper [11]. Here we will be content to present our results which may be loosely summarized as follows: (i) If u is generic (i.e., if all degrees of freedom are excited), then u is linearly stable, (ii) if u has a degree of freedom which is not excited, an exponential instability may occur. (iii) When degrees of freedom are not excited, we characterize all the exponentially unstable modes and derive explicit formulas for the rates of instability. (iv) The nonlinear state to which these instabilities saturate is clearly characterized by the geometric structure of the phase space.

II. Some Spectral Theory

We will now briefly review some facts concerning S.G. and its associated spectal theory which will enable us to state our results. For more details on these matters the reader is referred to [2,4].

The phase space here is $F = C_L^\infty \times C_L^\infty$ where C_L^∞ denotes the space of C^∞ functions which are L-periodic mod 2π (i.e. $(u,v) \in F$ only if $u(x+L) - u(x) + 2\pi n$, for some $n \in Z$ and $v(x+L) = v(x)$). The S.G. flow, as given by

$$\begin{aligned} u_t &= v \\ v_t &= u_{xx} - \sin(u). \end{aligned} \tag{II.1}$$

is integrated with the Fadeev-Takhtadjian eigenvalue problem,

$$\mathcal{L}\psi = \left[\begin{pmatrix} 0 & -1 \\ 1 & 0 \end{pmatrix}\frac{d}{dx} + \frac{u_x+u_t}{4}\begin{pmatrix} 0 & 1 \\ 1 & 0 \end{pmatrix} + \frac{1}{16\sqrt{E}}\begin{pmatrix} e^{iu} & 0 \\ 0 & e^{-iu} \end{pmatrix}\right]\psi = \sqrt{E}\ \psi. \tag{II.2}$$

If u flows in time according to (II.1), the Floquet spectrum of (II.2) remains invariant.

To display this spectrum we use the fundamental solution of (II.2), $M(x;E,\vec{u})$, which is defined by the initial conditions $M(x=0) = I$. One then defines the <u>transfer matrix</u>, $M(x=L;E,\vec{u})$, which maps $\vec{\psi}(x;E)$ across one period of $\vec{u}$. The <u>Floquet discriminant</u>, the fundamental object of the theory, is given by

$$\Delta(E;\vec{u}) = \mathrm{tr}[M(L;E,\vec{u})]. \tag{II.3}$$

The spectrum of (II.2) is characterized via Δ:

$$\text{spectrum (II.2)} = \{E \in \mathbb{C} \mid \Delta(E,\vec{u}) \text{ is real and } \Delta^2(E,\vec{u}) - 4 \leqq 0\}.$$

If $\vec{u}$ changes with time according to the sine-Gordon equation, $\frac{d}{dt}\Delta(E,\vec{u}(t)) = 0$, hence the spectrum of (II.2) remains invariant. Put differently, the isospectral level sets in F,

$$M_{\vec{u}} = \{\vec{w} \mid \Delta(E,\vec{w}) = \Delta(E,\vec{u})\},$$

are invariant level sets for the S.G. flow. These level sets are the tori near an N-phase solution to which we alluded in the introduction.

$\Delta(E,\vec{u})$ is an analytic function of E except at $E = 0$ and $E = \infty$ where it has essential singularities of exponential type. On the positive real axis, $\Delta(E,\vec{u})$ is (a) roughly sinusoidal, (b) real, (c) $-2 \leqslant \Delta(E,\vec{u}) \leqslant 2$, and (d) vanishes infinitely often. The periodic (anti-periodic) spectrum $\{E_k^+\}$ ($\{E_k^-\}$) are the roots of $\Delta - 2$ ($\Delta + 2$). These spectra satisfy the following list of properties: (II.4)

(i) If E_j is a root of $\Delta \pm 2$, then E_j^* is also a root.

(ii) If E_j is a simple root, then E_j is negative real or $\mathrm{Im}(E_j) \neq 0$.

(iii) If E_j is a positive real root, then E_j has algebraic multiplicity which is even and $\geqq 2$.

(iv) If E_j is a root, its algebraic multiplicity need not be $\leqq 2$.

(v) If $\vec{u}$ is an N-phase solution, there are only finitely many <u>non-positive</u> roots, E_j, having algebraic multiplicity $\geqq 2$. Furthermore, for all but a discrete set of values of L, the algebraic multiplicity of E_j is <u>always</u> $\leqq 2$.

(iv) $E_j = (\frac{j\pi}{L})^2 + O(1)$ as $j \to +\infty$

$E_j = (\frac{16j\pi}{L})^2 + O(1)$ as $j \to -\infty$.

One has infinite product representations of $\Delta'(E)$ and $\Delta^2(E) - 4$ [2]:

$$\Delta^2(E) - 4 = 4E \prod_{n>0} (1 - \frac{E}{E_n^-})(1 - \frac{E}{E_n^+}) \prod_{n<0} (1 - \frac{E_n^-}{E})(1 - \frac{E_n^+}{E})$$
$$\Delta'(E) = C_2(1 + \frac{1}{16E} (\prod \frac{E_n}{\Lambda_n^2})^{1/2}) \prod_{n>0} (1 - \frac{E}{\Lambda_n}) \prod_{n<0} (1 - \frac{\Lambda_n}{E}) \quad \text{(II.5)}$$

Here C_2 is a constant and $\{\Lambda_n\}$ are the roots of $\Delta'(E)$ which have an asymptotic distribution similiar to $\{E_n\}$.

The function $\sqrt{\Delta^2(E)-4}$ plays a central rôle in the theory. This function has branch points at the simple zeros of $\Delta^2(E) - 4$, $\{\mathcal{E}_k^+,\mathcal{E}_k^-\}$. Taking these simple zeros only, we construct a Riemann surface $\mathcal{R}$ over the E plane for a function $R(E)$:

$$R^2(E) = 4E \prod_{k>0} (1 - \frac{E}{\mathcal{E}_k^-})(1 - \frac{E}{\mathcal{E}_k^+}) \prod_{k<0} (1 - \frac{\mathcal{E}_k^-}{E})(1 - \frac{\mathcal{E}_k^-}{E}) \quad \text{(II.6)}$$

On this Riemann surface, with coordinates $(E,R(E))$, we lay down a canonical basis of homology cycles, $\{a_i,b_i\}$. Specifically, we make the following choice of a_j-cycles: Order those branch points on the negative real axis as $0 > E_1 > E_2 > \ldots > E_{2N}$. The remaining branch points occur in conjugate pairs E_{2N+1}, $E_{2N+2} = E_{2N+1}^*$, etc. Cycle a_1 encycles 0 and E_1 on one sheet. Cycle a_2 encycles E_2 and E_3 on the same sheet ..., cycle a_N encycles E_{2N-2} and E_{2N-1} on the same sheet. As for the conjugate pairs, cycle a_{N+1} begins at E_{2N+1} and travels to $E_{2N+2} = E_{2N+1}^*$ by crossing the positive real axis. At E_{2N+2} it changes sheets and returns, across the positive real axis, to E_{2N+1}. The remaining a_j cycles are constructed between conjugate pairs of branch points in an analogous fashion. The generic picture is depicted below.

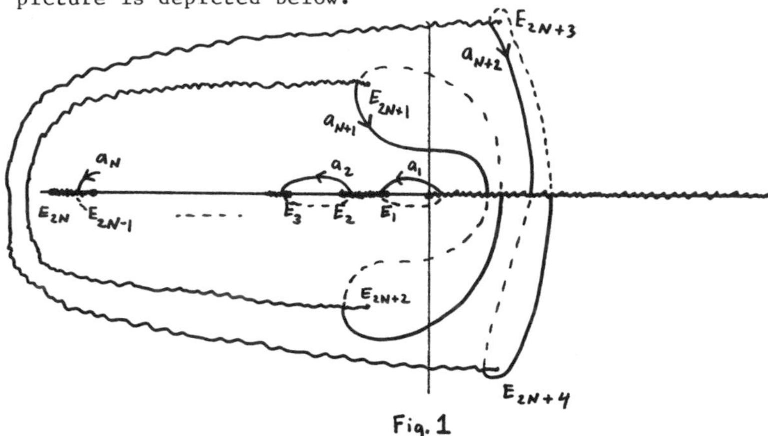

Fig. 1

This choice of cycles is the most natural one for implementing the constraints which insure that u is a real-valued function of x and t. For more on these reality constraints see [11].

Consider the differential

$$\Omega^{(-)} \equiv \frac{1}{L} \quad \frac{\Delta'}{E\sqrt{\Delta^2-4}} \, dE \tag{II.7}$$

where L denotes the spatial period. The infinite product expansions (II.5) show that this differential is well defined on the Riemann surface $(E,R(E))$. Moreover, it can be shown to satisfy the following properties: (we define a companion differential, $\Omega^{(+)}$, as well for later reference):

(i) $\oint_{a_j} \Omega^{(\pm)} = 0$

(ii) $\Omega^{(\pm)}$ is holomorphic except near $E = 0$ and $E = \infty$.

(iii) $\Omega^{(\pm)} \simeq (\frac{\xi^{-2}}{16} + \text{holo} \ldots)d\xi$ for $E = \xi^2$ near 0

$\Omega^{(\pm)} \simeq (\xi^2 + \text{holo} \ldots)d\xi$ for $E = \xi^{-2}$ near ∞

Indeed, these three properties uniquely characterize $\Omega^{(\pm)}$. Given $\Omega^{(-)}$, the "Hochstadt formula" yields Δ:

$$\cosh^{-1} \frac{\Delta}{2} = \int^{E} \Omega^{(-)} dE. \qquad \text{(II.9)}$$

The linear growth rates we present in the next section will be expressed explicitly in terms of the differentials $\Omega^{(\pm)}$.

III. The Identification of the Instability

In our approach to studying linear stability we seek to construct a basis for the L^2-completion of F which consists entirely of solutions to the linearized sine-Gordon equation (L.S.G.):

$$\begin{aligned} U_t &= V \\ V_t &= U_{xx} + (\cos u_0)U, \end{aligned} \qquad \text{(III.1)}$$

where u_0 is any N-phase solution of S.G. We construct almost all (N elements are missing and must be added for a complete set) of this basis from the following eigenvalue problem

$$\begin{aligned} \vec{\psi} &= \sqrt{E}\ \vec{\psi} \\ \vec{\psi}(x+L) &= \rho\vec{\psi}(\vec{x}) \end{aligned} \qquad \text{(III.2)}$$

where $\mathcal{L}$ is the linear operator (II.2) with N-phase potential $\vec{u}_0$, and the Floquet multiplier is an eigenvalue of the transfer matrix.

For fixed E,

$$\begin{aligned} \rho &= \frac{\Delta(E) \pm \sqrt{\Delta^2(E)-4}}{2} \\ &= \exp\left(i \int_{P_0}^{P} \Omega^{(-)}\right) \text{ (by Hochstadt's formula),} \end{aligned}$$

where $\qquad P_0 = (E_0^+, R(E_0^+))$ and $p = (E, \pm R(E))$.

Thus for each $p \in \mathcal{R}$ we have a unique (up to normalization) vector $\vec{\psi}(x)$ solving (III.2). We can regard $\vec{\psi}$ as a vector function

$$\vec{\psi}(p;x,t) = (\psi_1,\psi_2)^t$$

on $\mathcal{R}$. (The t-dependence is determined by the sine-Gordon flow acting on $\vec{u}_0$ in $\mathcal{L}$.) This function is called the <u>Baker vector</u>, and it can be constructed explicitly from hyperelliptic functions on $\mathcal{R}$ and the differentials $\Omega^{(\pm)}$ [12].

From $\vec{\psi}$ we form the squared eigenfunction

$$f(p,x,t) = i\psi_1\psi_2. \tag{III.3}$$

A straightforward calculation [3] shows that the vector

$$\vec{\Phi}(p;x,t) = \binom{f}{f_t} \tag{III.4}$$

solves L.S.G. (III.1). If $p_j = (E_j,R(E_j))$ then $\vec{\Phi}_j^{\pm} = \Phi(p_j^{\pm};x,t)$ is L-periodic in x and represents an element of F which solves (III.1).

We consider just those $\Phi_j^{\pm}$ associated to periodic or anti-periodic spectrum of algebraic multiplicity 2 (<u>double points</u>). For generic values of L, by (II.4.v), these vectors contain all the potentially unstable modes. Using the explicit representation of the Baker function $\vec{\psi}$, one can show that $\vec{\Phi}_j^{\pm}$ has the form

$$\vec{\Phi}_j^{\pm}(x,t) = \exp\left(ix \int_{P_0}^{p_j^{\pm}} \Omega^{(-)} + it \int_{P_0}^{p_j^{\pm}} \Omega^{(+)}\right)\vec{\phi}_j(x,t), \tag{III.5}$$

where the components of $\vec{\phi}_j$ are bounded quasi-periodic functions of both x and t. Hence, $\vec{\Phi}_j^{\pm}$ has the form of a Bloch wavefunction. The stability of this mode now is just a question of determining whether or not $\int_{P_0}^{P_j} \Omega^{(\pm)}$ has a non-zero imaginary part. For instance, we have the following result when u_0 is strictly periodic (i.e. $u_0(x+L) = u_0(x)$).

THEOREM If u_0 is strictly periodic, then

(i) For all double points, E_j, $\Phi_j^{\pm}$ is a bounded function of x.

(ii) If E_j is a positive real double point, $\Phi_j^{\pm}$ is uniformly bounded in t.

(iii) If E_j is a non-positive double point, $\Phi_j^{\pm}$ grows exponentially in t. The growth rate of this mode is given by the explicit formula

$$\Phi_j^{\pm} \sim \exp(-t \{\mathrm{Im} \int_{P_0}^{P_j^{\pm}} \Omega^{+}\}). \qquad \text{(III.6)}$$

When u_0 is not strictly periodic (for example, kink trains) an analogous theorem holds but is more complicated to state [11].

Thus we have found that a generic perturbation of an N-phase SG solution, u_0, is only unstable to a finite number of modes identified by the non-positive double points in the spectrum of $\mathcal{L}(\vec{u}_0)$. In addition, in [11] we construct a biorthogonal basis of F, each member of which solves linearized sine-Gordon. This basis will be useful for perturbation theory.

IV. The Saturation of the Instability

In the preceding section we used double points to identify degenerate tori, their associated instabilities and linear growth rates. These double points are (SG) constants of the motion which are related to action variables for the integrable (SG) Hamiltonian system. The origin of the instability and the states to which it saturates can be clearly and easily understood using dynamical variables, the $\vec{\mu}$'s, which are related to the angle variables. In this section we (i) define the $\vec{\mu}$ variables, (ii) describe difficulties with them such as the "reality problem", (iii) use them to show the origin of the instability, and (iv) clearly describe the saturated states.

IV.A. Definition of $\vec{\mu}(x,t)$

The transfer matrix, $M = M(L;E;\vec{u})$, yields the canonical variables for periodic (SG). While the trace of M, $\Delta(E;\vec{u})$, determines the E_j as the roots of $\Delta = \pm 2$, the conjugate variables, μ_j, are the roots of $M_{12}(L;E = \mu;\ \vec{u}) = 0$.

By a dynamical degree of freedom, we mean one variable μ_j which is free to move. Each μ_j can be associated to a pair of simple periodic spectra, (E_{2j-1}, E_{2j}). As long as $E_{2j-1} \neq E_{2j}$, μ_j is free to move, and this degree of freedom is said to be "excited". When $E_{2j-1} = E_{2j}$, μ_j becomes stationary and locked to the double point: $E_{2j-1} = \mu_j = E_{2j}$. In this case the degree of freedom is removed and said to be "closed".

When only N degrees of freedom are excited, the N-phase SG wave is given in terms of the variables $(\vec{E},\vec{\mu}) = (E_1,\ldots,E_{2N}, \mu_1,\ldots,\mu_N)$ by the formula

$$u = i\ell n\Big(\prod_1^N \mu_j \Big/ \sqrt{\prod_1^{2N} E_j}\Big), \tag{IV.1}$$

where $\mu_j(x,t)$ satisfy a system of ode's in x and t, whose precise form is not important here. We only remark that these ode's are well-defined on the Riemann surface of $\sqrt{\Delta^2(E;u)-4}$, Figure 1. (See [4,5] for a thorough discussion of the $\vec{\mu}$ ode's.)

IV.B The Sine-Gordon Reality Problem

Here we want to describe the SG reality problem and compare it to the simpler situation for KdV. Then in the rest of Section IV we will connect this reality discussion to a compelling geometric realization of both the origin and saturation of (SG) modulational instabilities.

The sine-Gordon equation as originally posed has a *real* phase space:

$$F = C_L^\infty \times C_L^\infty$$

where C_L^∞ consists of *real* functions of x. Through the function M we passed to variables E_j, μ_j on F which can take on complex values even though the point they label in F is real. This is different from KdV where, because of the self-adjointness of Hill's operator, E_j and μ_j must always be real for real potentials. For (SG) the reality conditions on E_j are elementary and given in (II.4). The *reality problem* is to determine what sequences of complex values $\{\mu_i\}_{i \varepsilon \mathbb{Z}}$ are roots of $M_{12}(L;E,\vec{u})$ for $\vec{u} \varepsilon F$. For KdV this problem could be treated one "μ_j" at a time: μ_j could take any value in the real gap between E_{2j-1} and E_{2j} independently of the position of $\mu_i (i \neq j)$ in the $i^{\underline{th}}$ gap (E_{2i-1}, E_{2i}). For sine-Gordon the situation is much more complicated. The complex coordinates $(\mu_1, \ldots, \mu_N)$ do live on an N-dimensional real manifold; however, the admissible μ_j locations are "coupled" by the reality conditions on $u(x)$, $u_t(x)$. The admissible locations of μ_1, for example, do not lie on a fixed curve which is independent of $\mu_2, \ldots, \mu_N$. Rather, the set of admissible locations of μ_1 depend upon $\mu_2, \ldots, \mu_N$ in a complicated way.

We provide an answer for the reality problem when $\vec{u}$ is an N-phase solution. We parametrize the *admissible sequences*

$\{\mu_j\}_{j\in\mathbb{Z}}$ by a real N dimensional torus. In this case, all but N degrees of freedom are closed, $\mu_j = E_{2j}$. For the N open degrees of freedom one has explicit expressions

$$\mu_j = \mu_j(\theta_1,\ldots,\theta_N), \tag{IV.2}$$

$\theta_i \in [0,2\pi)$, given in terms of theta functions. One finds that although the individual μ_j do not lie on fixed closed curves, they are constrained to lie in bounded regions of the complex plane. For details of this explicit parametrization see [5]. (In fact, in this reference, we parametrize the μ_j - sequences by a real "Kummer variety" rather than an N-torus. This may seem to complicate matters, but in fact it has the tremendous advantage that the μ_j are now fractional linear functions of the coordinates).

Although we have explicitly characterized the locations of the $\vec{\mu}$ variables, pointwise information about $\vec{\mu}(x,t)$ is very difficult to control. In every application to date, we have not needed pointwise information. Rather, it has been sufficient to characterize (as paths of integration) all closed cycles on the N-torus. Using the Kummer representation of [5], we can identify the homology class of any closed paths on the N-torus as a linear combination of $\vec{a}$ cycles on the Riemann surface (Fig. 1). This topological characterization of the $\vec{\mu}$ locus is sufficient for integration.

IV.C Saturated States

Here we use the topological characterization of the admissible $\vec{\mu}$-locus to see how the linear instabilities saturate.

When we introduce an order ε perturbation $\vec{u}_0 + \varepsilon\,\delta\vec{u}$ of a degenerate potential $\vec{u}_0$, a double point $(E_{2j-1} = E_{2j})$ in the spectrum of $\vec{u}_0$ splits into two simple points

($|E_{2j-1} - E_{2j}| = O(\varepsilon) \neq 0$) in the spectrum of $\vec{u}_0 + \vec{\delta}u$. When this happens, a degree of freedom, μ_j, is generated; the real N-torus which parametrized admissible $\vec{\mu}$, for $\vec{u}_0$, becomes a larger dimensional torus to accomodate the new degrees of freedom; and for each such new mode, another a-cycle, a_j, (see Fig. 1) must be introduced.

The saturation of the linear instabilities is most easily described in terms of these homology cycles. When $E_{2j-1} = E_{2j}$ is a positive real double point, the generation of a new a-cycle which encycles E_{2j-1} and $E_{2j}(= E^*_{2j-1})$ is described in Fig. 2.

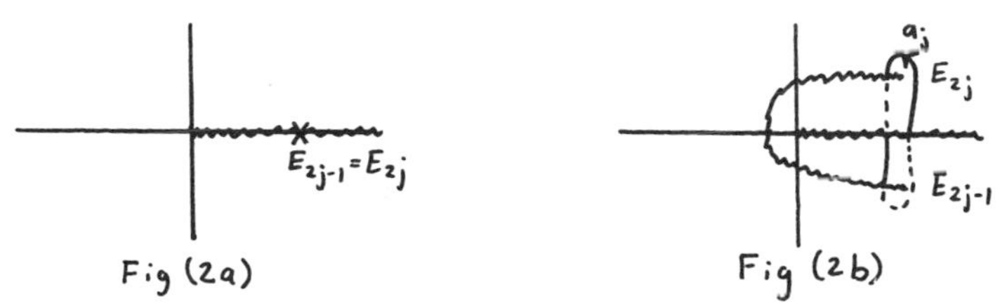

Fig (2a) Fig (2b)

When $E_{2j-1} = E_{2j}$ is a non-real (or real negative) double point, this unfolding is depicted in Fig. 3. In order to simplify our representation we have taken the liberty of choosing new cuts on $\mathcal{R}$ and a new basis of a-cycles.[1] For a more systematic topological discussion of the generation and degeneration of these a-cycles we refer the reader to [11].

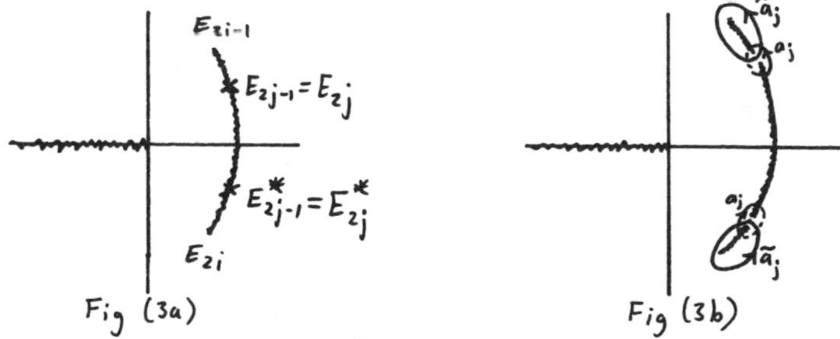

Fig (3a) Fig (3b)

[1]Note that a_j (similarly $\tilde{a}_j$) denotes two disjoint cycles a_j^+ and a_j^-; by this we mean that $a_j = a_j^+ + a_j^-$, a formal sum in the sense of integration theory: $\int_{a_j} = \int_{a_j^+} + \int_{a_j^-}$.

Observe that in Fig. 2b and 3b, the a_j cycle need only open $O(|E_{2j-1} - E_{2j}|) = O(\varepsilon)$. However, in Fig (3b), $\tilde{a}_j$ must encycle E_{2i-1} (or E_{2i}) and hence must open $O(1)$.

Reversing the perturbation by letting ε go to 0 we see the origin of the instability. The cycle a_j, in either figure, is a vanishing cycle as $\varepsilon \to 0$. However, $\tilde{a}_j$ is constrained, by its own topology, not to vanish. The pinched separatrix (Fig. 4) to which it limits provides a topological phase portrait of the instability.

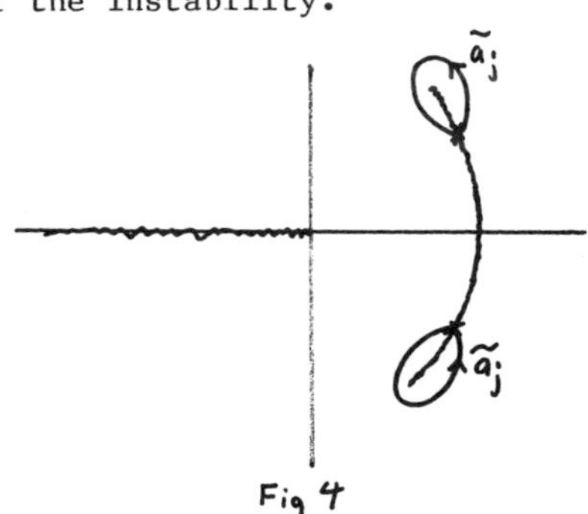

Fig 4

This description now makes it clear that in order to analytically capture the saturated state one must add to the dynamical variables $(\mu_1, \ldots, \mu_N)$ of $\vec{u}_0$, the finite set of variables $\{\tilde{u}_j\}$ associated to all double points which are not real and positive.

Having thus reduced the perturbation to finitely many degrees of freedom, one can proceed to study it analytically via Baker functions on Riemann surfaces (as we have begun to do [5]) or numerically through the $\vec{\mu}$-o.d.e.'s [10].

V. Conclusion

In this article we have established, for the periodic sine-Gordon equation, that the classical modulational instability is only a very special case of a general instability. Using the fundamental object of the periodic spectral theory, the Floquet discriminant, we (i) identify all linear instabilities of N phase solutions, (ii) compute explicit formulas for the linear growth rates, and (iii) represent the nonlinear state which emerges from the instability. It is this nonlinear state which must be used as the starting point in practical perturbation calculations such as those which arise in the physics of Josephson junction devices.

References

1. L. A. Takhtajian and L. D. Faddeev, "Essentially nonlinear one dimensional model of classical field theory", Theor. and Math. Phys. 21: 1046 (1974).

2. H. P. McKean, "The Sine-Gordon and Sinh-Gordon equations on the circle", C.P.A.M. 34: 197-257 (1981).

3. M. G. Forest and D. W. McLaughlin, "Modulations of Sinh-Gordon and Sine-Gordon wavetrains", Stud. Appl. Math., 68: 11-59 (1983).

4. M. G. Forest and D. W. McLaughlin, "Spectral theory for the periodic Sine-Gordon Equation: A concrete Viewpoint", J. Math. Phys. 23 (7): 1248-1277 (1982).

5. N. M. Ercolani and M. G. Forest, "The Geometry of Real Sine-Gordon Wavetrains", Comm. Math. Phys., to appear.

6. N. M. Ercolani, M. G. Forest and D. W. McLaughlin, "Modulations of Two-Phase Sine Gordon Wavetrains", Stud. Appl. Math, 69: 91-101 (1984).

7. T. B. Benjamin, "Instabilities of Periodic Wave Trains in Nonlinear Dispersive Systems", Proc. Royal Soc. A 299: 59-75 (1967).

8. D. J. Benney and A. Newell, "Propagation of Nonlinear Wave Envelopes", Jour. Math. and Phys. 46 (2): 133-139 (1967).

9. V. E. Zacharov, Dissertation, Inst. Nucl. Phys., Siberian Div., USSR Acad. Sci., 1966.

10. E. Tracy, Thesis, University of Maryland, 1984; (with H. H. Chen, and Y. C. Lee), "Study of Quasiperiodic Solutions of the Nonlinear Schrodinger Equation and the Nonlinear Modulational Instability", Physical Review Letters 53 (3): 218-221, 1984.

11. N. M. Ercolani, M. G. Forest and D. W. McLaughlin, "Geometry of the Benjamin-Feir Instability", preprint, June, 1984.

12. E. Date, "Multisoliton solutions and Quasi-periodic solutions of Nonlinear Equations of Sine-Gordon type", Osaka J. Math., 19: 125-158 (1982).

13. H. P. McKean and P. van Moerbeke, "The spectrum of Hill's equation", Invent. Math. 30 (11): 217-274 (1975).

[1]Department of Mathematics
Ohio State University
Columbus, Ohio 43210

[2]Program in Applied Mathematics
University of Arizona
Tucson, Arizona 85721

Lectures in Applied Mathematics
Volume 23, 1986

ON COMPLETELY INTEGRABLE SYSTEMS WITH HIGHER ORDER CORRECTIONS

Yuji Kodama[1]

ABSTRACT. The integrable Hamiltonian systems with certain class of perturbations such as higher order corrections are discussed.

0. INTRODUCTION. In many cases, integrable systems appearing in the physical problems are just the approximated equations of the original physical systems in an appropriate asymptotic sense. Among these systems, there are linearized equations for certain nonlinear problems (of finite or infinite degrees of freedom), the Korteweg-de Vries (KdV) equation and the nonlinear Schrödinger (NLS) equation describing weakly and strongly dispersive nonlinear wave phenomena, respectively, in the leading order approximations [1]. (Both KdV and NLS equations are known as the completely integrable Hamiltonian systems by means of the method of inverse scattering transformation [2].) Those integrable systems are valid for certain time length determined from the physical settings (e.g. the order of nonlinearity and/or the smoothness of the initial conditions). For the problem related to large time behavior of the solutions (such as the stability of the critical points), one needs to study the effects of the

---

[1]This work is partially supported by NSF.

higher order terms which are neglected in the derivation of the integrable systems. For finite dimensional systems, this problem has been studied extensivley [3], and the several methods for analyzing the problem have been also developed. A most successful one is the Binkhoff normal form thoery in which the perturbed Hamiltonian systems can be transformed into the integrable normal forms by successive canonical transformations under the non-resonant condition [3].

In recent years, there has been several discussions related to the problem for infinite dimensional systems, such as the higher order corrections of the KdV and the NLS equations [4,5,6,7]. Especially, reference [7] shows that the perturbed KdV and NLS equations can be transformed into the integrable systems in the same order by canonical transformations.

The main purpose in this lecture is to study the integrability of such perturbed Hamiltonian systems (i.e. the integrable systems with certain class of perturbations such as the higher order corrections). In the sections 1 and 2, we describe the type of perturbations considered here and give several physical examples which will be discussed through this lecture. The examples are (a) N-uncoupled harmonic oscillators (a well-known classical example [3]), (b) the linearized KdV equation (an example of the linear dispersive wave equation), and (c) the KdV equation. (The readers will find some similar properties among these examples, and the first example may help to understand the other examples.) In the section 3, we try to find the integrals for the perturbed systems for studying their integrability. Particularly, in the case of the linearized KdV equation, we show that the perturbed equation actually possesses infinitely many integrals up to second order by use of an algebra of the differential polynomials. It is however difficult to find all of the integrals of the perturbed equation in general. But for a certain class of perturbations, such as our case, the perturbed equations for given order can be characterized by only a few integrals. So, if we found several integrals (not necessary to be all) of the perturbed equation, one can show that the equation is actually integrable up to the same order. This is the main result in this lecture. In order to show this, we first define, in section 4, a normal form corresponding to the perturbed equation on a constant surface determined by some integrals of the unperturbed system. The normal form can be given in an integrable form if we found enough integrals to characterize the perturbed equation. Then, in section 5, we construct a map (a canonical transformation) between the perturbed equation and the corresponding normal form, and show that our examples (a)-(c) are integrable up to first order correction. In this lecture, we mainly consider the first order problem, but the higher order problems can be studied in the same manner used here.

1. INTEGRABLE SYSTEMS WITH PERTURBATIONS. We consider the following form of evolution equation as a perturbed integrable system for dynamical variable $u(t)$ on certain smooth manifold M,

$$\dot{u}(t) = \frac{d}{dt} u(t) = X(u(t);\epsilon) \tag{1}$$

$$= X^{(0)}(u) + \epsilon X^{(1)}(u) + \dots \quad \text{for} \quad |\epsilon| \ll 1,$$

which is also written in a Hamiltonian form, i.e. there exist a skew-symmetric Hamiltonian operator $\mathcal{J}$ and a Hamiltonian $H \in C^{\infty}(M)$ such that $X(u,\epsilon)$ is the Hamiltonian vector field given by

$$X(u;\epsilon) = \mathcal{J}(u;\epsilon)\,\nabla H[u;\epsilon] \,. \tag{2}$$

Here, $\mathcal{J}$ and H are also given in the power series of ϵ,

$$(3)\qquad \begin{cases} \mathcal{J}(u;\epsilon) = \mathcal{J}^{(0)}(u) + \epsilon\mathcal{J}^{(1)}(u) + \dots , \\ H[u;\epsilon] = H^{(0)}[u] + \epsilon H^{(1)}[u] + \dots , \end{cases}$$

and, the gradient of H, ∇H, is defined in the usual way, i.e. for any vector field Y,

$$(4)\qquad (Y \bullet \nabla H)[u] = \lim_{\delta \to 0} \frac{d}{d\delta} H[u + \delta Y] .$$

The poisson bracket generated by $\mathcal{J}$ for functions $F,G \in C^{\infty}(M)$ is given by

$$(5)\qquad \{F,G\}[u] = \lim_{\delta \to 0} \frac{d}{d\delta} F[u + \delta\mathcal{J}\nabla G] = (\nabla F \bullet \mathcal{J}\nabla G)[u]$$

Recall that $\mathcal{J}$ is a Hamiltonian operator if (5) forms a Lie algebra on $C^{\infty}(M)$ (i.e. the skew-symmetric bilinear form $\{\bullet,\bullet\}$ satisfies the Jacobi identity). The unperturbed equation $\dot{u} = X^{(0)} = \mathcal{J}^{(0)}\nabla H^{(0)}$ (i.e. (1) with $\epsilon = 0$) is an integrable Hamiltonian system in the usual sense: $\exists$ a set of $I_{\nu}^{(0)}[u] \in C^{\infty}(M)$ for $\nu \in \Gamma_0$ (where the number of elements of the index set Γ_0 is equal to the degree of freedom of the system), such that

$$\left.\begin{array}{ll} \text{I(a)} & X^{(0)}\bullet\nabla I_{\nu}^{(0)} = 0 \quad \text{for } \forall\, \nu \in \Gamma_0 , \\ \text{I(b)} & \{I_{\nu}^{(0)}, I_{\mu}^{(0)}\}^{(0)} = \nabla I_{\nu}^{(0)}\bullet\mathcal{J}^{(0)}\nabla I_{\mu}^{(0)} = 0 \quad \text{for } \forall\, \nu,\mu \in \Gamma_0 , \\ \text{I(c)} & \{\nabla I_{\nu}^{(0)}[u]\}_{\nu\in\Gamma_0} \text{ are independent.} \end{array}\right\} \quad \text{(I)}$$

(In the case of infinite dimensional system, one needs more careful definition. See examples below.)

For the perturbations $X^{(\ell)}(u)$, consisting of the polynomials or the differential polynomials in u, we define the degree of $X^{(\ell)}(u)$, say $\mathrm{Deg}(X^{(\ell)}(u))$, in the form,

$$\text{(6)} \qquad \mathrm{Deg}(X^{(\ell)}(u)) := (\#(u) \text{ in } X^{(\ell)}) \bullet \mathrm{Deg}(u) + (\#(\partial_x) \text{ in } X^{(\ell)}) \bullet \mathrm{Deg}(\partial_x),$$

where $\#(u)$ and $\#(\partial_x)$ denote the number of u's and the number of derivatives $\partial/\partial x$, respectively. Namely, $\mathrm{Deg}(X^{(\ell)})$ indicates the scales of nonlinearity and smoothness of the vector field. Here $\mathrm{Deg}(u)$ and $\mathrm{Deg}(\partial_x)$ are determined from the self-similar property of the unperturbed equation based on the scaling of the physical setting. In this lecture, we consider the perturbations satisfying,

$$\left.\begin{array}{lll} X(a) & \text{for each } \ell, \ \mathrm{Deg}(X^{(\ell)}(u)) \text{ is fixed,} \\ X(b) & \mathrm{Deg}(X^{(\ell+1)}(u)) = \mathrm{Deg}(X^{(\ell)}(u)) + \mathrm{Deg}(u), \\ X(c) & \#(u) \text{ in } X^{(\ell+1)} \le \#(u) \text{ in } X^{(\ell)} + 1. \end{array}\right\} \qquad (X)$$

It should be noted that these conditions (X) (i.e. the ordering of $X^{(\ell)}$'s) are naturally appeared in the method of asymptotic expansions used in the derivation of the equation (1) from the original physical problem [1].

2. EXAMPLES FOR THE PERTURBED SYSTEMS. Here we give three examples of the perturbed equations which will be studied through this lecture.

2a. N-uncoupled harmonic oscillators [3]. Our first example is a classical example of the weakly coupled nonlinear oscillators on $M = \mathbb{R}^{2N}$. The unperturbed vector field $X^{(0)} = \mathcal{J}^{(0)} \nabla H^{(0)}$ is given by the Hamiltonian describing the N-uncoupled harmonic oscillators, i.e. in terms of canonical coordinates $(x_1, \ldots, x_N, y_1, \ldots, y_N) = u \in \mathbb{R}^{2N}$,

$$\text{(7)} \qquad H^{(0)}[u] = \frac{1}{2} \sum_{\nu=1}^{N} \omega_\nu^{(0)} (x_\nu^2 + y_\nu^2).$$

Here the Hamiltonian operator $\mathcal{J}^{(0)}$ is given by $2N \times 2N$ antisymmetric matrix

$$(8) \qquad \mathcal{J}^{(0)} = \begin{pmatrix} 0 & -I \\ I & 0 \end{pmatrix}, \quad I = N \times N \text{ identity matrix.}$$

The perturbations are given by

$$(9) \qquad X^{(\ell)} = \mathcal{J}^{(0)} \nabla H^{(\ell)}$$

where $H^{(\ell)}$ is a homogeneous polynomial of degree $\ell + 2$,

$$(10) \qquad H^{(\ell)} = \sum_{|m| + |n| = \ell + 2} C_{mn} x^m y^n, \quad x^m y^n = \prod_{\nu=1}^{N} x_\nu^{m_\nu} y_\nu^{n_\nu},$$

$$\text{with } |m| = m_1 + \ldots + m_N, \quad m_\nu \in \mathbb{Z}^+.$$

Here $\mathrm{Deg}(X^{(\ell)})$ is given by $\#(u)$ in $X^{(\ell)}$ i.e. $\mathrm{Deg}(u) = 1$, so that $\mathrm{Deg}(X^{(\ell)}) = \ell + 1$. (We sometimes say $\mathrm{Deg}(H^{(\ell)}[u]) = \ell + 2$.)

2b. A linear dispersive wave (the linearized KdV equation). The second example is a dispersive wave equation where the unperturbed equation is the linearized KdV equation, i.e.

$$(11) \qquad \dot{u} - X^{(0)}(u) = u_{3x} := u_{xxx}.$$

This example may be considered as an infinite dimensional analogue of the previous example. The space M considered here for u is $M = C_\downarrow^\infty(\mathbb{R})$ defined by

$$C_\downarrow^\infty(\mathbb{R}) = \{u(x,\cdot) \mid u(x) \in C^\infty(\mathbb{R}) \text{ and, for any } m,n \in \mathbb{Z}^+, \ |x|^m \cdot |\partial_x^n u| \to 0, \text{ as } |x| \to \infty\}.$$

The first order perturbation satisfying the conditions (X) with the choice of degrees $\mathrm{Deg}(u) = 2$ and $\mathrm{Deg}(\partial_x) = 1$ (same as the case of the KdV equation) is given by

(12) $$X^{(1)}(u) = a_1^{(1)} u_{5x} + a_2^{(1)} uu_{3x} + a_3^{(1)} u_x u_{2x},$$

which can be put into a Hamiltonian form,

(13) $$X^{(1)} = \mathcal{J}^{(0)} \nabla H^{(1)} + \mathcal{J}^{(1)} \nabla H^{(0)},$$

where the Hamiltonian structure is given by

(14) $$\begin{cases} \mathcal{J}^{(0)} = \partial_x, & \mathcal{J}^{(1)} = b_1^{(1)} \partial_x^3 + b_2^{(1)} (\partial_x u + u\partial_x), \\ H^{(0)}[u] = -\frac{1}{2}\int_{-\infty}^{\infty} u_x^2 dx, & H^{(1)}[u] = b_3^{(1)} \int_{-\infty}^{\infty} uu_x^2 dx. \end{cases}$$

Here the sets of constants $\{a_\ell^{(1)}\}_{\ell=1}^3$ and $\{b_\ell^{(1)}\}_{\ell=1}^3$ are isomorphic, and $b_1^{(1)} = a_1^{(1)}$, $b_2^{(1)} = (2a_2^{(1)} - a_3^{(1)})/3$, $b_3^{(1)} = (a_2^{(1)} - 2a_3^{(1)})/6$. Note that the Hamiltonian structure is <u>not</u> unique, and in fact there is another choice given by

(14') $$\begin{cases} \mathcal{T}^{(0)} = \partial_x^3, & \mathcal{T}^{(1)} = c_1^{(1)} \partial_x^5 + c_2^{(1)} (\partial_x^3 u + u\partial_x^3), \\ \tilde{H}^{(0)}[u] = \frac{1}{2}\int u^2 dx, & \tilde{H}^{(1)}[u] = c_3^{(1)} \int u^3 dx \end{cases}$$

where $c_1^{(1)} = a_1^{(1)}$, $c_2^{(1)} = (3a_2^{(1)} - a_3^{(1)})/3$, $c_3^{(1)} = (-2a_2^{(1)} + a_3^{(1)})/6$. It should be also noted that the Hamiltonian operator for the infinite dimensional non-canonical systems (e.g. this example) generally depend on the coordinates on M unlike the finite dimensional case where the Darboux theorem holds (i.e. the symplectic structure is locally constant).

2c. The KdV equation. Let the KdV equation be in the following form,

(15) $$\dot{u} = X^{(0)}(u) = 6uu_x + u_{3x}, \quad \text{for } u \in C_\downarrow^\infty(\mathbb{R}),$$

where $X^{(0)} = \mathcal{J}^{(0)} \nabla H^{(0)}$, and

$$(16) \qquad \mathcal{J}^{(0)} = \partial_x, \quad H^{(0)}[u] = \int_{-\infty}^{\infty} (u^3 - \frac{1}{2} u_x^2)dx .$$

In this case, the degree of the perturbations are $\mathrm{Deg}(X^{(\ell)}) = 2\ell + 5$ with the choice of $\mathrm{Deg}(u) = 2$, $\mathrm{Deg}(\partial_x) = 1$, based on the self-similarity of the KdV equation (i.e. if $u(x,t)$ is a solution of (15), then $v(x,t) = \delta^2 u(\delta x, \delta^3 t)$ is also a solution). Thus, $X^{(1)}(u)$ satisfying $\mathrm{Deg}(X^{(1)}) = 7$ and the conditions (X) is

$$(17) \quad X^{(1)}(u) = a_1^{(1)} u_{5x} + a_2^{(1)} uu_{3x} + a_2^{(1)} u_x u_{2x} + a_4^{(1)} u^2 u_x$$

$$= \mathcal{J}^{(0)} \nabla H^{(1)}[u] + \mathcal{J}^{(1)}(u) \nabla H^{(0)}[u]$$

where the Hamiltonian structure is given by

$$(18) \quad \begin{cases} \mathcal{J}^{(1)}(u) = b_1^{(1)} \partial_x^3 + b_2^{(1)} (\partial_x u + u\partial_x) , \\ H^{(1)}[u] = \int_{-\infty}^{\infty} (b_3^{(1)} uu_x^2 + b_4^{(1)} u^4)dx . \end{cases}$$

Here the relations between the two sets of constants $\{a_\ell^{(1)}\}_{\ell=1}^4$ and $\{b_\ell^{(1)}\}_{\ell=1}^4$ are given by $b_1^{(1)} = a_1^{(1)}$, $b_2^{(1)} = (6a_1^{(1)} + 2a_2^{(1)} - a_3^{(1)})/3$, $b_3^{(1)} = (-30a_1^{(1)} - 10a_2^{(1)} + 5a_3^{(1)} + a_4^{(1)})/9$, $b_4^{(1)} = (30a_1^{(1)} + a_2^{(1)} - 2a_3^{(1)})/6$.

3. INTEGRALS OF THE PERTURBED EQUATIONS. The existence of the integrals for given equation is a key to its integrability. In this section, we look for the integrals for the perturbed equation (1) in the following formal power series,

$$(19) \quad I_\nu[u;\epsilon] = I_\nu^{(0)}[u] + \epsilon I_\nu^{(1)}[u] + \dots, \quad \text{for some} \quad \nu \in \Gamma_o ,$$

where $\{I_\nu^{(0)}[u]\}_{\nu \in \Gamma_o}$ are the integrals for the unperturbed equation satisfying (I). Let us define Γ_n, a subset of the index set Γ_0, if for each $\nu \in \Gamma_n \subset \Gamma_0$, there exist $I_\nu^{(\ell)}[u]$ in (19), for $1 \leq \ell \leq n$, satisfying $X \cdot \nabla I_\nu = O(\epsilon^{n+1})$, or equivalently,

$$(20) \qquad \sum_{m=0}^{\ell} X^{(\ell - m)} \cdot \nabla I_\nu^{(m)} = 0, \text{ for each } \nu \in \Gamma_n \text{ and } 1 \leq \ell \leq n,$$

(i.e. $I_\nu[u;\epsilon]$ for $\nu \in \Gamma_n$ is the integrals of (1) up to order ϵ^n). Note that $\Gamma_0 \supset \Gamma_1 \supset \ldots \supset \Gamma_n \supset \ldots \supset \Gamma_\infty$, and if $\Gamma_n = \Gamma_0$, then the perturbed system (1) is integrable up to ϵ^n, (it may be called "nearly integrable"). The main purpose in this section is to find Γ_n (i.e. find $I_\nu^{(\ell)}[u]$ by solving (20)). With this purpose, we study three examples presented in the previous section.

3a. N-uncoupled harmonic oscillators. The equation for $I_\nu^{(\ell)}[u]$ in this case can be expressed by

$$(21) \qquad L_{X^{(0)}} I_\nu^{(\ell)} = -X^{(0)} \cdot \nabla I_\nu^{(\ell)} = \sum_{m=0}^{\ell - 1} X^{(\ell - m)} \cdot \nabla I_\nu^{(m)}$$

where $L_{X^{(0)}}$ is the Lie derivative with respect to $X^{(0)} = \mathcal{J}^{(0)} \nabla H^{(0)}$, and given by

$$(22) \qquad L_{X^{(0)}} = \sum_{\ell=1}^{N} \omega_\ell^{(0)} (x_\ell \frac{\partial}{\partial y_\ell} - y_\ell \frac{\partial}{\partial x_\ell}) .$$

In terms of the action-angle variables defined by $\rho_\ell = (x_\ell^2 + y_\ell^2)/2$, $\theta_\ell = \arctan(x_\ell/y_\ell)$, $\ell = 1,\ldots,N$, (21) can be written in

$$(23) \qquad L_{X^{(0)}} I_\nu^{(\ell)} = \sum_{\ell=1}^{N} \omega_\ell^{(0)} \frac{\partial}{\partial \theta_\ell} I_\nu^{(\ell)} = G_\nu^{(\ell)} ,$$

where $G_\nu^{(\ell)} = \sum_{m=0}^{\ell - 1} X^{(\ell - m)} \cdot \nabla I_\nu^{(m)}$. For the case $\ell = 1$, choosing $I_\nu^{(0)} = \rho_\nu$, we have, from $X^{(1)} = \mathcal{J}^{(0)} \nabla H^{(1)}$,

(24) $$L_{X^{(0)}} I_\nu^{(1)} = -\frac{\partial}{\partial\theta_\nu} H^{(1)} .$$

Here $H^{(1)}$ in terms of (ρ,θ) is given by

(25) $$H^{(1)}[u] = \sum_{|r|+|s|=3} c_{rs} x^r y^s = \sum_{|r|+|s|=3} \tilde{c}_{rs}(\rho) e^{i(r-s)\cdot\theta},$$

where $\tilde{c}_{rs} = \tilde{c}^*_{sr}$ (complex conjugate), and $(r-s)\cdot\theta = \sum_{\ell=1}^{N} (r_\ell - s_\ell)\theta_\ell$. Thus, if $(r-s)\cdot\omega^{(0)} = \sum_{\ell=1}^{N} (r_\ell - s_\ell)\omega_\ell^{(0)} \neq 0$ for $|r| + |s| = 3$ (i.e. the non-resonant case), the particular solution of $I_\nu^{(1)}$ can be obtained by

(26) $$I_\nu^{(1)}[u] = \sum_{|r|+|s|=3} \tilde{c}_{rs}(\rho) \frac{r_\nu - s_\nu}{(r-s)\omega^{(0)}} e^{i(r-s)\cdot\theta} .$$

Moreover, one can prove that if $\{\omega_\ell^{(0)}\}_{\ell=1}^{N}$ are rationally independent (i.e. for $\forall\, r \in \mathbb{Z}^N$, if $r\cdot\omega^{(0)} = 0$ then $r = 0$), then the solution $I_\nu^{(\ell)}$ of (23) exists for any $\ell \geq 1$ and $\nu \in \Gamma_0$ (i.e. $\Gamma_\infty = \Gamma_0$). (In the resonant case, $\{I_\nu[u;\epsilon]\}_{\nu=1}^{N}$ do not exist in general.) This is a consequence of the theorem for the normal form expansion of (1) (see [3] and the following sections).

3b. The linearized KdV equation. The equation for $I_\nu^{(0)}$ is given by the following form similar to (23),

(27) $$L_{X^{(0)}} I_\nu^{(\ell)} = \int_0^\infty dk\, \omega_k^{(0)} \frac{\partial}{\partial\theta_k} I_\nu^{(\ell)} = G_\nu^{(\ell)}$$

where $\omega_k^{(0)} = k^3$ (the linear dispersion relation), and the action-angle variables (ρ,θ) are defined by

(28) $$\rho_k = \frac{|\hat{u}(k)|^2}{\omega_k^{(0)}}, \quad \theta_k = \arg \hat{u}(k), \quad \text{for } k \geq 0 .$$

Here $\hat{u}(k)$ is the Fourier component of $u(x)$,

$$(29)\quad \begin{cases} \hat{u}(k) = \frac{1}{\sqrt{2\pi}} \int_{-\infty}^{\infty} u(x) e^{-ikx} dx, \quad \text{for} \quad k \geq 0, \\ u(x) = \sqrt{\frac{2}{\pi}} \int_0^{\infty} \sqrt{k\rho(k)} \cos(kx + \theta_k) dk . \end{cases}$$

Let $I_\nu^{(0)}[u]$ be the integrals for (11) given by

$$(30)\quad I_{\nu+1}^{(0)}[u] = \frac{(-1)^\nu}{2} \int_{-\infty}^{\infty} (u_{\nu x})^2 dx, \quad \text{for} \quad \nu = 0, 1, \ldots .$$

Then $G_\nu^{(1)}(\rho,\theta)$ in (27) can be written by

$$(31)\quad G_\nu^{(1)}(\rho,\theta) = \int_0^{\infty}\int_0^{\infty} \Phi_\nu^{(1)}(k,\rho) \sin(\theta_{k_1} + \theta_{k_2} - \theta_{k_1+k_2}) dk_1 dk_2 ,$$

where $\Phi_\nu^{(1)}(k,\rho) = \Phi_\nu^{(1)}(k_1,k_2,\rho_{k_1},\rho_{k_2})$ does not depend on θ, and $|\Phi_\nu^{(1)}(k,\rho)| \propto k^{2\nu+1}$ for small k. (See (33) below for the explicit form of $G_\nu^{(1)}$ in terms of u.) By virtue of the fact that $\omega_{k_1}^{(0)} + \omega_{k_2}^{(0)} \neq \omega_{k_1+k_2}^{(0)}$ for $k_1,k_2 > 0$ (the <u>non-resonant</u> condition), one can find a solution of (31) for $\ell = 1$ in the similar form of (26),

$$(32)\quad I_\nu^{(1)}[u] = - \int_0^{\infty}\int_0^{\infty} \frac{\Phi_\nu^{(1)}(k,\rho)}{\omega_{k_1}^{(0)} + \omega_{k_2}^{(0)} - \omega_{k_1+k_2}^{(0)}} \cdot \cos(\theta_{k_1} + \theta_{k_2} - \theta_{k_1+k_2}) dk_1 dk_2 .$$

Thus if there is no-resonance, one can find $I_\nu^{(\ell)}$. However, the higher order equations ($\ell \geq 2$) generally contain the resonant terms (e.g. $\omega_{k_1}^{(0)} + \omega_{k_2}^{(0)} - \omega_{k_3}^{(0)} - \omega_{k_1+k_2-k_3}^{(0)}$) and in these cases, we need to study the singular integral equations having the poles at the resonant points, and to give the analytical estimates for the existence of the solution $I_{\nu}^{(\ell)}$.

Fortunately, there is another way to find the solution for (27) based on an algebra of the functionals consisting of the differential polynomials (FDP), instead of the analytical way discussed above. This algebraic method is particularly useful for the case of the KdV equation where the higher order equation does not have simple forms like those in the previous case (e.g. (31)). Noticing that the linear part of $X^{(1)}$, i.e. u_{5x}, is perpendicular to $\nabla I_{\nu}^{(0)}$ (this implies $\#(u)$ in $(X^{(1)} \cdot \nabla I_{\nu}^{(0)}) = 3$), equation (27) for $\ell = 1$ can be written in

$$(33) \qquad (X^{(0)} \cdot I_{\nu}^{(1)})[u] = \sum_{\substack{\ell_1 + 2\ell_2 = 2\nu+1 \\ 1 \le \ell_1 \le \ell_2}} a_{\ell_1 \ell_2}^{(1)} \int_{-\infty}^{\infty} u_{\ell_1 x} u_{\ell_2 x}^2 dx ,$$

where the constants $a_{\ell_1 \ell_2}^{(1)}$ are determined by $\{a_{\ell}^{(1)}\}_{\ell=1}^{3}$ in $X^{(1)}$. From the relation $\mathrm{Deg}(I_{\nu}^{(1)}[u]) = 2\nu + 4$, one can find the solution $I_{\nu}^{(1)}$ in the following form,

$$(34) \qquad I_{\nu}^{(1)}[u] = \Sigma\, b_{\ell_1 \ell_2}^{(1)} \int_{-\infty}^{\infty} u_{(\ell_1 - 1)x} u_{(\ell_2 - 1)x}^2 dx ,$$

where the sum is taken over the same way as (33), $c_{\nu}^{(1)}$ is an arbitrary constant, and $b_{\ell_1 \ell_2}^{(1)}$ are determined uniquely from $a_{\ell_1 \ell_2}^{(1)}$ by comparing the coefficients of the independent FDPs, $\int (u_{\ell_1 x})(u_{\ell_2 x})^2 dx$ for each ℓ_1, ℓ_2, and $\ell_1 + 2\ell_2 = 2\nu + 1$ on both sides of (33). The explicit forms of $I^{(1)}$ for $\nu = 1, 2$ are given by

$$(35) \qquad \begin{cases} I_1^{(1)}[u] = b_{11}^{(1)} \int u^3 dx + c_1^{(1)} I_2^{(0)}[u] , \\ \\ I_2^{(1)}[u] = b_{12}^{(1)} \int u u_x^2 dx + c_2^{(1)} I_3^{(0)}[u] , \end{cases}$$

where $b_{11}^{(1)} = (-2a_2^{(1)} + a_3^{(1)})/6$, $b_{12}^{(1)} = (-a_2^{(1)} + 2a_3^{(1)})/6$. It is important to note that the first order vector field $X^{(1)}$ can be determined conversely by giving the set of integrals (35), except the linear term $a_1^{(1)} u_{5x}$ which can be determined by fixing the linear dispersion relation i.e. $\omega_k = k^3 - \epsilon a_1^{(1)} k^5$. Namely, the perturbed equation up to order ϵ can be characterized by two integrals (35) and the linear dispersion relation. This fact will be important in the following sections.

The next order solution $I_\nu^{(2)}$ can be also constructed easily in the same way. Equation (27) for $\ell = 2$ is expressed by

$$(36)\quad X^{(0)} \cdot \nabla I_\nu^{(2)} = \sum_{\substack{\ell_1 + \ell_2 + 2\ell_2 = 2\nu + 1 \\ 0 \le \ell_1 < \ell_2 \le \ell_3}} a^{(2)}_{\ell_1 \ell_2 \ell_3} \int u_{\ell_1 x} u_{\ell_2 x} u^2_{\ell_3 x} dx$$

$$+ \sum_{\substack{m_1 + 2m_2 = 2\nu + 3 \\ 1 \le m_1 \le m_2}} a^{(2)}_{m_1 m_2} \int u_{m_1 x} u^2_{m_2 x} dx .$$

It is easy to see that the solution of (36) is given by

$$(37)\quad I_\nu^{(2)}[u] = \Sigma\, b^{(2)}_{\ell_1 \ell_2 \ell_3} \int u_{\ell_1 x} u_{(\ell_2 - 1)x} u^2_{(\ell_3 - 1)x} dx$$

$$+ \Sigma\, b^{(2)}_{m_1 m_2} \int u_{(m_1 - 1)x} u^2_{(m_2 - 1)x} dx$$

$$+ c_\nu^{(2)} I^{(0)}_{\nu + 2}[u] ,$$

where $c_\nu^{(\ )}$ is an arbitrary constant, and the constants $b^{(2)}_{\ell_1 \ell_2 \ell_3}$, $b^{(2)}_{m_1 m_2}$ are determined by the same way as before. Thus, even for the case $\ell = 2$ containing the resonant terms, this method can derive the solution explicitly in terms of the FDPs. However, at the next order $\ell = 3$, the solution may not be obtained directly from the form of the FDP on the right hand side of (27), because of the increasing nonlinearity, and the method should be modified.

3c. The KdV equation [7]. The set of integrals $\{I_\nu^{(0)}[u]\}_{\nu=0}^{\infty}$ of the KdV equation is given by the following recursion formula [2],

$$\mathcal{J}_0^{(0)}\nabla I_{\nu+1}^{(0)} = \mathcal{J}_1^{(0)}\nabla I_\nu^{(0)} \quad \text{with} \quad I_0^{(0)}[u] = \frac{1}{2}\int_{-\infty}^{\infty} u\,dx \,, \tag{38}$$

where the Hamiltonian operators $\mathcal{J}_0^{(0)}$ and $\mathcal{J}_1^{(0)}$ are defined by

$$\mathcal{J}_0^{(0)} = \mathcal{J}^{(0)} = \partial_x, \quad \mathcal{J}_1^{(0)} := \mathcal{J}_0^{(1)} = \partial_x^3 + 2(\partial_x u + u\partial_x) \,. \tag{39}$$

The first three integrals are as follows:

$$\begin{cases} I_1^{(0)}[u] = \frac{1}{2}\int u^2 dx \,, \\ I_2^{(0)}[u] = H^{(0)}[u] = \int (u^3 - \frac{1}{2}u_x^2)dx \,, \\ I_3^{(0)}[u] = \int (\frac{5}{2}u^4 - 2uu_x^2 + \frac{1}{2}u_{2x}^2)dx \,. \end{cases} \tag{40}$$

In this case, the equation for $I_\nu^{(1)}$ has the following form,

$$X^{(0)} \bullet \nabla I_\nu^{(1)} = \Sigma\, a^{(1)}_{\ell_0 \cdots \ell_N} \int (u^{\ell_0} u_x^{\ell_1} \cdots u_{Nx}^{\ell_N})dx \,, \tag{41}$$

where the sum is taken over $(\ell_0,\ldots,\ell_N)$ with the constraints $2\ell_0 + 3\ell_1 + \ldots + (N+2)\ell_N = 2\nu + 7$, and $\ell_N \geq 2$, and the constants $a^{(1)}_{\ell_0 \cdots \ell_N}$ are determined from $\{a_\ell^{(1)}\}_{\ell=1}^4$ in $X^{(1)}$. The general solution of (41) may not be obtained directly. However, for the first two cases $(\nu = 1, 2)$ where the degree of nonlinearity $(\ell_0 + \ell_1 + \ldots + \ell_N \leq \nu + 2)$ is 4 or less, the solutions can be found explicitly by the same way as the previous case. Namely,

$$(42)\qquad \begin{cases} (X^{(0)}\cdot\nabla I_1^{(1)})[u] = a_{01}^{(1)}\int u_x^3 dx\,, \\ (X^{(0)}\cdot\nabla I_2^{(1)})[u] = \int (a_{11}^{(1)} uu_x^3 + a_{012}^{(1)} u_x u_{2x}^2)dx\,, \end{cases}$$

which lead to the solutions,

$$(43)\qquad \begin{cases} I_1^{(1)}[u] = b_{01}^{(1)}\int u^3 dx + c_1^{(1)} I_2^{(0)}[u]\,, \\ I_2^{(1)}[u] = H^{(1)}[u] + c_2^{(1)} I_3^{(0)}[u]\,, \end{cases}$$

where $H^{(1)}[u]$ is the higher order Hamiltonian given in (18), $c_\nu^{(1)}$ are arbitrary constants, and $b_{01}^{(1)} = (-2a_2^{(1)} + a_3^{(1)})/6$. The next integral $I_3^{(1)}$ can be also found in the form $I_3^{(1)}[u] = \int (b_5^{(1)} uu_{2x}^2 + b_6^{(1)} u^2 u_x^2 + b_7^{(1)} u^5)dx + c_3^{(1)} I_4^{(0)}[u]$ in the same way. While in the cases $\nu \geq 4$, it is not obvious that the forms $I_\nu^{(1)}$ assumed in the similar manner as above satisfy (41), since the possible number of dimension for $X^{(1)}\cdot\nabla I_\nu^{(0)} \geq$ the maximum dimension of $I_\nu^{(1)}$, (the dimension of FDP with fixed degree is defined by the number of independent FDPs there). However, one can actually show that the solutions $I_\nu^{(1)}$ exist for all $\nu \in \Gamma_0$ $(= \{0,1,2,\ldots,\infty\})$, i.e. $\Gamma_1 = \Gamma_0$. We will discuss this in the last section where we construct a transformation between the perturbed equation (1) with (17) and an integrable system from the fact that the perturbed equation has two integrals (43).

4. NORMAL FORMS. If the perturbed equation has several integrals, then by changing the coordinates on M, one can expect to transform the equation into a simpler equation having the same number of integrals. The transformed equation may help us to find more integrals (which we could not find easily from the perturbed equation), because of its (simple) form (it it sometimes linear or integrable). We call the resulting

simple equation "a normal form" of the perturbed equation. In the next section, we will construct such transformations.

The normal forms may be defined as follows: Let $\dot{v} = X_0^{(0)}(v) = \mathcal{J}_0^{(0)} \nabla H_0^{(0)}[v]$ be an integrable Hamiltonian system in the sense of (I), and $X_0^{(\ell)}(v)$, $\ell \geq 1$, be the vector fields satisfying the conditions (X). A perturbed Hamiltonian system

$$\dot{v} = \sum_{\ell=0}^{n} \epsilon^{\ell} X_0^{(\ell)}(v) + O(\epsilon^{n+1}) \tag{44}$$

$$= X_0(v;\epsilon) = \mathcal{J}_0(v;\epsilon) \nabla H_0[v;\epsilon]$$

where $\mathcal{J}_0 = \mathcal{J}_0^{(0)} + \epsilon \mathcal{J}_0^{(1)} + \ldots$, $H_0 = H_0^{(0)} + \epsilon H_0^{(1)} + \ldots$, is said to be a normal form on $\{I_\nu^{(0)}[v] = \text{const.}\}_{\nu \in \Gamma_n}$ of the perturbed equation (1) (sometimes called a normal form on Γ_n), if the vector fields $X_0^{(\ell)}(v)$ satisfy

$$X_0^{(\ell)} \cdot \nabla I_\nu^{(0)} = 0 \quad \text{for} \quad 1 \leq \ell \leq n \quad \text{and} \quad \nu \in \Gamma_n . \tag{45}$$

(i.e. $\{I_\nu^{(0)}[v]\}_{\nu \in \Gamma_n}$ are the integrals of (44) up to order ϵ^n). Here, Γ_n is determined from the integrals of the perturbed equation (1) by (20), i.e. $\Gamma_n = \{\nu \in \Gamma_0 \mid \exists\ I_\nu[u;\epsilon]$ such that $X \cdot \nabla I_\nu = O(\epsilon^{n+1})\}$. Namely, the normal form on Γ_n is a perturbed equation in which $X_0^{(\ell)}$, $1 \leq \ell \leq n$, are on the tangent space of the integral surface given by $I_\nu^{(0)} = \text{const.}$ for $\nu \in \Gamma_n$. Note that if $\Gamma_n = \Gamma_0$ then $X_0^{(\ell)}$, $1 \leq \ell \leq n$, are the Hamiltonian vector fields given by the integrals $I_\nu^{(0)}$, and (44) is integrable up to ϵ^n. In the rest of this section, we give the normal forms for the examples in the previous section.

4a. N-uncoupled harmonic oscillators. As an example of the normal forms of (1) with (9), we consider the classical one which is defined on the constant energy surface, i.e.

(46) $$X_0 \cdot \nabla H^{(0)} = 0 .$$

If $X_0 = \mathcal{J}^{(0)} \nabla K_0$ where $\mathcal{J}^{(0)}$ is defined in (8) and K_0 a Hamiltonian, then (46) can be written in

(47) $$L_{X^{(0)}} K_0 = \sum_{\mu=1}^{N} \cdot \omega_\mu^{(0)} \frac{\partial}{\partial \theta_\mu} K_0 = 0 .$$

In the case where $\{\omega_\mu^{(0)}\}_{\mu=1}^N$ are rationally independent (non-resonant case), the solution K_0 can be expressed as a function of the action variables only, i.e. $K_0 = K_0[\rho_1, \ldots, \rho_N]$, and the normal form is an integrable system given by

(48) $$\begin{cases} \dot{\theta}_\mu = \dfrac{\partial K_0}{\partial \rho_\mu} = \omega_\mu^{(0)} + \epsilon \omega_\mu^{(1)}(\rho) + \ldots , \\ \dot{\rho}_\mu = - \dfrac{\partial K_0}{\partial \theta_\mu} = 0 , \end{cases}$$

where $\omega_\mu^{(\ell)} = \partial K_0^{(\ell)} / \partial \rho_\mu$ with the formal series $K_0 = K_0^{(0)} + \epsilon K_0^{(1)} + \ldots$. In the resonant case, i.e. there exists a non-empty set of integers $\Delta = \{ s \in \mathbb{Z}^N \mid \omega^{(0)} \cdot s = \sum_{\mu=1}^{N} \omega_\mu^{(0)} s_\mu = 0 \}$, and the Hamiltonian K_0 is given by

(49) $$K_0 = \sum_{\ell - m \in \Delta} c_{\ell m} z^\ell \bar{z}^m , \quad z^\ell \bar{z}^m = \prod_{\mu=1}^{N} z_\mu^{\ell_\mu} \bar{z}_\mu^{m_\mu} ,$$

where $z_\mu = x_\mu + iy_\mu$, $\bar{z}_\mu = x_\mu - iy_\mu$. The normal form in this case is <u>not</u> integrable, in general (see [3]).

4b. The linearized KdV equation. It is easy to see that a linear dispersive wave equation given by

(50) $$\dot{v} = v_{3x} + \sum_{n \geq 1} \epsilon^n a_1^{(n)} v_{(2n+3)x} ,$$

where $a_1^{(n)}$'s are constants, is the integrable normal form on Γ_0 (i.e. (50) has the same set of integrals $\{I_\nu^{(0)}[u]\}_{\nu=1}^{\infty}$ in (30) of (11)). Let us see the normal forms on the constant surfaces given by the finite sets of integrals $\{I_\nu^{(0)}[u] = \text{const.}\}_{\nu=\Gamma_1}$ for $\Gamma_1 = \{1\}$ and $\Gamma_1 = \{1,2\}$. The normal form on $\Gamma_1 = \{1\}$ is given by the vector field (12) satisfying

$$(51) \qquad (X_0^{(1)} \cdot \nabla I_1^{(0)})[v] = \int_{-\infty}^{\infty} X_0^{(1)}(v) v dx = 0 ,$$

which implies that $a_3^{(1)} = 2a_2^{(1)}$ and $a_1^{(1)}$, $a_2^{(1)}$ are arbitrary, i.e.

$$(52) \qquad X_0^{(1)}(v) = a_1^{(1)} v_{5x} + a_2^{(1)} (vv_{3x} + 2v_x v_{2x}) .$$

Similarly, the normal form on $\Gamma_1 = \{1,2\}$ is given by the conditions (51) and

$$(53) \qquad (X_0^{(1)} \cdot \nabla I_2^{(0)})[v] = \int_{-\infty}^{\infty} X_0^{(1)}(v) v_{2x} dx = 0 ,$$

which lead to $a_2^{(1)} = a_3^{(1)} = 0$ and $a_1^{(1)}$ is arbitrary. Consequently, the normal form on the constant surface determined by the <u>two</u> constants $I_\nu^{(0)}[v]$, $\nu = 1,2$, is nothing but the <u>linear</u> equation (50) up to order ϵ, i.e.

$$(54) \quad X_0^{(1)}(v) = a_1^{(1)} v_{5x} = \mathcal{J}_0^{(0)} \nabla H_0^{(1)}[v] + \mathcal{J}_0^{(1)} \nabla H_0^{(0)}[v] ,$$

where the Hamiltonian structure is given by

$$(55) \quad \begin{cases} \mathcal{J}_0^{(0)} = \partial_x, \quad \mathcal{J}_0^{(1)} = b_1^{(1)} \partial_x^3 \\ H_0^{(0)}[v] = I_2^{(0)}[v], \quad H_0^{(1)}[v] = c_2^{(1)} I_3^{(0)}[v] , \end{cases}$$

with $I_\nu^{(0)}[v]$ in (30), and $b_1^{(1)} + c_2^{(1)} = a_1^{(1)}$ (i.e. (51) and (53) give $b_2^{(1)} = b_3^{(1)} = 0$ in (14)). Thus, the normal form (54) is actually the normal form on Γ_0, i.e. $X_0^{(1)} \cdot \nabla I_\nu^{(0)} = 0$ for any $\nu \in \Gamma_0$.

4c. The KdV equation. Since the results for the KdV equation are similar to those of the previous example 4b, we just state the results. (The calculations are left for the readers as an exercise.) The normal form on the constant surface given by $I_\nu^{(0)}[v] = \text{const.}$ for $\nu = 1, 2$ (defined in (40)) is

$$
(56) \qquad \begin{aligned} X_0^{(1)}(v) &= a_1^{(1)}(v_{5x} + 10vv_{3x} + 20v_xv_{2x} + 30v^2v_x) \\ &= \mathcal{J}_0^{(0)} \nabla H_0^{(1)}[v] + \mathcal{J}_0^{(1)}(v) \nabla H_0^{(0)}[v] , \end{aligned}
$$

where the Hamiltonian structure is given by

$$
(57) \qquad \begin{cases} \mathcal{J}_0^{(0)} = \partial_x, \quad \mathcal{J}_0^{(1)}(v) = b_1^{(1)}(\partial_x^3 + 2(\partial_x v + v\partial_x)) , \\ H_0^{(0)}[v] = I_2^{(0)}[v], \quad H_0^{(1)}[v] = c_2^{(1)} I_3^{(0)}[v] , \end{cases}
$$

with $I_\nu^{(0)}[v]$ in (40) and $b_1^{(1)} + c_2^{(1)} = a_1^{(1)}$ (arbitrary). Note that the normal form given by (56) is also an <u>integrable</u> system (known as the Lax hierarchy of the KdV equation, and its set of integrals is the same as that of the KdV equation, i.e. $\Gamma_1 = \Gamma_0$ again).

5. TRANSFORMATIONS. As we have seen in the previous section, it is not practical to find the integrals $I_\nu[u;\epsilon]$ of the perturbed system for <u>all</u> $\nu \in \Gamma_0$. But practically, this is not an easy problem (i.e. solve (20) for all $\nu \in \Gamma_0$). However, we have noticed in the examples 4.b and 4.c that the perturbed equation and the corresponding normal form are characterized by

the few integrals (not all of them). So that if there is a transformation between the perturbed equation and the normal form, it may be constructed from those integrals only, and if the normal form is integrable, so is the perturbed equation.

We consider such transformation denoted φ in the following power series,

$$u = \varphi(v;\epsilon) = v + \epsilon\varphi^{(1)}(v) + \dots , \tag{58}$$

where u and v are the solutions of the perturbed equation (1) and its normal form (44), respectively. Then the equations for $\varphi^{(\ell)}$'s are given by

$$\sum_{m=0}^{\ell} \{X_0^{(\ell-m)} \cdot \nabla\varphi^{(m)} - \lim_{\delta\to 0} \frac{1}{m!} \frac{d^m}{d\delta^m} X^{(\ell-m)}(v + \delta\varphi^{(1)} + \dots)\} \tag{59}$$

$$= 0, \quad \text{for} \quad 1 \leq \ell \leq n ,$$

where we have assumed $\varphi^{(0)}(v) = v$. For $\ell = 1$, we have the following equation for $\varphi^{(1)}$,

$$[X^{(0)}, \varphi^{(1)}] := X^{(0)} \cdot \nabla\varphi^{(1)} - \varphi^{(1)} \cdot \nabla X^{(0)} \tag{60}$$

$$= X^{(1)} - X_0^{(1)} ,$$

where $[\cdot,\cdot]$ is called the Lie bracket. Under the transformation φ, the integrals $I_\nu[u;\epsilon]$ of the pertrubed equation (1) are transformed into those of the normal form, i.e.

$$I_\nu[u;\epsilon] = (\varphi^* I_\nu)[v;\epsilon] \tag{61}$$

$$= I_\nu^{(0)}[v] + \sum_{\ell=1}^{n} \epsilon^\ell J_\nu^{(\ell)}[v] + O(\epsilon^{n+1}) ,$$

$$\text{for} \quad \nu \in \Gamma_n ,$$

where $J_\nu^{(\ell)}[v]$ are the integrals of the normal form i.e.

$X_0^{(0)}\cdot\nabla J_\nu^{(\ell)} = 0$, and φ^* is the pull-back map. Especially, for the Hamiltonians H and H_0, we have

$$(62)\qquad H[u;\epsilon] = (\varphi^* H)[v;\epsilon] = H_0[v;\epsilon]\,.$$

For each order of ϵ, (61) can be written in the form

$$(63)\quad \lim_{\delta\to 0}\ \sum_{\ell=0} \frac{1}{\ell!}\frac{d^\ell}{d\delta^\ell} I_\nu^{(m-\ell)}[v+\delta\varphi^{(1)}(v)+\dots] = J_\nu^{(m)}[v],$$

$$\text{for}\quad 1\le m\le n,\quad \nu\in\Gamma_n\,.$$

Thus, the equation for $\varphi^{(1)}$ in terms of the integrals (instead of the vector fields as (60)) is

$$(64)\qquad \varphi^{(1)}\cdot\nabla I_\nu^{(0)} + I_\nu^{(1)} = J_\nu^{(1)}\,.$$

Note that (63) can be derived directly from (59). In order to solve $\varphi^{(1)}$, we would rather use (64) than (60) which is more difficult to solve (because the dimension of differential polynomials (DP) is much greater than that of functional DP, and as we noticed that we deal with (64) for only the few number of ν's (e.g. $\nu = 1, 2$ for the examples 2b, 2c)).

In connection with Hamiltonian formalism, let us discuss how the Hamiltonian structure changes under the transformation. We note that by the transformation φ in (58), the gradient of a functional, $\nabla_u K]u]$, becomes, for any vector field Y,

$$\begin{aligned}(65)\quad Y\cdot\nabla_u K[u] &= \lim_{\delta\to 0}\frac{d}{d\delta}K[u+\delta Y]\\ &= Y\cdot\nabla_v K_0[v] - \epsilon(Y\cdot\nabla_v\varphi^{(1)}(v))\cdot\nabla_v K_0[v] + 0(\epsilon^2)\end{aligned}$$

where $K_0[v] = (\varphi^* K)[v] = K[u]$. From (1) and (44) in the

Hamiltonian forms, and using (65), we obtain the relation between the Hamiltonian operators $\mathcal{J}$ in (1) and $\mathcal{J}_0$ in (44) at order ϵ,

$$(66) \quad (\mathcal{J}^{(1)}(v) - \mathcal{J}_0^{(1)}(v))\nabla K_0[v]$$
$$= (\mathcal{J}_0^{(0)}\nabla K_0[v]) \bullet \nabla\varphi^{(1)}(v) + \mathcal{J}^{(0)}(\ \bullet\nabla\varphi^{(1)}(v)) \bullet \nabla K_0[v],$$

where $(\ \bullet\nabla\varphi^{(1)})\bullet\nabla K_0$ is defined by, for any vector field Y,

$$(67) \qquad (Y\bullet\nabla\varphi(v))\bullet\nabla K[v] := \lim_{\delta\to 0}\frac{d}{d\delta}K[v+\delta Y\bullet\nabla\varphi(v)] .$$

In the sense (66) (i.e. $\varphi^{(1)}$ connects the Hamiltonian structures for the systems (1) and (44)), the transformation φ may be considered as "a canonical transformation". It is interesting to note that in finite dimensional case, the canonical transformation $\varphi^{(1)}$, which is given by $\varphi^{(1)} = \mathcal{J}^{(0)}\nabla S^{(1)}$ with $\mathcal{J}^{(0)}$ in (8) and a functional $S^{(1)}$ (so-called the generator of the transformation), does not change the Hamiltonian structure (i.e. $\mathcal{J}^{(1)} = \mathcal{J}_0^{(1)}$). On the other hand, in infinite dimensional case, $\varphi^{(1)}$ generally changes the Hamiltonian structure, and the generator for $\varphi^{(1)}$ may not exist (except for the auto-canonical transformations, (see [8,9])).

We now construct $\varphi^{(1)}$ for the examples given in the previous sections. Here we consider only the first order problem, but again the higher order problems can be discussed in the same way.

5a. N-uncoupled harmonic oscillators. Here we recover the classical result [3] in which the perturbed equation,

$$(68) \qquad \dot{u} = \mathcal{J}^{(0)}\nabla(H^{(0)}[u] + \epsilon H^{(1)}[u] + \dots) ,$$

with $\mathcal{J}^{(0)}$ in (8) and $H^{(\ell)}$ in (10), can be canonical transformed into the normal form

(69) $$\dot{v} = \mathscr{J}_0^{(0)} \nabla (H_0^{(0)}[v] + \epsilon H_0^{(1)}[v] + \ldots),$$

where $\mathscr{J}_0^{(0)} = \mathscr{J}^{(0)}$ and $\{H_0^{(0)}, H_0^{(\ell)}\}^{(0)} = \nabla H_0^{(0)} \bullet \mathscr{J}^{(0)} \nabla H_0^{(\ell)} = 0.$

In the equation (62) at order ϵ, i.e.

(70) $$\varphi^{(1)} \bullet \nabla H^{(0)} + H^{(1)} = H_0^{(1)},$$

we look for the solution $\varphi^{(1)}$ in the form given by the generator $S^{(1)}$,

(71) $$\varphi^{(1)} = \mathscr{J}^{(0)} \nabla S^{(1)}.$$

Then, (70) can be written by

(72) $$L_{X^{(0)}} S^{(1)} = \sum_{\mu=1}^{N} \omega_\mu \frac{\partial}{\partial \theta_\mu} S^{(1)} = H^{(1)} - H_0^{(1)}.$$

The solution $S^{(1)}$ can be found by choosing $H_0^{(1)}$ as the part of the kernel of $L_{X^{(0)}}$ in $H^{(1)}$, i.e.

(73) $$H_0^{(1)} = (\ker L_{X^{(0)}}) \cap H^{(1)},$$

which is nothing but the <u>resonant</u> term in $H^{(1)}$.

The higher order problems, which is given by

(74) $$L_{X^{(0)}} S^{(\ell)} = P^{(\ell)} - H_0^{(\ell)}$$

with $P^{(\ell)} = P^{(\ell)}[H_0^{(0)}, \ldots, H_0^{(\ell-1)}, S^{(1)}, \ldots, S^{(\ell-1)}]$, can be solved in the same way. Namely, choose $H_0^{(\ell)}$ as

(75) $$H_0^{(\ell)} = (\ker L_{X^{(0)}}) \cap P^{(\ell)}.$$

(Note that the decomposition of $P^{(\ell)}$ into two parts, ker $L_{X^{(0)}}$ and im $L_{X^{(0)}}$ (the image of $L_{X^{(0)}}$) is unique.) This is the result of the Binkhoff normal form theorem (see the section 3 and reference [3]).

5b. The linearized KdV equation. For the perturbed equation,

$$\dot{u} = \mathcal{J}^{(0)}\nabla H^{(0)}[u] + \epsilon(\mathcal{J}^{(0)}\nabla H^{(1)}[u] + \mathcal{J}^{(1)}\nabla H^{(0)}[u]) + O(\epsilon^2), \tag{76}$$

with the Hamiltonian structure given by (14), we have shown in the previous sections that up to order ϵ, (76) has at least two integrals given by (35), and the corresponding normal form is

$$\dot{v} = \mathcal{J}_0^{(0)}\nabla H^{(0)}[u] + \epsilon(\mathcal{J}_0^{(0)}\nabla H_0^{(1)}[u] + \mathcal{J}_0^{(1)}\nabla H_0^{(0)}[u]) + O(\epsilon^2), \tag{77}$$

with (55). The transformation $\varphi^{(1)}$ between (76) and (77) can be constructed by the equations (64) for $\nu = 1$ and 2, i.e.

$$\left\{\begin{aligned} (\varphi^{(1)}\cdot\nabla I_1^{(0)})[v] &= \int_{-\infty}^{\infty}\varphi^{(1)}(v)v\,dx \\ &= (\gamma_1^{(1)} - c_1^{(1)})I_2^{(0)}[v] - b_{11}^{(1)}\int_{-\infty}^{\infty}v^3dx, \\ (\varphi^{(1)}\cdot\nabla I_2^{(0)})[v] &= \int_{-\infty}^{\infty}\varphi^{(1)}(v)u_{2x}dx \\ &= (\gamma_2^{(1)} - c_2^{(1)})I_3^{(0)}[v] - b_{12}^{(1)}\int_{-\infty}^{\infty}vv_x^2dx, \end{aligned}\right. \tag{78}$$

where $J_\nu^{(1)} = \gamma_\nu^{(1)}I_{\nu+1}^{(0)}$ with constants $\gamma_\nu^{(1)}$ (since $\operatorname{Deg} F_\nu^{(1)} = \operatorname{Deg} I_{\nu+1}^{(0)}$), and $I_\nu^{(0)}$ are given by (30). In order to solve (78), we make an ansatz [7,10] for $\varphi^{(1)}$ having $\operatorname{Deg}(\varphi^{(1)}) = 4$,

$$\varphi^{(1)}(v) = \alpha_1^{(1)}v^2 + \alpha_2^{(1)}v_x\int_{-\infty}^{x}v\,dx + \alpha_3^{(1)}v_{2x}, \tag{79}$$

where $\{\alpha_\ell^{(1)}\}_{\ell=1}^3$ are constants determined from (78). Substituting (79) into (78), we obtain the constants from the coefficient of the independent FDPs, i.e. $\int u^3 dx$, $\int v_x^2 dx$, $\int vv_x^2 dx$ and $\int v_{2x}^2 dx$,

$$(80)\quad \begin{cases} \alpha_1^{(1)} = \frac{1}{3}(-b_{11}^{(1)} + b_{12}^{(1)}) = \frac{1}{3}(b_2^{(1)} - b_3^{(1)}), \\ \alpha_2^{(1)} = \frac{2}{3}(2b_{11}^{(1)} + b_{12}^{(1)}) = -\frac{2}{3}(2b_2^{(1)} + b_3^{(1)}), \\ \alpha_3^{(1)} = \frac{1}{2}(\gamma_1^{(1)} - c_1^{(1)}) = \frac{1}{2}(\gamma_2^{(1)} - c_2^{(1)}). \end{cases}$$

Also, it is easily checked from (66) that under this transformation the Hamiltonian operator $\mathcal{J}^{(1)} = b_1^{(1)}\partial_x^3 + b_2^{(1)}(\partial_x u + u\partial_x)$ is transformed into $\mathcal{J}_0^{(1)} = b_1^{(1)}\partial_x^3$.

Before ending this section, we comment on the ansatz (79). This ansatz can be constructed from an operator $\mathcal{L}$ which is a map on the space of the Hamiltonian vector fields of different degree, i.e. $\mathcal{L}$ is given by

$$(81)\qquad \mathcal{J}_2 = \mathcal{L}\mathcal{J}_1 \quad \text{or} \quad \mathcal{L} = \mathcal{J}_2\mathcal{J}_1^{-1}$$

where $\mathcal{J}_1$, $\mathcal{J}_2$ are the Hamiltonian operators such that $\mathcal{J}_2\mathcal{J}_1^{-1}\mathcal{J}_2$ is also a Hamiltonian operator. A non-trivial example of such $\mathcal{L} = \mathcal{L}^{(1)}$ with $\mathrm{Deg}(\mathcal{L}^{(1)}) = 2$ is given by

$$(82)\qquad \begin{aligned} \mathcal{J}_1 &= \partial_x, \\ \mathcal{J}_2 &= \partial_x^3 + 2(\partial_x u + u\partial_x). \end{aligned}$$

(it is not difficult to check that $\mathcal{J}_2\mathcal{J}_1^{-1}\mathcal{J}_2$ in (82) is a Hamiltonian operator, i.e. it satisfies the Jacobi identity for the Poisson bracket $\{F,G\} = \nabla F \cdot \mathcal{J}_2\mathcal{J}_1^{-1}\mathcal{J}_2\nabla G$.) By using $\mathcal{L}^{(1)}$ given by (82), $\varphi^{(1)}$ can be written by

$$(83)\quad \varphi^{(1)}(v) = \frac{\alpha_2^{(1)}}{2}\,\mathcal{L}^{(1)}\cdot v + (\alpha_1^{(1)} - 2\alpha_2^{(1)})u^2 + (\alpha_3^{(1)} - \frac{\alpha_2^{(1)}}{2})u_{xx}\,.$$

(The geometrical meaning of $\varphi^{(1)}$ may be an interesting subjust, and will be discussed in the future.)

5c. The KdV equation. In the similar discussions as the previous example, we obtain the result where the perturbed equation (17) is transformed into the integrable normal form (56) by a transformation having the same form as (76). We leave the calculation for the readers as an exercise [7].

BIBLIOGRAPHY

1. For example, T. Taniuti, "Reductive perturbation method and for fields of wave equations," Prog. Theor. Phys. Suppl. 55 (1974) 1.

2. For example, M. J. Ablowitz and H. Segur, Solitons and the inverse scattering transform, (1981) SIAM Studies in Applied Mathematics.

3. J. Moser, "Lectures on Hamiltonian systems," Memoirs, AMS (1968).

4. For review, Y. H. Ichikawa, "Topics on solitons in plasmas," Physica Scripta, 20 (1979) 296.

5. Y. Kodama and T. Taniuti, "Higher order approximation in the reductive perturbation method I. The weakly dispersive system," J. Phys. Soc. Jpn. 45 (1978) 298.

6. Y. Kodama, "Higher order approximation in the reductive perturbation method II. The strongly dispersive system," J. Phys. Soc. Jpn. 45 (1978) 311.

7. Y. Kodama, "Nearly integrable systems," (preprint DPNU-11-83, Nagoya University, 1983), to be published in Physica D.

8. Y. Kodama and M. Wadati, "Theory of canonical transformations for nonlinear evolution equations I," Prog. Theor. Phys. 56 (1976) 1740.

9. Y. Kodama, "Theory of canonical transformations for nonlinear evolution equations II," Prog. Theory Phys. 57 (1977) 1900.

10. J. Gibbons, Private communications (1982).

DEPARTMENT OF MATHEMATICS
THE OHIO STATE UNIVERSITY
231 WEST 18TH AVENUE
COLUMBUS, OHIO 43210

Current Address:
Department of Physics
Nagoya University
Chikusa-Ku Nagoya 464
Japan

Lectures in Applied Mathematics
Volume 23, 1986

FORCED INTEGRABLE SYSTEMS - AN OVERVIEW

D. J. Kaup

ABSTRACT. Forced integrable systems are discussed and the forced sine-Gordon equation is used as an example of a trivial forced integrable system. Then the forced nonlinear Schroedinger equation is used as an example of a nontrivial forced integrable system. The difficulties with solving such forced systems are outlined and the prospects for eventually obtaining a complete solution are discussed. Finally, a method for obtaining approximate solutions is given, and this method is then used to predict what the soliton spectrum would be if the nonlinear Schroedinger field were forced to follow a square-pulse at $x = 0$.

1. INTRODUCTION

Currently it is well accepted and understood how one may use the method of the Inverse Scattering Transform (IST) for solving a broad class of integrable evolution equations. This method was originated by Gardner, Greene, Kruskal, and Muira[1] in 1967 when they demonstrated that the Korteweg-de Vries (KdV) equation could be solved by this method. Then in 1971, Zakharov and Shabat[2] demonstrated that the method could also be used for solving the 'nonlinear Schroedinger equation' (NLS). Shortly after that, the modified KdV equation (MKdV) was solved by Wadati[3], the sine-Gordon equation (SG) by Ablowitz, Kaup, Newell, and Segur (AKNS)[4], the Toda lattice by Flaschka[5] and a host of other integrable evolution equations of various natures, origins and importances, but now really too numerous to mention.

The major properties of this IST were collected and described

1980 Math. Subj. Class. 58F07, 35Q20

in the well-known work by AKNS[6]. AKNS also demonstrated how one could determine these classes of integrable evolution equations as well as how this IST was very similar to the well-known Fourier Transform (FT). This similarity between the IST and the FT is one which I have always found to be extremely useful and valuable. I don't really remember now which one of us in the AKNS group did originate this concept. However, it is certainly true that we all had a part in the development of this concept. I remember quite well that after I saw and understood this connection between these two transforms, I, myself, as well as the others in the AKNS group, realized that this was a very key point for understanding and utilizing the IST. This was one of those key items which could be used in innumerable manners. It could be used to test and verify a calculation or a result. It could be used as a guidepost for developing an inversion procedure. It could be used for predicting what additional and new ideas might be worthwhile attempting to develop, etc.

I have always held to the premise that 'if something can be done with an FT, then there is a corresponding result to be had with the IST.' Some may wish to dispute this contention of mine. But I do not recognize any exception to this rule. However, I do acknowledge that currently there are some 'corresponding IST results' that are either missing or totally absent. But, instead of considering these as exceptions to the rule, I would prefer to consider them as either 'incomplete' or as 'unfinished.' I maintain that instead of being an exception, it is much more likely to be simply a matter of putting in the necessary effort along with the right viewpoint than it is to be a matter of being an actual exception and thereby impossible to obtain. I will say this because I have gained a tremendous respect for this transform. I have seen it produce results which I (and many others) consider to be singular and remarkable. And these results were obtained only after what seemed to be the purest accidental mathematical cancellation and/or reduction. And these purest mathematical accidents are indeed the essence of this method of solution. Because, without them, there could never have been an IST. So this is the very standard routine with the IST. You do a calculation and again find that another mathematical miracle has occurred. And after a while you do begin to believe in miracles, at least mathematical ones.

II. FORCED INTEGRABLE SYSTEMS

So much for generalities. Let us now get down to specifics. There are several areas of the FT-IST connection which I could discuss here, but the one that I wish to emphasize now is 'forced integrable systems.' Or, in other words, integrable systems with nontrivial boundaries. In these systems, there are sources and/or sinks and/or reflections at the edges or boundaries. In between the edges the system is integrable. A simple example is the standard sine-Gordon chain constructed from a rubber band and a box of straight pins. Neglecting air resistance, the motion of the pins which are stuck in the rubber band will closely model the sine-Gordon equation. However, the solution of the mathematical model is only known for the infinite interval[4,7,8]. How can one solve the finite interval case where one has sources and sinks (the fingers holding the stretched rubber band) at the edges? With a little reflection, one will undoubtedly agree that this and similar problems are indeed more practical problems than the (simple) infinite interval case and are an important set of problems in need of a solution. One knows quite well that the analogous linear problems are solvable by the FT. Thus the analogous nonlinear integrable problem may be expected to be also solvable.

Now in some cases such a forced system is almost trivial to solve. Consider the SG equation in light-cone coordinates[4]

$$\partial_x \partial_t u = \sin(u). \tag{1}$$

Let us say that we are to specify

$$\gamma(x) = u(x,t=0), \tag{2a}$$

$$\mu(t) = u(x=0,t). \tag{2b}$$

The solution of this forced system is quite straightforward and follows from the already known solution of the self-induced transparency (SIT) equations[9,10] upon taking the sine-Gordon limit of these equations. But that is not of importance here since I want to use this system as an almost trivial example of a forced system which later will be contrasted against a very definite nontrivial example.

The method of solution for this system is given in terms of two operators, loosely referred to as the 'Lax pair'[11]. The first operator is the well-known Zakharov-Shabat eigenvalue problem[2]

$$\partial_x v_1 + i\, z\, v_1 = q\, v_2, \tag{3a}$$

$$\partial_x v_2 - i\, z\, v_2 = -q\, v_1 \tag{3b}$$

where z is the eigenvalue and q is a potential. The Zakharov-Shabat potentials[2,6] for the sine-Gordon equation are

$$r = -q^*, \tag{4a}$$

$$q = -(1/2)\partial_x u. \tag{4b}$$

The other operator is

$$\partial_t v_1 = [i/(4z)][v_1 \cos(u) + v_2 \sin(u)], \tag{5a}$$

$$\partial_t v_2 = [i/(4z)][v_1 \sin(u) - v_2 \cos(u)], \tag{5b}$$

which governs the time evolution of the ZS eigenfunctions.

The method of solution of this system is as follows. From Eq.(2a) one knows what q is at $t = 0$. Without loss of generality we may take $q = 0$ for x negative. Then it follows that we may solve the ZS eigenvalue problem [Eqs.(3)] for the scattering data, providing we require $\gamma(x)$ to vanish sufficiently rapidly as x approaches positive infinity. Here I shall not define this scattering data since by now it is well documented in many places, as in Refs. 2 and 6. It simply suffices here to say that this scattering data consist of a reflection coefficient as a function of z for z real and also information about each bound state such as its eigenvalue and its normalization coefficient.

Next we outline how one determines the time evolution of this scattering data. Consider now the ψ-solution for which x approaching + ∞ is defined by

$$\psi_1 \sim 0, \tag{6a}$$

$$\psi_2 \sim \exp[izx - t/(4z)]. \tag{6b}$$

Do note that this definition of the ψ-solution differs from that used in Refs. 2 and 6 by the inclusion of the time in the phase. The scattering coefficients (by this I mean the usual coefficients a, b and the same quantities barred) may be determined from the values of the ψ-solution at x = 0

$$\psi_1(0,t) = \bar{b}(t) \exp[-it/(4z)], \tag{7a}$$

$$\psi_2(0,t) = a(t) \exp[-it/(4z)], \tag{7b}$$

Now this ψ-solution must also be a solution for Eqs.(5). And we also note that Eqs.(5) evaluated at x = 0 only require the knowledge of $\mu(t)$ in order to construct the ψ-solution at x = 0 for all time. Thus it follows that the time evolution of a(t) and $\bar{b}(t)$, due to the relations in Eqs.(7), may be obtained by integrating Eqs.(5) at x = 0. Of course, this will give time dependences for a and $\bar{b}$ that will be considerably different from that which one usually finds in any unforced system. In the unforced cases, a is time independent while $\bar{b}$ varies in time only through a phase factor. Thus it is obvious that the time evolution of $\mu(t)$ will force a and $\bar{b}$ to vary considerably from the unforced case. And by varying $\mu(t)$ in various manners one could inject, reflect, destroy, etc., solitons and/or radiation into the system at various times.

At this point, let us look at the linear limit and the FT-IST connection for this forced sine-Gordon equation. There are some connections that I wish to bring out before we go ahead into a nontrivial example. First we shall work out the solution by the FT and then the linear limit of the IST. In the linear limit, Eq.(1) becomes

$$\partial_x \partial_t u = u, \tag{8}$$

We can represent the general solution of Eq.(8) as

$$u(x,t) = \int dk\ A(k)\ \exp[i(kx - t/k)], \tag{9}$$

where A(k) and the contour for k are to be determined from the initial conditions and the forcing term. There are two possible contours because we have two essential singularities, one at $k = \infty$ and the other at $k = 0$. One contour will pass through the singularity at $k = \infty$ as shown in Fig. 1a and the other will pass through the singularity at $k = 0$ as shown in Fig. 1b.

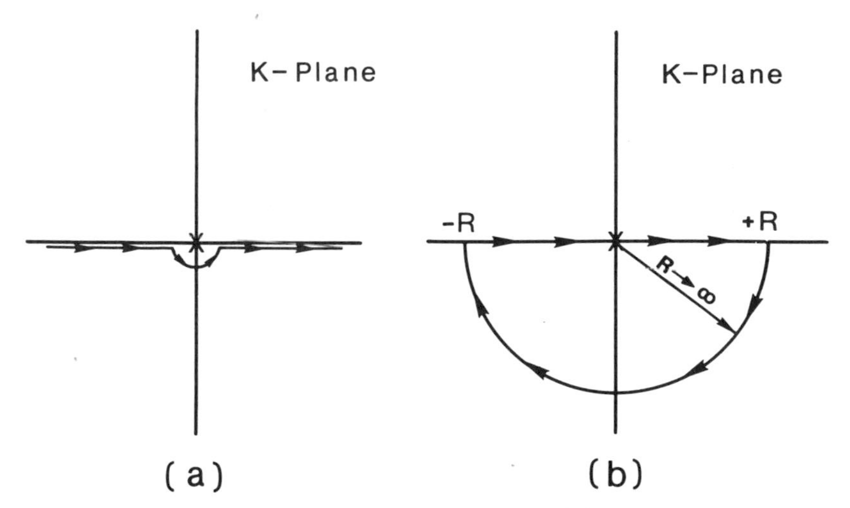

Figure 1. The contours c_1(a) and c_2(b) in the complex k-plane

And since we have in reality <u>two</u> contours, we also have two A(k)'s, $A_1(k)$ for contour #1 and $A_2(k)$ along contour #2.

At t = 0, the essential singularity at k = 0 becomes removed and contour #2 can then be collapsed to a point, giving no contribution to u. Thus $A_1(k)$ is then determined from

$$A_1(k) = [1/(2\pi)] \int dx\ u(x,0) \exp(-ikx). \tag{10a}$$

Similarly one finds at x = 0 that the essential singularity at $k = \infty$ becomes removed, allowing one to collapse coutour #1 to a point. Whence

$$A_2(k) = [1/(2\pi)] \int dt\ u(0,t) \exp(it/k). \tag{10b}$$

As one can see, there is no difficulty in solving the linear problem with the FT. The general solution is just the sum of the contributions from each contour.

Now let us see how this corresponds to the linear limit of the IST method. What I want to do here is simply to demonstrate that each result from the IST does correspond to a similar result with the FT, although there may be disparity in the forms. In the linear limit, the ψ-solution is given by

$$\psi_1 = (1/2)\exp[izx - it/(4z)] \int_x^\infty ds\ \partial_s u(s,t) \exp[2iz(s-x)] \tag{11a}$$

$$\psi_2 = \exp[izx - it/(4z)]. \tag{11b}$$

Let's consider this solution at t = 0. From the definition of $\overline{b}$ in Eq.(7a), we can construct the solution

$$\overline{b}(z,0) = -(1/2)[u(0,0) + 4iz\pi A_1(-2z)], \tag{12}$$

This is the linear limit of the solution that would have been obtained from the IST via Eq.(3) and is what we would call the 'initial scattering data.' As one can see, it also corresponds to the FT of the initial data.

From Eq.(5), one may determine how this $\overline{b}$ will evolve in time. This is

$$\bar{b}(z,t) = \bar{b}(z,0)\exp[it/(2z)] + [i/(4z)]\int_0^t ds\, u(0,s)\exp[i(t-s)/(2z)] \quad (13)$$

One should note that $\bar{b}$ does indeed consist of two parts. The first part is $\bar{b}(z,0)$ which is proportional to A_1 and therefore would generate the contribution to u from contour #1 in the FT solution. Similarly, the last part which is the integral over t would generate the contribution from contour #2. (It is true that this part does not appear to be the same as the result with the FT. But such can be equated after some contours have been distorted.) Thus, upon using the inverse scattering equations, one would find that this linear limit of the IST would give the same solution as the FT would.

Now let us spend a short while considering the analytical properties of these various solutions. We shall find that there are some nontrivial consequences here. First let us note that each eigenvalue problem [Eq.(3) and Eq.(5)] can have its own characteristic set of eigenfunctions. In other words, the natural Jost eigenfunctions of the ZS-eigenvalue problem are not necessarily the natural Jost eigenfunctions of the time-evolution operator. Due to integrability, an eigenfunction of one is automatically an eigenfunction of the other. But, in general, these two sets of Jost functions will differ although in the forced sine-Gordon case they are the same. In the above, we have used the ψ-solution which is a right-eigenfunction (i.e.-defined by its value on the right where x approaches $+\infty$) and which has a definite analytical property. This analytical property is that

$$\psi(x,t;z)\exp[-izx + it/(4z)]$$

is analytic in the upper-half z-plane. Whence by Eq.(7a), so is $\bar{b}(z,t)$. This is all true providing that q remains integrable in x[12].

So far we have used only the analytical properties which follow from the ZS eigenvalue problem, Eqs.(3). But $\bar{b}(z,t)$ must also satisfy the time-evolution equations, Eqs.(5). These equations are an eigenvalue problem in their own right. But since by a simple gauge transformation[8] one can transform Eqs.(5) into Eqs.(3), it follows that the natural eigenfunctions of the time-evolution eigenvalue problem have exactly the same analytical properties as those of the ZS eigenvalue problem. Note that this need not always be the case. In fact, this is the exception rather than the rule. And because it is an exception, it also becomes trivial to find the solution for $\overline{b}$ in terms of the eigenfunctions of the time-evolution operator. We may find this solution as follows. First, the solution for $\overline{b}$ must be proportional to some linear combination of the two linearly independent eigenfunctions of the time-evolution operator and must be of such a combination that up to a phase factor, as in Eq.(7a), the combination is analytic in the upper-half z-plane. This is a stringent requirement, particularly when one considers that it must hold for all time $t > 0$. And when one also considers the asymptotics of these eigenfunctions, one can then exactly determine what the required combination must be. And one concludes that only <u>one</u> eigenfunction is required, and that one which is needed is that right-eigenfunction of the time-evolution operator which is analytic in the upper-half z-plane (the ψ-solution). In hindsight this all becomes very obvious. We need a function analytic in the upper-half z-plane whose asymptotics match those of the ψ- solution of the time-evolution operator. Then upon factoring out the appropriate phase factor, the ψ-solution is also analytic in the same region. So we have that $\overline{b}$ must be proportional to the ψ-solution and that <u>the required analytic properties have determined the necessary admixture of the eigenstates</u>. [This last point does not seem to agree with Eq.(13). That is true because Eq.(13) gives $\overline{b}$ in terms of the left-eigenfunctions (also referred to as the ϕ- and the $\overline{\phi}$ solutions) and not the right-eigenfunctions. If one did re-express $\overline{b}$ in terms of the right-eigenfunctions of the time-evolution operator, one would then find that $\bar{b}$ was indeed analytic in the upper-half z-plane whenever q was integrable.]

Let me state this another way, and at the same time try to summarize what I am trying to emphasize. From the initial-value problem we can determine that $\bar{b}$ has a certain analytical property. Up to a phase factor, $\bar{b}$ must also be a solution of the time-evolution operator. By expanding $\bar{b}$ in eigenfunctions of the time-evolution operator and also requiring this expansion to have the correct analytical properties, one can then determine what the required expansion coefficients are. For the forced sine-Gordon equation, this is really a trivial exercise because the regions of analyticity for the eigenfunctions of the time-evolution operator and the regions of analyticity for the ZS eigenfunctions are the same. Thus, an expansion in one set of eigenfunctions can be readily converted into an expansion in the other set of eigenfunctions.

III. THE FORCED NONLINEAR SCHROEDINGER EQUATION

Let us now consider a less trivial example. And in order to give a physical motivation, let us consider a possible application for the forced NLS equation. The forced NLS could be used to model the generation of Langmuir waves by an intense rf-wave which was being directed directly straight up into the ionosphere[13]. The excitation of Langmuir waves would occur because the rf-wave would penetrate up until it reaches its WKB turning-point, at which point the rf-wave energy would accumulate due to the WKB swelling. This intense collection of rf-energy could then be expected to excite Langmuir waves in the direction of the electric-field polarization. These Langmuir waves would then propagate away from this source, and their propagation would be governed by the NLS equation. So the important questions which one would like to answer are: what type of Langmuir waves may we expect to see propagating away from the source? solitons? continuous spectrum?

and how would this all be dependent on the rf-source? Well, we will not be able to answer these questions here because I still do not have the final answer. But I can outline the major considerations necessary for their resolution. Of course, these considerations are preliminary and as such are subject to change. But I am fairly confident that these are all of the essential considerations necessary for the resolution of these systems.

Let us take the NLS in the form

$$i\partial_t q = -\partial_x^2 q + 2(rq)q \,, \tag{14a}$$

$$i\partial_t r = \partial_x^2 r - 2(rq)r \,, \tag{14b}$$

where

$$r = \pm q^* . \tag{15}$$

The eigenvalue problem is the generalized ZS eigenvalue problem

$$\partial_x v_1 + i\, z\, v_1 = q\, v_2 \,, \tag{16a}$$

$$\partial_x v_2 - i\, z\, v_2 = r\, v_1 \,, \tag{16b}$$

while the time-evolution operator is given by the ABC equations[6]

$$i\partial_t v_1 = A\, v_1 + B\, v_2 \,, \tag{17a}$$

$$i\partial_t v_2 = C\, v_1 - A\, v_2 \,, \tag{17b}$$

where for the NLS

$$A = 2\, z^2 + rq \,, \tag{18a}$$

$$B = 2\, i\, z\, q - \partial_x q \,, \tag{18b}$$

$$C = 2\, i\, z\, r + \partial_x r \,. \tag{18c}$$

Let us specify that the forced NLS has the boundary conditions

$$S(x) = q(x,t=0) \,, \tag{19a}$$

$$Q(t) = q(x=0,t) \,, \tag{19b}$$

$$P(t) = \partial_x q(x=0,t) \,. \tag{19c}$$

From the linear problem one may show that P(t) and Q(t) are <u>not</u>

linearly independent and that only one may be specified or else the problem would be overdetermined. Since it is fairly straightforward to demonstrate this with the FT, I shall leave this as an exercise for the reader and go directly into the fully nonlinear problem.

Proceeding just as with the forced SG case, we use the ZS eigenvalue problem, Eq.(16), at $t = 0$ to determine the initial scattering data. When r is related to q as in Eq.(15), only two of the four scattering coefficients are required. The two necessary scattering coefficients can be obtained from either the ψ or the $\overline{\psi}$ solution[2,6]. These are the right-eigenfunctions of Eq.(16) and for x approaching $+\infty$ are defined by

$$\psi_1 \sim 0 \tag{20a}$$

$$\psi_2 \sim \exp(izx + 2iz^2t) , \tag{20b}$$

$$\overline{\psi}_1 \sim \exp(-izx - 2iz^2t) , \tag{20c}$$

$$\overline{\psi}_2 \sim 0 . \tag{20d}$$

The scattering coefficients can then be obtained from the values at $x = 0$. We shall use only the ψ-solution for which

$$a(z,t) = \psi_2(x=0,t;z) \exp(-2iz^2t) , \tag{21a}$$

$$\overline{b}(z,t) = \psi_1(x=0,t;z) \exp(-2iz^2t) . \tag{21b}$$

Clearly one can use Eqs.(16) and (19a) to obtain the ψ-solution at $t = 0$. Then from Eqs.(21) one can obtain the initial scattering data.

Once we have the initial scattering data, we would like to know how it evolves in time. So we now want to integrate the ABC equations at $x = 0$. This gives us the ψ-solution at $x = 0$ from which we can obtain a and $\overline{b}$ according to Eqs.(21). If we rewrite these ABC equations, we can find

$$\partial_t v_1 + i(2z^2 + rq)v_1 = (2zq + i\partial_x q)v_2, \tag{22a}$$

$$\partial_t v_2 - i(2z^2 + rq)v_2 = (2zr - i\partial_x r)v_1. \tag{22b}$$

This is a familiar set of equations and is an eigenvalue problem which is somewhat different from the ZS problem. This eigenvalue problem has been used to study the Thirring model[14,15] and the derivative-NLS[16]. [Equations (22) do differ from those eigenvalue problems by the term 'rq' which has been added to the $2z^2$ term. But this term could easily be eliminated by simply rotating the phases of v_1 and v_2 in opposite directions.] Thus we understand quite well the analytical properties of its eigenfunctions[17]. But first let us note that we are not able to integrate Eqs.(22) even at $x = 0$ because we do not know what $P(t) = \partial_x q(x=0,t)$ is. In the solution of the linear problem by the FT, P(t) can be determined from S(x) and Q(t). Presumably the same is true for the nonlinear problem. But how is one to determine what P(t) is to be in the nonlinear case?

I don't have a final answer to this problem yet, but I can indicate the direction in which one has to go in order to resolve it. And that is from the analytical properties of the desired solution. I pointed out this feature in the forced SG case above. It seems to be a key consideration in these problems because with it I can solve the problem initially around $t = 0$. To demonstrate this, let us note that if we are to use the ψ-solution, then we need a solution of Eq.(22) which is analytic in the upper half z-plane and whose values at $t = 0$ are the appropriate scattering coefficients. I claim that given this and $Q(t) = q(x=0,t)$, then the solution becomes unique. This claim is certainly reasonable; it does agree with the linear limit, and it can be verified to be true for infinitesimal instants beyond $t = 0$. But a complete solution remains to be given.

This problem of having to determine a second or dependent potential will occur in almost all forced integrable systems. The only exceptions are those like the forced sine-Gordon equation, SIT[11], stimulated Raman-scattering (SRS)[18,19], simple harmonic generation[20], etc. These are all trivial forced systems. But having to determine a dependent potential is not the only complication in forced integrable systems. Another difficulty is the nonsimilarity between the regions of analyticity for the ZS eigenfunctions and the eigenfunctions of the time-evolution operator. The ZS eigenfunctions are naturally analytic in the upper- or lower-half z-planes. But the eigenfunctions of Eqs.(22) are naturally analytic in the upper- or the lower-half z^2-plane, not the z-plane. In the forced SG case, we were able to construct the solution so easily in part because the regions of analyticity for both these eigenvalue problems were identical. In the forced NLS case, this is definitely not so and it becomes necessary to determine how to linearly combine two independent functions so as to create a function with a definite analytic property. In other words, let Ω and $\overline{\Omega}$ represent the two independent solutions of Eqs.(22), with Ω ($\overline{\Omega}$) being analytic in the upper- (lower-) half z^2-plane. Then the precise problem that we must solve is what are the required values for α and β such that

$$\alpha\,\Omega + \beta\,\overline{\Omega}$$

will be analytic in the upper-half z-plane (not the z^2-plane). This is simply the well-known 'Riemann-Hilbert problem' stated in another form. And it does have a solution. I shall not try to describe here how I can solve it because it would be long and not necessarily enlightening at this time. But simply stated, one can use Cauchy's theorem and the asymptotics of the solutions to construct such a function.

But if that was not enough, then one is also plagued with the problem of determining an acceptable and legitimate form of the 'transformation kernels.' Let me illustrate this problem for you in a simple case by using the Schroedinger eigenvalue problem as an example. Let $Y(x,k)$ be an eigenfunction of this Schroedinger eigenvalue problem (where k is the eigenvalue) for some potential $V(x)$. Let $Y\exp(-ikx)$ also be analytic in the upper-half k-plane and approach 1 as x approaches $+\infty$. Then, as is well known, we may represent Y(x,k) by

$$Y(x,k) = \exp(ikx) + \int_x^\infty ds\ K(x,s)\ \exp(iks)\ , \tag{23}$$

where K(x,s) is known as the 'transformation kernel.' One may prove the existence of this kernel by substituting this expression for Y(x,k) into the Schroedinger equation, obtaining the partial differential equations and boundary conditions which K must satisfy, then verifying that a solution exists. For the K in Eq.(23) above, it is obvious that we will have two boundary conditions, one at $s = x$ and one at $s = +\infty$. At these boundaries, K and $\partial_s K$ will have to satisfy certain conditions. Now these conditions at $s = +\infty$ reduce in effect to essentially trivial conditions, due to the structure of the partial differential equations that K satisfies. But now consider what happens if the integral in Eq.(23) above would have been from $s = 0$ to $s = x$ instead as it would have to be if we were considering a forced system. In this case, neither of these two conditions at either boundary value of s will collapse into trivial conditions. And, in fact, the total of these now four conditions can even overdetermine Y in certain classes of eigenvalue problems. Whether or not this has any implications on the existence of a limit beyond which transformation kernels just do not exist is not known. However, in those cases which I have worked on, in spite of this, I still have found that certain linear combinations of the linear independent solutions do have transformation kernels. So, no such limit has yet been found.

Well, where does this leave us? At the best, I have been able to give you an inkling of what is involved in trying to solve these forced integrable systems. But the exact solution is indeed complex and it does cause one to at least hesitate and ponder whether or not there might be an easier way to obtain some information on these systems. And that is exactly what I wish to discuss next.

IV. APPROXIMATIONS WITH THE IST

Although it is certainly interesting and amusing for one to spend his time in attempting to solve more and more integrable systems with the IST, one always will reach a point beyond which there will be diminishing returns. One illustration of this was the analysis of the experimental data of the internal waves found in the Andaman Sea. These data were analyzed by Osborne[21] by using simply the spectral transform for the Schroedinger eigenvalue problem, not the spectral transform for the 'nonlinear intermediate long wave equation'[22] which more correctly describes the propagation of intermediate depth internal waves. Rather he found that the Schroedinger spectral transform (which is the spectral transform for decomposing the generic KdV equation into its normal modes) was indeed quite adequate for analyzing these data, and that the more complicated spectral transform of the 'nonlinear intermediate long wave equation' just was not required. The experimental data just did not require this more complicated spectral transform for its explanation. This does not mean that the 'nonlinear intermediate long wave equation' is in error by any means. Rather, this simply means that the experimental data could not have distinguished between the two.

The moral of this is that for the present we well may have an adequate collection of integrable systems for describing many physical phenomena. And instead of exerting our efforts in the direction of attempting to solve newer and more exotic integrable systems, (which by now all of us and many others well know that given the time, we can indeed solve) we may be better off to instead seek more applications of what we now have. And it was in this spirit that I started this investigation of forced integrable systems. I have known for several years that these systems were an important set of problems. And I finally decided to just simply force my way through one of the cases and find some method to glean some information from it. My first result was with the forced Toda lattice which has been used as a model in molecular dynamics studies[23,24]. The same problem existed there as in the forced NLS, manely, one had two potentials appearing in the eigenvalue problem for $\bar{b}$ and only one potential could be specified without overdetermining the problem. The way I then bypassed that was to approximate the unknown potential, and then determine the time-evolution of the scattering data using that approximation[25]. It turned out that in this case this approximation worked quite well and that one could use it to predict fairly accurately the soliton birth-rate in the molecular dynamics model[26].

Let me close now by applying these same ideas again to the forced NLS and make a prediction concerning its soliton birth-rate. Let's consider the forced NLS when $Q(t)$ is a square-pulse of amplitude Q_0 and of width τ_0 which is turned on at $t = 0$. And let's also take $q(x,0) = 0$ for simplicity. Thus initially the scattering data are empty and $a = 1$ and $\bar{b} = 0$. Now we wish to solve Eq.(22) subject to the initial conditions for the ψ-solution given in Eq.(21). Obviously we need to know what $P(t)$ is. I shall approximate by taking $P(t) = 0$. This may not be as drastic an approximation as one may at first suspect. As a justification of this approximation, I have some preliminary numerical results which show $P(t)$ to be usually small and oscillatory, with an average value of zero. In any case, with this approximation, Eq.(22) becomes solvable and the solution for $a(z,t)$ is given by

$$a(z,t) = [(2z^2 + R_oQ_o + \Omega)/(2\Omega)] \exp[i(\Omega - 2z^2)t]$$
$$-[(2z^2 + R_oQ_o - \Omega)/(2\Omega)] \exp[-i(\Omega + 2z^2)t] , \quad (24)$$

where

$$\Omega^2 = 4z^4 + (R_oQ_o)^2. \quad (25)$$

Equation (24) is valid for $0 < t < \tau_o$. For $\tau_o < t$, $a(z,t)$ remains constant and with the value that it had at $t = \tau_o$. Also

$$R_o = \pm Q_o^* , \quad (26)$$

according to Eq.(15). Now if this is a reasonable approximation to the actual $a(z,t)$, then we would expect to be able to use this approximation of $a(z,t)$ for predicting the soliton spectrum as a function of time. The soliton eigenvalues occur at the zeros of $a(z,t)$, which according to Eq.(24) are at

$$\exp[2i\Omega t] = (2z^2 + R_o Q_o - \Omega)/(2z^2 + R_o Q_o + \Omega). \qquad (27)$$

This is a rather simple equation and, in fact, its solutions have already been analyzed by Mjølhus[27]. The spectrum is quite complicated with all its roots being complex in general, except for the one at $\Omega^2 = 0$. This one is the only one which is time-independent, and is created with the initial turning-on of the pulse. The other roots are a time-dependent spectrum whose motion can be predicted from Mjølhus' results.

Thus the prediction I make is that the spectrum of solitons generated in the forced NLS equation for $Q(t)$ constrained to be a square-pulse, should closely follow the soliton spectrum which would arise from the derivative NLS with the same $Q(t)$, however with $P(t) = 0$.

V. SUMMARY

In this paper I have discussed the major considerations required for solving forced integrable systems. At the present time only 'trivial' forced integrable systems (such as the forced SG equation, SIT, SRS, simple harmonic generation, etc.) have been solved. However, I strongly suspect that it shall only be a short time before nontrivial forced integrable systems (e.g. the forced NLS, the forced Toda lattice, etc.) will be completely solved and thereby proven to be completely integrable. Currently we can only solve these systems as 'almost integrable systems,' by which I mean that once a small additional set of information about the solution is given, then the system becomes completely integrable. For the forced NLS, this additional set of information would be the value of P(t) [or the value of Q(t) if one would specify P(t) instead]. Because once P(t) was known, then the rest of the solution becomes known. Of course, there is a consistency problem here. Whatever one takes for P(t) will determine the future evolution of the scattering data. And from that scattering data, one can use the inverse scattering equations to calculate what P(t) is for that scattering data which were evolved from that initial choice of P(t). In general, this calculated P(t) will then differ from the initial choice for P(t) <u>unless</u> the initial choice just happened to be the exact solution. But if the initial choice would be close to the exact solution, then we may expect that the time evolution of the scattering data as predicted from the (judiciously) chosen value for P(t) should be close, in some sense, to the true solution. I have already demonstrated that this approach can work for the forced Toda lattice, and we have every reason to expect that it can also work for other forced systems.

VI. ACKNOWLEDGEMENTS

I wish to thank Dr. P. J. Hansen for valuable discussions and collaboations on this problem. This research has been supported in part by the AFOSR and the NSF.

BIBLIOGRAPHY

1. C. S. Gardner, J. M. Greene, M. D. Kruskal, and R. M. Miura, "Korteweg-de Vries Equation and Generalizations. VI. Methods for Exact Solution," Phys. Rev. Lett. <u>19</u> (1967) 1095-1097; Comm. Pure Appl. Math. <u>27</u> (1974) 97-133.

2. V. E. Zakharov and A. B. Shabat, "Exact Theory of Two-Dimensional Self-Focusing and One-Dimensional Self-Modulation of Waves in Nonlinear Media," Zh. Eksp. Teor. Fiz. <u>61</u> (1971) 118-134; [Sov. Phys.-JETP <u>34</u> (1972) 62-69].

3. M. Wadati, "The Exact Solution of the Modified Korteweg-de Vries Equation,"J. Phys. Soc. Japan <u>32</u> (1972) 1681.

4. M. J. Ablowitz, A. C. Newell, D. J. Kaup, and H. Segur "Method for Solving the Sine-Gordon Equation," Phys. Rev. Lett. <u>30</u> (1973) 1262-1264.

5. H. Flaschka, "On the Toda Lattice. II -Inverse Scattering Solution-," Prog. Theor. Phys. 51 (1974) 703-716.

6. M. J. Ablowitz, D. J. Kaup, A. C. Newell, and H. Segur, "The Inverse Scattering Transform-Fourier Analysis for Nonlinear Problems," Stud. Appl. Math. 53 (1974) 249-315.

7. L. A. Takhtadzhyan and L. C. Faddeev, "Essentially Nonlinear One-Dimensional Model of Classical Field Theory " Teor. Math. Fiz. 21 (1974) 160-174.

8. D. J. Kaup, "Method for Solving the Sine-Gordon Equation in Laboratory Coordinates," Stud. Appl. Math. 54 (1975) 165-179.

9. G. L. Lamb, Jr., "Coherent-Optical-Pulse Propagation as an Inverse Problem," Phys. Rev. A 9 (1974) 422-430.

10. M. J. Ablowitz, D. J. Kaup, and A. C. Newell, "Coherent Pulse Propagation, a Dispersive, Irreversible Phenomenon," J. Math. Phys. 15 (1974) 1852-1858.

11. P. D. Lax, "Integrals of Nonlinear Equations of Evolution and Solitary Waves," Comm. Pure Appl. Math. 21 (1968) 467-490.

12. The reader who is experienced with some of the more subtle features of the SG equation in light-cone coordinates will recognize that certain smoothness conditions are also necessary, in addition to the integrability requirement. These points are discussed in: D. J. Kaup and A. C. Newell, "The Goursat and Cauchy Problems for the Sine-Gordon Equation," SIAM J. Appl. Math. 34 (1978) 37-54.

13. D. R. Nicholson (private communication).

14. A. V. Mikhailov, "Integrability of the Two-Dimensional Thirring Model," Pis'ma Zh. Eksp. Teor. Fiz. 23 (1976) 356-358; [Sov. Phys.-JETP Lett. 23 (1976) 320 323].

15. D. J. Kaup and A. C. Newell, "On the Coleman Correspondence and the Solution of the Massive Thirring Model," Lett. al Nuovo Cimento 20 (1977) 325-331.

16. D. J. Kaup and A. C. Newell, "An Exact Solution for a Derivative Nonlinear Schroedinger Equation," J. Math. Phys. 19 (1978) 798-801.

17. V. S. Gerdzhikov, V. S. Ivanov, and P. P. Kulish, "Quadratic Bundle and Nonlinear Equations," Teor. Math. Fiz.(USSR) 44 (1980) 342-357; [Theor. Math. Phys. (USA) 44 (1980) 784-795].

18. D. J. Kaup, "The Method of Solution for Stimulated Raman Scattering and Two-Photon Propagation," Physica 6D (1983) 143-154.

19. H. Steudel, "Solitons in Stimulated Raman Scattering and Resonant Two-Photon Propagation," Physica 6D (1983) 155-178.

20. D. J. Kaup, "Simple Harmonic Generation: an Exact Method of Solution," Stud. Appl. Math. 59 (1978) 25-35.

21. A. R. Osborne, "Experimental Observations of Large Amplitude Solitons in the Ocean: The Nonlinear Deformable 'On-Site' Potential," Poster session at Solitons '82, Scott Russell Centenary Conference, Edinburgh, Scotland, August 1982, 23.

22. Y. Kodama, J. Satsuma, and M. J. Ablowitz, "Direct and Inverse Scattering Problems of the Nonlinear Intermediate Long Wave Equation," J. Math. Phys. 23 (1982) 564-576.

23. B. L. Holian and G. K. Straub, "Molecular Dynamics of Shock Waves in One-Dimensional Chains," Phys. Rev. B 18 (1978) 1593-1608.

24. T. G. Hill and L. Knopoff, "Propagation of Shock Waves in One-Dimensional Crystal Lattices," J. Geophys. Res. 85 (1980) 7025-7030.

25. D. J. Kaup, "The Forced Toda Lattice: An Example of an Almost Integrable System," J. Math. Phys. 25 (1984) 277-281.

26. D. J. Kaup and D. H. Neuberger, "The Soliton Birth Rate in the Forced Toda Lattice," J. Math. Phys. 25 (1984) 282-284.

27. E. Mjølhus, "A Note on the Modulational Instability of Long Alfven Waves Parallel to the Magnetic Field," J. Plasma Phys. 19 (1978) 437-447.

DEPARTMENT OF PHYSICS
CLARKSON UNIVERSITY
POTSDAM, NEW YORK 13676

Lectures in Applied Mathematics
Volume 23, 1986

NOTE ON THE INVERSE PROBLEM FOR A CLASS OF FIRST ORDER MULTIDIMENSIONAL SYSTEMS

A.I. Nachman*, A.S. Fokas[1] and M.J. Ablowitz[1]

ABSTRACT. The inverse problem for a multidimensional system of first order differential equations is considered. The $\bar{\partial}$ methodology is employed and integral equations are developed for which the potential may be reconstructed.

In recent years there has been substantial interest in the study of: (a) inverse scattering problems for appropriately decaying potentials (i.e. given suitable scattering data reconstruct the potential $q(x)$); (b) the initial value problem of certain physically important nonlinear evolution equations (i.e. given $q(x,0)$ find $q(x,t)$). In this note we shall consider the inverse problem associated with

$$\psi_{x_0} + i \sum_{\ell=1}^{n} J_\ell \psi_{x_\ell} = q\psi, \tag{1}$$

where $q(x_0,x)$ is an $N \times N$ matrix-valued off-diagonal function in $\mathbb{R}^{n+1}$ and J_ℓ are constant real diagonal $N \times N$ matrices (we denote the diagonal entries of J_ℓ by $J_\ell^1,\dots,J_\ell^N$). We note that the methods presented here can be easily extended to the system $\psi_{x_0} + \sigma \sum_{\ell=1}^{n} J_\ell \psi_{x_\ell} = q\psi$, $\sigma = \sigma_R + i\sigma_I$, which as $\sigma_I \to 0$ becomes the linear eigenvalue problem associated with the so called N wave-interaction equation in $n+1$ spatial dimensions (see [1]). Associated with (1) is a nonlinear evolution equation (a complexified form of the N-wave equation) which is in a sense illposed. Nevertheless (1) provides a natural scattering system to study with the methods at our disposal.

Using the transformation $\psi(x_0,x,k)=\mu(x_0,x,k)\exp(i\Sigma k_\ell(x_\ell - ix_0J_\ell))$; $k \in \mathbb{C}^n$, we may alternatively consider the system

$$\psi_{x_0} + \sum_{\ell=1}^{n} (iJ_\ell \mu_{x_\ell} - k_\ell[J_\ell,\mu]) = q\mu. \tag{2}$$

Equations (1), (2) are natural extensions of well known problems:

1980 Mathematics Subject Classification. 35R30
[1]Supported by NSF Foundation under grant number MCS-8202117, ONF under grant number N00014-76-C09867, and AFOSR under grant number 84-005.

i) In one spatial dimension, i.e. $n=0$, (1) would correspond to

$$\psi_{x_0} + ikJ\psi = q(x_0)\psi, \qquad -\infty < x_0 < \infty. \tag{3}$$

The transformation $\psi(x_0,k) = \mu(x_0,k)e^{-ikJx_0}$, $k \in \mathbb{C}$ leads to the system of differential equations

$$\mu_{x_0} + ik[J,\mu] = q\mu. \tag{4}$$

The function $\mu(x_0,k)$ has desirable analytic properties in k, provided that q is in an appropriate space. Utilization of these analytic properties leads to the formulation of a Riemann-Hilbert (RH) problem for the solution of the inverse problem associated with (4). The 2 x 2 case has been studied in [2], [3]; it can be used to solve the initial value problem of the nonlinear Schrödinger, Sine-Gordon, and modified Korteweg-deVries equation. The 3 x 3 case was studied in [4]; it can be used to solve the initial value problem of the 3 wave interaction (a review of the above work appears in [5]). Recently the N x N case was studied by a number of authors and in a completely rigorous manner by Beals and Coifman [6,7].

ii) In two spatial dimensions (i.e. $n=1$) equations (1) and (2) were studied in [8]. The inverse problem was formulated and formally solved in terms of a DBAR ($\bar{\partial}$) problem (a $\bar{\partial}$ problem generalizes the notion of a RH problem). The 2 x 2 case of this inverse problem was used to solve the initial value problem of certain nonlinear evolution equations in two spatial and one temporal dimension: the Modified Kadomtsev-Petviashvili II (MKPII), and Davey-Stewartson II (DSII) equation. The hyperbolic analogs of (1), (2) (i.e. $J_\ell \to iJ_\ell$)) in two spatial dimensions (i.e. $n=1$) was studied in [9]. The inverse problem in this case was adequately treated via a RH problem; it was used to solve the initial value problem of the N wave interaction, MKPI and DSI.

The solution of the inverse problem associated with (2) has two aspects: (a) develop a formalism such that given appropriate inverse data $T^{ij}(k,\lambda)$ one may reconstruct the potential $q(x_0,x)$. (b) It turns out that $T^{ij}(k,\lambda)$ depends on $3n-1$ parameters while the potential depends only on $n+1$. Thus one needs a characterization equation that restricts the scattering data. In this note we only consider (a) above by extending the method of [8,10], question (b) is considered in [1].

In component form equation (4) is written as:

$$(L\mu)^{ij} = \mu^{ij}_{x_0} + \sum_{\ell=1}^{n} ij^i_\ell \mu^{ij}_{x_\ell} - k_\ell(J^i_\ell - J^j_\ell)\mu^{ij} = (q\mu)^{ij}\,. \tag{5}$$

The specific eigenfunctions we shall work with are defined by the integral equations:

$$\mu = I + \tilde{G}(q\mu) \tag{6a}$$

or

$$\mu^{ij}(x,_0x,k)=\delta_{ij} + \int\int_{\mathbb{R}^{n+1}} G^{ij}(x_0-y_0,x-y,k)(q(y_0,y)\mu(y_0,y,k))^{ij}dy_0dy \tag{6b}$$

where δ_{ij} is the usual Kronecker delta function. The Green's function satisfies: $(LG)^{ij} = \delta(x_0-y_0)\delta(x-y)$ and is given by:

$$G^{ij}(x_0,y_0,k) = \frac{\operatorname{sgn}(J_1^i)}{2\pi i(x_1-iJ_1^i x_0)} e^{i\alpha^{ij}(x_0,x,k)} \prod_{\ell=2}^{n} \delta(x_\ell - \frac{J_\ell^i}{J_1^i} x_1),$$

where

$$\alpha^{ij}(x_0,x,k) = \sum_{\ell=1}^{n} (J_\ell^i - J_\ell^j)(x_0 k_{\ell_I} - \frac{x_\ell k_{\ell_R}}{J_\ell^i}) . \tag{7}$$

(7) is obtained by looking for a Fourier representation of the Green's function whereby one finds:

$$G^{ij}(x_0,x,k) = \frac{-i}{(2\pi)^{n+1}} \int \frac{e^{i(x_0\xi_0+x\cdot\xi)}}{\xi_0+i\sum_{\ell=1}^{n}[(J_\ell^i\xi_\ell + k_\ell(J_\ell^i - J_\ell^j)]} d\xi_0 d\xi. \tag{8}$$

(7) is then calculated by using:

$$\int_{-\infty}^{\infty} \frac{e^{i\xi x}}{\xi+a+ib} d\xi = 2\pi i \operatorname{sgn}(x)e^{-i(a+ib)x}\,\theta(-xb)$$

$$\int_{c\xi<A} e^{(c+id)\xi}d\xi = \frac{\operatorname{sgn}(c)}{c+id} e^{(c+id)A/c}, \quad c \neq 0$$

where the heaviside function is defined by: $\theta(x) = \{1, x>0;\ x, x<0\}$. We next show that $\partial\mu/\partial\bar{k}_p$ (where $\partial/\partial\bar{k}_p = \frac{1}{2}(\partial/\partial k_{pR}+i\,\partial/\partial k_{pI})$) can be expressed in terms of μ. From (6a) we have

$$\frac{\partial\mu}{\partial\bar{k}_p} = \frac{\partial\tilde{G}}{\partial\bar{k}_p}(q\mu) + \tilde{G}(q\frac{\partial\mu}{\partial\bar{k}_p}) \tag{9}$$

and by direct calculation

$$\frac{\partial G^{ij}}{\partial\bar{k}_p} = i(\frac{J_p^i - J_p^j}{2})(x_0 + i\frac{x_p}{J_p^i})G^{ij} \tag{10a}$$

$$= -\frac{1}{2}\frac{(J_p^i - J_p^j)}{(2\pi)^n}\int_{\mathbb{R}^n} \delta(\sum_{\ell=1}^{n} J_\ell^i\lambda_\ell)e^{i\beta^{ij}(x_0,x,k,\lambda)} d\lambda \tag{10b}$$

where

$$\beta^{ij}(x_0,x,k,\lambda) = \alpha^{ij}(x_0,x,k) + \sum_{\ell=1}^{n} x_\ell\lambda_\ell .$$

Defining the scattering data:

$$T^{ij}(k,\lambda) = \int_{\mathbb{R}^{n+1}} e^{-i\beta^{ij}(y_0,y,k,\lambda)}(q\mu)^{ij}(y_0,y,k)dy_0dy \tag{11}$$

(9) may be written in the form:

$$(\frac{\partial\mu}{\partial\bar{k}_p})^{ij}(x_0,x,k) = -\frac{1}{2}\frac{(J_p^i - J_p^j)}{(2\pi)^n}\int_{\mathbb{R}^n}\delta(\sum_{\ell=1}^n J_\ell^i\lambda_\ell)e^{i\beta^{ij}(x_0,x,k,\lambda)}T^{ij}(k,\lambda)d\lambda$$

$$+ \int_{\mathbb{R}^{n+1}} G^{ij}(x_0-y_0,x-y,k)(q\frac{\partial\mu}{\partial\bar{k}_p})^{ij}(y_0,y,k)dy_0dy. \tag{12}$$

In order to express $\partial\mu/\partial\bar{k}_p$ in terms of μ we decompose $\partial\mu/\partial\bar{k}_p$ into fundamental matrices $M_{\nu\nu'}(x_0,x,k,\lambda)$ on $\sum_{\ell=1}^n J_\ell^\nu\lambda_\ell = 0$:

$$M_{\nu\nu'}(x_0,x,k,\lambda) = e^{i\beta^{\nu\nu'}(x_0,x,k,\lambda)}E_{\nu\nu'} + \tilde{G}(qM_{\nu\nu'})(x_0,x,k) \tag{13}$$

where the elementary matrix $E_{\mu\nu}$ has components:

$$(E_{\nu\nu'})^{ij} = \begin{Bmatrix} 1 & \nu=i,\ \nu'=j \\ 0 & \text{otherwise} \end{Bmatrix}.$$

Hence once we have $M_{\nu\nu'}$ then we have $\partial\mu/\partial\bar{k}_p$ via:

$$\frac{\partial\mu}{\partial\bar{k}_p}(x_0,x,k) = \sum_{\nu,\nu'=1}^n(-\frac{1}{2})\frac{(J_p^\nu - J_p^{\nu'})}{(2\pi)^n}\int_{\mathbb{R}^n}\delta(\sum_{\ell=1}^n J_\ell^\nu\lambda_\ell)e^{i\beta^{\nu\nu'}(x_0,x,k,\lambda)}T^{\nu\nu'}(k,\lambda)E_{\nu\nu'}d\lambda$$

$$+ \tilde{G}(q\frac{\partial\mu}{\partial\bar{k}_p})(x_0,x,k) \tag{14a}$$

and hence

$$\frac{\partial\mu}{\partial\bar{k}_p} = \sum_{\nu,\nu'=1}^n -(\frac{1}{2})\frac{(J_p^\nu\ J_p^{\nu'})}{(2\pi)^n}\int_{\mathbb{R}^n}\delta(\sum_1^n J_\ell^\nu\lambda_\ell)T^{\nu\nu'}(k,\lambda)M_{\nu\nu'}(x_0,x,k,\lambda)d\lambda. \tag{14b}$$

From (14) it is clear that the only nontrivial combinations come from columns of $(\partial\mu/\partial\bar{k}_p)^{ij}$ such that $j=\nu'$. Letting $(\eta_{\nu j})^{rj} = (M_{\nu j})^{rj}e^{-i\beta^{\nu j}}$, we have

$$(\eta_{\nu j})^{rj}(x_0,x,k) = \delta_{r\nu} + \int_{\mathbb{R}^{n+1}} e^{-i\beta^{\nu j}(x_0,x,k,\lambda)} \times$$

$$\times\ G^{rj}(x_0-y_0,x-y,k)e^{i\beta^{\nu j}(y_0,y,k,\lambda)}(q\eta_{\nu j})^{rj}(y_0,y,k)dy_0dy \tag{15}$$

The fact that the Greens function admits the following symmetry condition:

$$e^{-i\beta^{\nu j}(x_0,x,k,\lambda)}G^{rj}(x_0,x,k) = G^{r\nu}(x_0,x,\hat{k}^{\nu j}(k,\lambda)),\ \text{on}\ \sum_{\ell=1}^n J_\ell^\nu\lambda_\ell=0, \tag{16a}$$

where

$$\hat{k}^{\nu n}(k,\lambda)=(\frac{J_\ell^j}{J_\ell^\nu}k_{\ell R}+\lambda_\ell,k_{\ell I})_{\ell=1,\ldots,n}, \text{ with } \lambda_\ell \text{ satisfying } \sum_{\ell=1}^{n} J_\ell^\nu\lambda_\ell=0. \tag{16b}$$

(16b) immediately gives:

$$(\eta_{\nu j})^{rj}(x_0,x,k) = \mu^{r\nu}(x_0,x,\hat{k}^{\nu j}(k,\lambda))$$

whereupon from (14) we have

$$\frac{\partial\mu}{\partial\bar{k}_p}(x_0,x,k) = \sum_{\nu,\nu'}(-\frac{1}{2})\frac{(J_p^\nu-J_p^{\nu'})}{(2\pi)^n}\int_{\mathbb{R}^n}\delta(\sum_{\ell=1}^{n}J_\ell^\nu\lambda_\ell)T^{\nu\nu'}(k,\lambda)e^{i\beta^{\nu\nu'}(x_0,x,k,\lambda)} \times$$

$$\times\ \mu(x_0,x,\hat{k}^{\nu\nu'}(k,\lambda))E_{\nu\nu'}\,d\lambda. \tag{17}$$

It should be noted that (16a) is suggested by the transformation between bounded eigenfunctions of (3). To see this explicitly, note that if ψ is a solution of (3) then so is $\psi E_{\nu\nu'}$, and therefore the function $\nu(x_0,x,k) = \psi E_{\nu\nu'}\exp(i\Sigma k_\ell(x_\ell-ix_0J_\ell))$ satisfies (4). But since the function $\mu(x_0,x,h)\exp(i\Sigma h_\ell(x_\ell-ix_0J_\ell))$ also satisfies (4) we have the transformation law:

$$\nu(x_0,x,k) = \mu(x_0,x,h)e^{i\sum_{\ell=1}^{n}h_\ell(x_\ell-ix_0J_\ell)}E_{\nu\nu'}e^{-i\sum_{\ell=1}^{n}k_\ell(x_\ell-ix_0J_\ell)} \tag{18}$$

For boundedness we require:

$$\mathrm{Re}\{i\sum_{\ell=1}^{n}[(h_\ell-k_\ell)x_\ell-(h_\ell J_\ell^\nu-k_\ell J_\ell^{\nu'})ix_0]\} = 0. \tag{19}$$

hence:

$$h = \hat{k}^{\nu\nu'}(k,\lambda) = (\frac{J_\ell^{\nu'}}{J_\ell^\nu}k_{\ell R}+\lambda_\ell,\ k_{\ell I})_{\ell=1,\ldots,n} \tag{20a}$$

for any λ_ℓ on

$$\sum_{\ell=1}^{n}\lambda_\ell J_\ell^\nu = 0. \tag{20b}$$

Finally the reconstruction is effected by inverting $\bar{\partial}\mu$ one variable at a time:

$$\mu(x_0,x,k) = I + \frac{1}{\pi}\iint_{\mathbb{R}^2}\frac{\frac{\partial\mu}{\partial\bar{k}_p}(x_0,x,k_1,\ldots k_p',\ldots k_n)}{k_p - k_p'}dk_{p_R}'dk_{p_I}' \tag{21}$$

and using (17) to obtain a linear integral equation for μ. Asymptotically, as $k_p \to \infty$, (21) yields

$$\mu \sim I + \frac{1}{\pi k_p}\iint\frac{\partial\mu}{\partial\bar{k}_p}(x_0,x,k_1,\ldots,k_p',\ldots,k_n)dk_{p_R}'dk_{p_I}'\ . \tag{22}$$

On the other hand substituting the asymptotic expansion

$$\mu \sim I + \frac{1}{k_p}\mu^{(1)} + \ldots$$

into (4) gives the relation:

$$q = -[J_p,\mu^{(1)}]$$

from which we have the formula:

$$q(x_0,x) = \frac{1}{\pi}\left[\iint \frac{\partial\mu}{\partial\bar{k}_p}(x_0,x,k_1,\ldots,k_p',\ldots,k_n)dk'_{p_R}dk'_{p_I},J_p\right], \tag{23}$$

with (17) used in (23).

Formulae (17), (21), (23) can be used for the reconstruction of $q(x_0,x)$. At this point one needs to show that: (a) $q(x_0,x)$ given by (23) is independent of $k_1,\ldots,k_{p-1}, k_{p+1},\ldots,k_n$; (b) the same $q(x_0,x)$ is found regardless of which inversion formula is used ($p=1,\ldots,n$); (c) there exists a restriction on the scattering data $T^{ij}(k,\lambda)$, which has 3n-1 parameters whereas $q(x_0,x)$ has only n+1. It can be easily shown that (i), (ii) are equivalent. Furthermore there exists a characterization equation restricting the scattering data T^{ij}, this equation is given in [1].

BIBLIOGRAPHY

1. A.I. Nachman and M.J. Ablowitz, "A Multidimensional Inverse Scattering Method", Stud. Appl. Math, 71 (1984) 243-250.

2. V.E. Zakharov and P.B. Shabat, "Exact Theory of Two-Dimensional Self-Focusing and One-Dimensional Self-Modulation of Waves in Nonlinear Media", Sov. Phys. JETP, 34 (1972) 62-69.

3. M.J. Ablowitz, D.J. Kaup, A.C. Newell and H. Segur, "The Inverse Scattering Transform-Fourier Analysis for Nonlinear Problems", Stud. in Appl. Math., 53 (1974) 249-315.

4. D.J. Kaup, "The Three-Wave Interaction - A Nondispersive Phenomenon", Stud. Appl. Math. 55 (1976) 9-44.

5. M.J. Ablowitz and H. Segur, Solitons and the Inverse Scattering Transform, SIAM, Philadelphia, 1981.

6. i) A.V. Mikhailov, "The Reduction Problem and the Inverse Scattering Method", Physica D, 3 (1981) 73-111.

ii) P. Caudrey, "The Inverse Problem for a General N x N Spectral Equation", Physica D, 1 (1982) 51-66.

7. R. Beals and R. Coifman, "Scattering and Inverse Scattering for First Order Systems", Commun. Pure Appl. Math. 37 (1984) 39-90.

8. A.S. Fokas and M.J. Ablowitz, "On the Inverse Scattering Transform of Multidimensional Nonlinear Equations Related to First Order Systems in the Plane", J. Math. Phys. 25 (1984) 2494.

9. A.S. Fokas, "On the Inverse Scattering of First Order Systems in the Plane Related to Nonlinear Multidimensional Equations", Phys. Rev. Lett. 51 (1983) 3.

10. A.S. Fokas and M.J. Ablowitz, "On a Method of Solution for a Class of Multidimensional Nonlinear Evolution Equations", Phys. Rev. Lett. 51, (1983) 7.

*DEPARTMENT OF MATHEMATICS AND COMPUTER SCIENCE
CLARKSON UNIVERSITY
POTSDAM, NEW YORK 13676

DEPARTMENT OF MATHEMATICS
UNIVERSITY OF ROCHESTER
ROCHESTER, NEW YORK 14627

Lectures in Applied Mathematics
Volume 23, 1986

THE DIAGONALIZATION OF NONLINEAR DIFFERENTIAL OPERATORS - A NEW APPROACH TO COMPLETE INTEGRABILITY

by M. S. Berger[1]

ABSTRACT. The functional analysis methods are used to define an idea of diagonalization of a nonlinear operator acting between Hilbert spaces H_1 and H_2. This idea is then applied to various examples, including the Riccati equation with periodic data, and certain nonlinear elliptic boundary value problems. This notion leads to explicit solutions of nonlinear boundary value problems up to smooth coordinate transformations, and thus a new notion of integrability is produced. This idea has the virtues of being stable under perturbation in the cases we discuss, and in addition, global invariants measuring diagonalizability are determined.

INTRODUCTION

One goal in studying problems of nonlinear science consists of formulating a given problem in terms of systems of nonlinear differential equations (ordinary or partial) supplemented by boundary conditions. One then attempts to integrate these equations as explicitly as possible using all the resources of mathematical analysis, geometry, algebra, and current large scale computers. In fact, the goal of explicitly integrating nonlinear differential equations goes back to the origins of calculus with

1980 Mathematics Subject Classification.
[1]Partially supported by NSF and AFOSR grants.

Newton, and has proved to be of crucial importance in subsequent centuries. In this article, we wish to exhibit some new ideas for integrating nonlinear differential equations supplemented by boundary conditions by using modern mathematical analysis and topology, but with enough explicit detail so that the problems studied can be calculated explicitly on present-day computers.

Our discussion is split into several parts. First we discuss the goals and conceptual details of my new theory of integrability, especially as it differs from the contemporary inverse scattering approach to complete integrability. Then we review some of the history of the subject of the explicit integration of nonlinear differential equations. In the third section we discuss details concerning the notion of diagonalizing a nonlinear operator, and compute some invariants for this problem in terms of explicit examples. Then we show how this work leads to a systematic procedure for explicitly solving nonlinear differential equations of varying types.

SECTION 1. A NEW THEORY OF COMPLETE INTEGRABILITY

There are a number of desirable properties of our new theory of integrability for nonlinear differential equations. They include:

(1) INDEPENDENCE OF SPACE DIMENSION. (The usual theories have been limited to partial differential equations of one-space dimension.)

(2) STABILITY. Under a small perturbation a differential equation integrable in our sense should be stable in that the methods used for explicit computation are not totally destroyed, but on the other hand should lead once more to the explicit solution of the perturbed problem.

(3) INCLUSION OF CLASSICAL EXAMPLES (not integrable by quadrature). The theory we describe applies to both ordinary and partial differential equations, and we wish to rethink the classical notion

of integrability by quadrature in terms of our new point of view. In doing this, examples that were not integrable by quadrature will turn out to be integrable by our new methods.

(4) GENUINELY NONLINEAR PHENOMENON. The current examples of complete integrability do not involve bifurcation, and in fact, are based on "global linearizations". The theory we shall describe will include bifurcation phenomenon, and at the same time not involve the notion of global linearization, but rather use linearization of a differential equation in a more intrinsic manner.

(5) A SYSTEMATIC PROCEDURE. The theory we describe can be checked for complete integrability by computing certain "invariants". These invariants are intrinsic, and are necessary conditions for our notion to work. In addition, we shall use the ideas of functional analysis as a unifying feature for our problems.

(6) CONNECTION WITH EXESTENCE THEORY. The methods we shall describe build on and sharpen the existence theory for nonlinear differential systems based on topological invariance such as the degree of a mapping, minimax determination of saddle points of calculus of variations, etc. In so doing we show that the integrable systems that we describe cannot be treated by these existential arguments.

The basic idea of our notion of complete integrability begins with the following question.

(Q) Suppose a differential equation of the form

$$Au = g \tag{1}$$

is regarded abstractly as an operator equation between two Banach spaces, that is the mapping A has domain X, and range Y. Suppose all the singular points*, S, and singular values, $A(S)$,

* We assume A to be a Fredholm operator of index 0 in the sequel so $S = \{x : A'(x)\}$ is not invertible and $A(S)$ is the image under A of the set S.

can be computed explicitly for the mapping A. Then what more need be said to integrate to solve equation (1) explicitly?

It turns out that in the theory of differential equations we shall be able to answer this question in a number of cases, and the ideas going into the theory are an extension of bifurcation theory and linearized stability theory of applied mathematics. Our idea is that the mapping A is equivalent to a canonical mapping C called "the global normal form" for A after global changes of coordinates. The question then becomes one of determining how to compute the global coordinate changes, and what are the appropriate normal forms for the operator A? The simplest case of this situation occurs when A is an n by n matrix (singular or nonsingular) and the spaces X and Y are identified with $\mathbb{R}^n$. In this case, using linear, but different, changes of coordinates on the domain X and the range Y, we find first that every matrix can be diagonalized with 0's and 1's on the diagonal. Secondly, the number of 1's = the rank of matrix A, and in fact this rank is the only invariant of the problem. In matrix analysis, a variant of this result, as in the recent book of Golub and Van Loan, is called the "singular value decomposition".

We ask the question, what <u>nonlinear</u> differential operators A can be diagonalized by similar notions? To extend our notions to nonlinear differential operators, supplemented by boundary conditions, the following picture is helpful.

The diagram (2)

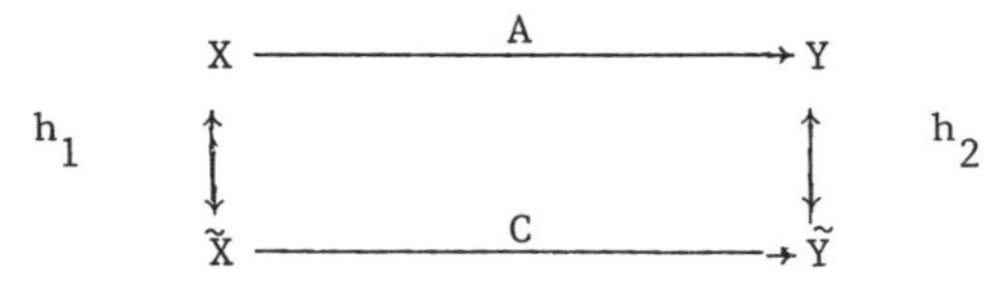

This picture signifies that A is an operator between the spaces X and Y, h_1 and h_2 are global coordinate changes between the Banach spaces X and $\tilde{X}$, and Y and $\tilde{Y}$, and C is the canonical map between X and Y. The simplest canonical map would be a diagonal operator which can be described as follows. Suppose $\tilde{X}$ and Y are Hilbert spaces and an orthonormal basis for $\tilde{X}$ and $\tilde{Y}$ are denoted $(x_1, x_2, x_3, \ldots)$, $(y_1, y_2, y_3, \ldots)$, then the diagonal map D would be defined as $D(x_1, x_2, x_3, \ldots) = (f_1(x_1), f_2(x_2), f_3(x_3), \ldots)$.

Now the diagram (2) clearly identifies what will be the invariants for diagonalizing a nonlinear operator exactly in the same manner as the rank of a matrix is the invariant in the linear case. The invariants for diagonalizing the operator A will be the singular points and singular values of the operator A as a mapping between the Banach spaces X and Y. Indeed, these points and their structure are preserved under the global changes of coordinates h_1 and h_2. In the case of diagonalizability, the canonical map C will be a diagonal map as defined above.

Thus the questions leading to diagonalizability of a nonlinear operator A can be listed as follows.

(1) Compute all singular points S for the mapping A and determine the geometric structure of S as a submanifold of X.

(2) Compute all singular values for the operator $A(S)$ and determine the geometric structure of $A(S)$ as a submanifold of Y.

(3) Compute the same singular points, Σ, and singular values, $D(\Sigma)$, for the desired diagonal map D and determine their geometric structures.

(4) Find global changes of coordinates that map S onto Σ and $A(S)$ onto $D(\Sigma)$.

Thus the question of diagonalizability becomes a combination of geometry and analysis. By this I mean analysis is needed in the selection of the Banach spaces X and Y, and the determination of the singular points S and $A(S)$. These turn out to be interesting questions of boundary value problems for differential operators. The geometric part of the problem consists of determining the geometric structures of the singular sets in question and determining the appropriate changes of coordinates.

Perhaps this is a good place to compare our ideas with those of the inverse scattering method of complete integrability of differential equations. This theory has been immensely successful in discussing evolution equations in one-space and one-time dimension of the form

$$u_t = Lu + Nu \ . \tag{3}$$

However, this theory has had little success with problems involving higher dimensions partially because, I believe, the theory requires an infinite number of independent nontrivial conservation laws for its applicability. The theory is based on the equivalence of the system (3) to a global linearization of the form (4) below.

$$u_t = Lu \ . \tag{4}$$

Here L is a linear differential operator involving only the spacial derivatives, and N is the nonlinear part of the equation (3). This approach avoids bifurcation phenomenon in integrable nonlinear problems. This contrasts with our goals since bifurcation phenomenon are the intrinsic invariants that we utilize for diagonalizing a nonlinear differential operator. Indeed, the singular points that we utilize in our research are precisely the bifurcation points of the associated nonlinear differential operator. This theory of inverse scattering and the invariants associated with it, namely the conservation laws, are generally destroyed under a perturbation. Thus any theory of stability associated with complete integrability in this sense requires an entirely new analysis. As mentioned at the very beginning of this section, our approach includes a stability analysis in its intrinsic formulation. In addition, as we shall point out in later sections, our theory relates to nonlinear differential equations in all space dimensions.

In fact, the usual arguments of applied mathematics is relevant here. Due to experimental error, the nonlinear differential equations defining a physical problem are not known precisely. Thus since we suppose diagonalizability of A is an intrinsic property of the equation (1), and the physical problems associated with it, some such stability property is a neccessary point of our new theory.

SECTION 2. A SHORT HISTORY OF EXPLICIT INTEGRATION OF DIFFERENTIAL EQUATIONS

In the 18th century the notion of "integrability by quadrature" was the key approach to explicit integration. In this theory, one begins with a differential equation

$$\frac{dy}{dt} = f(y, t) \tag{5}$$

and attempts, by changes of coordinates of the form $y = y(Y, T)$ and $t = t(Y, T)$, to reduce (5) to the inhomogeneous linear equation

$$\frac{dY}{dT} = g(T) \ . \tag{6}$$

This method proved immensely successful. However, not all ordinary differential equations could be analyzed by this means. A simple example, such as Riccati's equation

$$\frac{dy}{dt} = y^2 + t^n \tag{7}$$

could be integrated by quadrature for a countable number of n's, but could not be integrated by quadrature for other values of n (see Watson, Theory of Bessel Functions, for details [1]). This is also mentioned in Arnold's recent book ([2] p.32). This theory was amplified by Ritt forty years ago, and became the subject of differential algebra.

Another approach in the 19th century was due to Liouville. In this approach one considered canonical Hamiltonian systems of dimension $2N$, and showed that if such systems had N independent conservation laws in involution, then the system could be integrated by quadrature. The method of inverse scattering can be viewed as an extension of this method to the case when $N \to \infty$. E. Cartan studied the problem of complete integrability locally by using differential geometry of connections and Lie theory (see Arnold [3] for recent survey). However, this method was purely local and requires a great deal of additional research. In fact, earlier, Sophus Lie studied integrability of ordinary differential equations as an application of his ideas on symmetry. In all these cases coordinate transformations to a canonical form are utilized. In fact, in the physics of nonlinear Hamiltonian system canonical coordinate transformations preserving the Hamiltonian structure are a well-developed and useful aspect of nonlinear science. Our methods are slightly different because they use notion of functional analysis and Fourier analysis in their formulation.

SECTION 3. DETERMINATION OF SINGULAR POINTS AND SINGULAR VALUES AND THEIR CLASSIFICATION

A key problem in our approach to integrability is the determination of singular points and singular values of a differential operator. Here we outline a general method for this determination and illustrate it by examples.

Suppose $g(u)$ is a C' function of its argument and set

$$Au = Lu + g(u) \tag{8}$$

where L is a linear (partial) differential operator supplemented by linear boundary conditions (assumed to be a linear Fredholm operator of index 0 acting between Banach spaces X and Y). To find the singular points S of A we attempt to find the nontrivial solutions (v) of the linearized equation $A'(u)v = 0$ at $u = \bar{u}$, i.e.,

$$Lv + g'(\bar{u})v = 0 . \tag{9}$$

Letting $\bar{u}$ is such a value, then $\bar{u} \varepsilon S$ and the singular set S is the set of such point $\bar{u}$ where (9) has a nontrivial solution. Moreover, assuming the real-valued function $g'(t)$ has a single-valued inverse, G, we have the following formula for $\bar{u}$ from (9).

$$\bar{u} = G(- \frac{Lv}{v}) . \tag{10}$$

Here we suppose $\dim\ker(L + g'(\bar{u})) = 1$. Then the singular values of the differential operator A defined by (10) are given by the formula

$$A(\bar{u}) = L\bar{u} + g(\bar{u}) , \tag{11}$$

i.e.,

$$A(u) = LG(- \frac{Lv}{v}) + gG(- \frac{Lv}{v}) \quad \text{from (10).}$$

EXAMPLE 1. RICCATI'S EQUATION

As an example of this approach we consider Riccati's equation

$$u' + u^2 = g(t) \quad \text{with} \quad u(0) = u(T) \tag{12}$$

where $g(t)$ is a L_2 T-periodic function, for fixed T. We consider only T-periodic solutions and so regard $Au = u' + u^2$ and a mapping between the Sobolev space $H_1(0, T)$ into $L_2(0,T)$. One easily shows A is a smooth Fredholm operator of index 0 between these Hilbert spaces and

$$A'(u)v = v' + 2uv . \tag{13}$$

Nontrivial T-periodic solutions for the equation $A'(u)v = 0$ occur (as is easily shown) exactly when the mean-value

$$\int_0^T u(t)dt = 0 . \tag{14}$$

Thus the set of singular points for the operator Au is a hyperplane of codimension one on H_1 (the set of all Fourier series without constant term).

To find the singular values we follow the prescription given above with $Lu = u'$ and $N(u) = u^2$ so that $Gu = \frac{1}{2}u$. Then, by (11), and setting $v = w^{-2}$ so $\bar{u} = w'/w$ we find

$$A(\bar{u}) = (\frac{w'}{w})' + (\frac{w'}{w})^2 = \frac{w''}{w} \tag{15}$$

(after simplification). Thus $A(\bar{u})$ satisfies the Hill's equation

$$w'' - A(\bar{u})w = 0 \tag{16}$$

with $w > 0$, and so we can apply the elementary spectral theory for this situation so that w is the positive eigen-function of $Lw = w'' - A(\bar{u})w$ with associated eigenvalues $\lambda_1(A(\bar{u})) = 0$ and so the singular values of A consist of all $g \in L_2(0, T)$ for which $\lambda_1(g) = 0$. Symbolically we suppose

$$Lg(w) = w'' - g(t)w \tag{16'}$$

has first eigenvalue $\lambda_1(g)$ and we write

$$A(S) = \{g \in L_2, \lambda_1(g) = 0\} \quad . \tag{17}$$

Thus for this example we have reached the situation mentioned in question (2) of SECTION 1.

In order to proceed further, it is necessary to classify the singular points that arise in the above examples. In fact, we shall find a local normal form for each singular point that we have computed, and moreover obtain a local diagonalization of the operators involved near each singular point. to this end we shall extend the classification of singular points as folds and cusps due to Whitney. In particular, we shall extend the notion of a Whitney fold to infinite dimensions. To this end we make the following Definition.

Fredholm operator A of index 0 acting between two Banach spaces X and Y, has a fold at $\bar{u}$ if

(i) $\dim\ker A'(\bar{u}) = 1$

(ii) $(A''(u)(e_0, e_0), h^*) \neq 0$ $\qquad$ (18)

where e_0 is the nonzero element in the null space of $A'(\bar{u})$, and h^* is the element in the kernel of the adjoint of $A'(\bar{u})$. With this definition we can prove the following result.

THEOREM: (Berger-Church [5]) If u is a fold for the operator A near a point U, then locally we can introduce coordinates t and v with $t \in \mathbb{R}'$ and v in the associated orthogonal complement such that A can be written $(t, v) \to (t^2, v)$.

In the above examples it is interesting to see in just what cases the singular points are folds. In fact it turns out that A is a global fold [4], and so A can be globally diagonalized.

We state this result formally in the following terms.

THEOREM: (Mc Keen-Scovil [4]) Let the Riccati operator $Au = u' + u^2$ (subject to T-periodic boundary conditions) acting between the Hilbert spaces $H_1(0, T)$ and $L_2(0, T)$. Then there are explicit smooth global coordinate changes h_1 and h_2 such that (relative to appropriate Fourier series decompositions of H_1 and L_2)

$$h_2^{-1} A h_1(t, w) = (t^2, w) \tag{19}$$

(i.e. A is a global fold).

Here explicit formulae for the global coordinate transformations can be written:

$$h_1(t, w) = u(w) + t\exp(-2 \int_0^x u(s)ds) . \tag{20}$$

Here $u(w)$ denotes a function in S with coordinates $(0, w)$ [notice that as required when $t = 0$ h_1 map the singular points of A into the singlar point of the canonical global fold map. It is easily shown that h_1 is a global homeomorphism of H_1 into itself].

$$h_2(t, w) = \begin{cases} A(u(w)) + t\exp(-4\int_0^t u(w)) , & \text{for } t > 0 \\ A(u(w)) + t , & \text{for } t < 0 . \end{cases} \tag{21}$$

EXAMPLE 2. NONLINEAR ELLIPTIC BOUNDARY VALUE PROBLEMS

In order to fulfill the goal of finding integrable nonlinear systems independent of space dimensions, we turn to the boundary value problems for nonlinear elliptic partial differential operators first studied by Ambrosetti and Prodi and subsequently by many others. Consider the oeprator $\Delta Au = u + f(u)$ subject to zero Dirichlet boundary conditions on a bounded domain $\Omega \varepsilon \mathbb{R}^N$ with N arbitrary. Assuming the eigenvalues of Δ on Ω, with the same boundary conditions are written in ascending order $0 < \lambda_1 < \lambda_2 < \lambda_3 \cdots$ then we restrict consideration to those C^2 convex functions f with $f''(0) > 0$ with normalization $f(0) = 0$ satistying the asymptotic conditions

$$\lambda_2 > \lim_{t\to\infty} f(t)/t > \lambda_1 > \lim_{t\to-\infty} f(t)/t > 0 . \tag{22}$$

In [5], we prove that as a Fredholm operator of index 0 from $H_1^0(\Omega) \to H_1^0(\Omega)$ (defined by duality), Au can be continuously globally diagonalized with the global normal form $D:(t, v) \to (t^2, v)$. Thus A is a global fold exactly as in the example of Riccati's equation mentioned above. To achieve this we prove the Cartesian coordinate representation

$$A[tu_1 + w(t)] = h(t)u_1 + g_1 \tag{23}$$

where u_1 is the positive normalized first eigenvector of Δ, g_1 and $w(t) \perp u_1$. Setting $u(t) = tu_1 + w(t)$, we find by differentiating that at a singular point t

$$\Delta u'(t_0) + f'(u(t_0))u'(t_0) = 0 . \tag{24}$$

Since $h'(t_0) = 0$, $u'(t_0)$ is a nontrivial solution of (24) and by the asymptotic conditions (22) we may suppose $u'(t_0) > 0$ in (24). Differentiating again and evaluating at $t = t_0$ we find

$$A''(u(t_0))[u'(t_0)]^2 + A'(u(t_0))u''(t_0) = h''(t_0)u_1 . \tag{25}$$

Taking the inner product of both sides with $u'(t_0)$ and using the self-adjointness of $A'(u(t_0))$ to insure $(A'(u(t_0)u''(t_0), u'(t_0)) = 0$ we find for (18) $e_0 = u'(t_0)$, $h^* = e_0$ and

$$(A''(\bar{u})(e_0, e_0), h^*) = (A''(u(t_0)(u'(t_0))))^2, \quad (26)$$

$$u'(t_0)) = \int_\Omega f''(u(t_0)[u'(t_0)]^3 > 0 .$$

Here we have used the convexity of $f(t)$ and the fact that $f''(0) > 0$. By virtue of the above Theorem this result implies all singular points S of A are Whitney folds, and moreover, a direct analysis of the relation (2) shows that both S and $A(S)$ are manifolds of codimension 1 in H_1 (see [5]). Moreover, in [5] we construct diffeomorphisms mapping the singular points of A onto the singular points of the global Whitney fold map D; ditto for the singular values between A and D. Using these diffeomorphisms we show A is actually globally diagonalizable with global normal form the Whitney fold map: $(t, v) \to (t^2, v)$ with t a coordinate in the u_1 direction and v orthogonal to u_1. For the details and the added smoothness of the global equivalence see [5].

SECTION 4. A SYSTEMATIC PROCEDURE FOR DIAGONALIZABILITY

We now state a theorem applicable to both EXAMPLES 1. and 2. showing that in each case the nonlinear equations are the simplest possible nonlinear diagonalizable operators. In the case of EXAMPLE 1., the theorem shows why complete integrability by quadrature must fail in the generality we discuss.

THEOREM: The mapping A is C^0 completely integrable in the sense that there are canonical homeomorphisms such that A is equivalent to the mapping $D: H_1 \to H_2$ defined by $D(x_1, x_2, x_3, \ldots) = (x_1^2, x_2^2, x_3^2, \ldots)$.

COROLLARY: The mapping A is a proper mapping.

COROLLARY: The solutions of (11) can be found explicitly (provided they exist), in terms of the "canonical coordinate changes" and the eigenfunctions of the Laplace operator (Fourier series, in special cases).

COROLLARY: All the singular points of the mapping A are "infinite dimensional" folds in the sense of Whitney.

Thus the mapping A is _the_ _"simplest"_ _nonlinear_ _operator_ that is associated with a nonlinear Dirichlet problem and exhibits bifurcation phenomena _independent_ of the domain Ω and the dimension A.

IDEA OF THE PROOF OF THE THEOREM (for EXAMPLE 2)

The proof divides into 2 distinct parts:

Part I - An _analytical_ _part_ consisting of 4 steps:

Step 1: Reduction to a finite dimensional problem;

Step 2: Explicit cartesian representation for the singular points of A;

Step 3: Explicit cartesian representation for the singular values of A;

Step 4: Coerciveness estimates for the mapping A.

Thus in EXAMPLE 2. we proceed as follows to show $h(t) \to \infty$ as $t \to \infty$. This fact follows from the representation (23), we find in fact

$$h(t) = -\lambda_1 t + \int_\Omega f(tu_1 + \omega(t))u_1 \tag{27}$$

the asymptotic relation and the fact that as $t \to \infty$ the contribution due to $\omega(t)$ is negligible via the a priori estimate

$$\|\omega'(t, g_1)\|_H \leq c \text{ (independent of } t \text{ and } g_1). \tag{28}$$

The following picture illustrates the behaviour of the function $h(t)$.

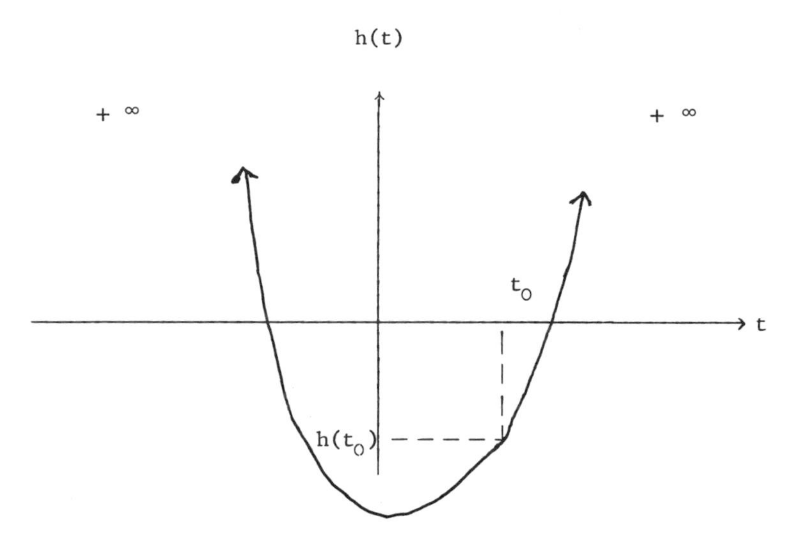

From this picture we read off the cartesian representation of the singular points and singular values of A.

Part II - The second part of the proof is geometric, namely construction of the diffeomorphisms α and β using the fact of Part I. This part consists of 4 steps also.

Step 1: Layering of the mapping A in accord with Step 1 of Part I by a diffeomorphism α_1;

Step 2: "Translation" of the Singular Points of the Mapping A to those of B by a diffeomorphism α_2;

Step 3: Translation of the Singular Value of A to those of B by a diffeomorphism α_3;

Step 4: The final homeomorphism.

Indeed after Step 3 we find

$$\alpha_3 A \alpha_1 \alpha_2 = (\alpha(t, \omega), \omega) \ . \tag{29}$$

Using Step 4 of Part I we represent the right hand side of (29) by the composition $\beta\phi$ where ϕ is a diffeomorphism $H \to H$.

Thus

$$\alpha_3 A \alpha_1 \alpha_2 = \beta\phi \tag{30}$$

which is the desired equation.

SECTION 5. STABILITY OF THE METHODS USED UNDER PERTURBATION OF A

Under a C^1 perturbation of A in the sense of the metric in H, our analytical results of Step 1 carry over to study perturbation problems. For details see [5] and [6].

SECTION 6. EXTENSIONS FOR CUSP SINGULARITIES

In recent work [7] we have studied cusp singularities as generalizations of our work on folds mentioned above. Moreover, we can prove that certain differential operators can be globally diagonalized to cusp maps. An example is

$$\Lambda y = \frac{dy}{dt} + \lambda y - y^3 , \quad y(0) = y(T) \tag{31}$$

acting between the spaces $H_1(0, T)$ and $L_2(0, T)$.

BIBLIOGRAPHY

1. G. N. Watson, Theory of Bessel Functions, Cambridge University Press

2. V. I. Arnold, Theory of Ordinary Differential Equation, MIT Press.

3. Geometrical Aspects of Ordinary Differential Equations, Springer Verlag, N. Y.

4. H. McKean and C. Scovil (forthcoming paper).

5. M. S. Berger and P. T. Church, Complete Integrability and Perturbation of a Nonlinear Dirichlet Problem I and II, Indiana Journal of Mathematics, 1979-1980.

6. M. S. Berger, Stability of Diagonalizability of Certain Nonlinear Differential Operators Under Perturbation.

7. M. S. Berger and P. T. Church and J. Tinnouvian, Folds and Cusps in Banach spaces with Applications I, (Indiana Journal (to appear 1985).

DEPARTMENT OF MATHEMATICS
UNIVERSITY OF MASSACHUSETTS
AMHERST , MA 01003

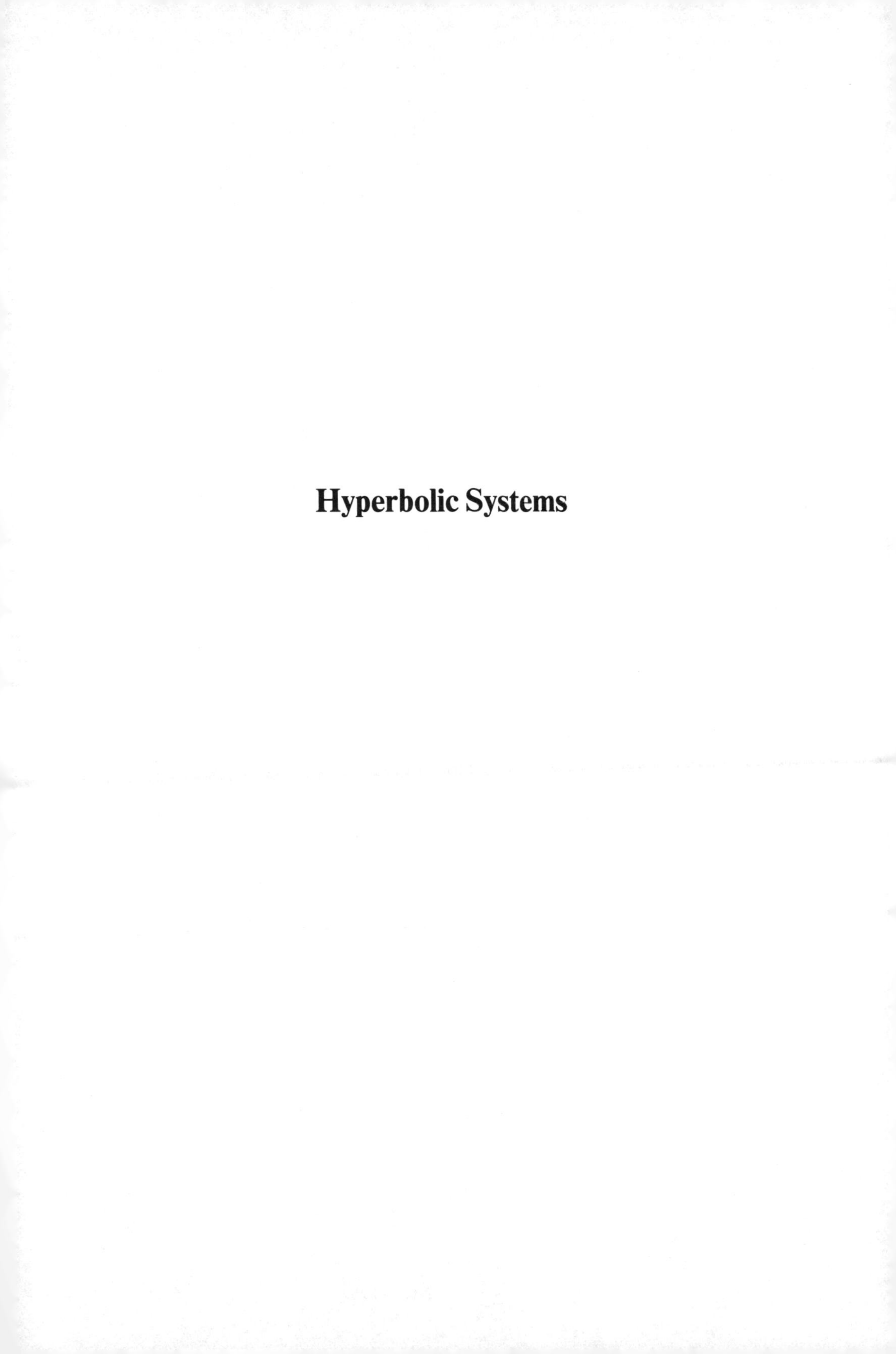

Hyperbolic Systems

Lectures in Applied Mathematics
Volume 23, 1986

OSCILLATIONS IN NONLINEAR PARTIAL DIFFERENTIAL EQUATIONS : COMPENSATED COMPACTNESS AND HOMOGENIZATION

Luc TARTAR*

I will present here some of the results of a program of research carried in the last ten years, partly in collaboration with François Murat, whose aim is to understand the relations between microscopic oscillations and their macroscopic averages and develop new mathematical tools which are expected to give then a rigorous basis to answer some of the important open problems in continuum mechanics and physics and replace some of the strange rules followed by physicists. Very little of the results has been written down completely, although some of the ideas have spread considerably often without any mention of our names. My purpose here is to comment on the ideas and give some examples of the results as I did in [1] which I do suggest as a basic reference to the field (a good occasion to read something in french !, and to work at guessing what the two pages of introduction, suppressed by the publisher, were saying).

Compensated compactness is a tool which has generalized some special result in homogenization theory (found at a time where the word itself was not yet used) and whose goal is to describe the constraints imposed on oscillations of some functions because of the differential equations they satisfy.

* Centre d'Etudes de Limeil-Valenton and Ecole Polytechnique

The actual framework has been developped in order to apply to situations arising in continuum mechanics, emphasizing the different roles of balance equations (which I will only consider here) and constitutive relations and showing how the entropy conditions, so important for understanding the creation of discontinuities in hyperbolic systems, could also be handled (I cannot resist the temptation to harass here the champions of the hamiltonian ideology who, also they have known for years that their framework can only describe the time, often short, where the solution stays smooth, continue to present their methods as the unifying concept in physics).

In the actual compensated compactness theory oscillations are being studied through relations between different weak limits but some new kind of weak topologies has to be introduced in order to describe homogenization ; there is no doubt that the next improvement of the theory, as I already foresee it, will encompass these two aspects. With this in mind I will then describe homogenization as the problem of understanding how oscillations of coefficients of partial differential equations create oscillations in its solution.

In nonlinear equations, the solution itself playing the role of a coefficient, it is then natural to study the existence of oscillating solutions and their behaviour. Such phenomena as propagation, interaction and creation of oscillations can then be observed for simple semilinear hyperbolic equations. Although it should be obvious to everyone I want to emphasize once more the fact that studying these questions is quite different than looking at propagation and interaction of singularities.

Although this was not the original goal of this program, as the motivation was coming from mathematics and continuum mechanics, it then became obvious to me that it could explain some of the rules of quantum physics through the mathematical question of homogenization and oscillations of solutions of partial differential equations.

Propagation of singularities of a system of partial differential equations exhibit some kind of hamiltonian framework and, as singularities in a physical system are more easily noticed, they were first observed by physicists and the laws governing their behaviour were quickly found ; but the partial differential equations were not always written : if you have understood how geometrical optics comes from the wave equation and you are told about experimental evidence of polarized light, would you deduce that one should study Maxwell's system of equations ?

But then physicists discovered particles and soon observed strange facts about them.

I believe that particles are but oscillatory solutions of the equation whose singularities they were studying before. But how could you describe the oscillations of the solution of a system that you do not know except perhaps through some information on the behaviour of its singularities ? and, even if you knew that this was the question to answer, how could you have done it knowing that the corresponding mathematical theory is only now in its infancy ?

The set of rules of quantum mechanics seems to be the answer given more than fifty years ago ; in a few years the answer may become different when the mathematical theory of oscillations in nonlinear partial differential equations will have developped.

I. COMPENSATED COMPACTNESS

The basic framework deals with a sequence of real functions u_j^ε, defined on an open set Ω of R^N which converge weakly as the sequence ε goes to 0 :

H1 : $u_j^\varepsilon \rightarrow u_j^0$ in $L^2(\Omega)$ weak $j = 1,\dots p$.

On the other hand some information on derivatives on the u^ε is given :

H2 : $\sum_{j,k} a_{ijk} \frac{\partial u_j^\varepsilon}{\partial x_k} \rightarrow v_i$ in $H^{-1}_{loc}(\Omega)$ strong $i=1,\dots q$

where a_{ijk} are real constants.

This information restricts the possible oscillations of the sequence u^ε and thus restricts the possible weak limits of the functions of u^ε ; the basic theorem characterizes the possible limits of quadratic functions of u^ε under hypotheses H1 + H2. Let us take a quadratic function Q on R^p and extract a subsequence such that

H3 : $Q(U^\varepsilon) \rightarrow f$ in $D'(\Omega)$ weak

(Notice that f can be a measure and H3 means also a weak convergence in the sense of measures), then we have the

Theorem 1 : H1 + H2 + H3 implies

C_+ : $f \geqslant Q(u^0)$ in the sense of measures
if and only if Q satisfies

H_+ : $Q(\lambda) \geqslant 0 \;\forall \lambda \in \Lambda$

where $\Lambda = \{\lambda \in R^p : \exists\, \xi \in R^N\setminus\{0\} : \sum_{jk} a_{ijk}\, \lambda_j\, \xi_k = 0 \quad i=1,\dots q\}$

A useful corollary is then :

<u>Corollary</u> : H1 + H2 + H3 implies

$$C_0 : f = Q(u^0)$$

if and only if Q satisfies :

$$H_0 : Q(\lambda) = 0 \quad \forall \lambda \in \Lambda$$

Obviously the set V of (λ, ξ) appearing in the definition of Λ plays the role of a wave front set ; it contains more information about the list H2 than Λ, but for describing limits of quadratic funtions of u^ε, knowing Λ is enough.

The description of Q satisfying H_+ or H_0 is not always easy ; it is more easy to check if a given Q satisfies one of these conditions and this is often how the corollary is used.

Quadratic functions Q satisfying H_0 are robust quantities with respect to oscillations and so play an important role ; some of them appear in questions apparently not related to oscillations and often have some algebraic or geometric interpretation.

The point of view of using "weak" topologies is radically different from the one using "strong" topologies (use of singularity theory, implicit function theorem, bifurcation analysis) although, after having understood oscillations and introduced the right macroscopic quantities one will be led to use classical tools (with probably a new insight of the problem).

We will now describe some typical examples of the corollary. We will leave the computation of Λ to the reader and replace the functions v_i appearing in H2 everywhere by * (and also deal with vector valued functions).

Example 1 : (div-curl lemma)

$E^{\varepsilon} \rightarrow E^{0}$; $D^{\varepsilon} \rightarrow D^{0}$ with $\Omega \subset R^{N}$, E, D being N-vectors
curl $E^{\varepsilon} \rightarrow *$; div $D^{\varepsilon} \rightarrow *$. Then $E^{\varepsilon}.D^{\varepsilon} \rightarrow E^{0}.D^{0}$

This is the basic result for understanding homogenization of second order elliptic equations. The physical interpretation is that in electrostatics no internal energy need be introduced because the macroscopic fields E^{0} and D^{0} give the right value for the macroscopic density of energy.

Example 2 : (wave equation)

$u^{\varepsilon} \rightarrow u^{0}$ in $H^{1}(\Omega)$ weak with $(x,t) \in \Omega \subset R^{N}$.

$$\frac{\partial^{2} u^{\varepsilon}}{\partial t^{2}} - \Delta u^{\varepsilon} \rightarrow *. \text{ Then } \left|\frac{\partial u^{\varepsilon}}{\partial t}\right|^{2} - |\text{grad } u^{\varepsilon}|^{2} \rightarrow \left|\frac{\partial u^{0}}{\partial t}\right|^{2} - |\text{grad } u^{0}|^{2}$$

This is a simple consequence of example 1. One should notice that if u^{ε} is a solution of the wave equation the integral of $\left|\frac{\partial u^{\varepsilon}}{\partial t}\right|^{2} + |\text{grad } u^{\varepsilon}|^{2}$ is conserved ; here we find that the action is a nice quantity. Although the energy is a conserved quantity, part of it can be trapped at a microscopic level ; small wiggles on u^{ε} can carry a finite amount of energy which, of course will propagate. What the above result says is that there is a macroscopic equipartition of this energy between $\left|\frac{\partial u}{\partial t}\right|^{2}$ and $|\text{grad } u|^{2}$

Example 3 : (Maxwell's system)

$E^{\varepsilon} \rightarrow E^{0}$, $B^{\varepsilon} \rightarrow B^{0}$, $H^{\varepsilon} \rightarrow H^{0}$, $D^{\varepsilon} \rightarrow D^{0}$ $(x,t) \in R^{4}$,E,B,D,H being 3 vectors $\operatorname{div} B^{\varepsilon} \to *$; $\frac{\partial B^{\varepsilon}}{\partial t} + \operatorname{curl} E^{\varepsilon} \to *$; $\operatorname{div} D^{\varepsilon} \to *$; $\frac{\partial D^{\varepsilon}}{\partial t} - \operatorname{curl} H^{\varepsilon} \to *$

Then $E^{\varepsilon}.B^{\varepsilon} \to E^{0}.B^{0}$; $H^{\varepsilon}.D^{\varepsilon} \to H^{0}.D^{0}$; $E^{\varepsilon}.D^{\varepsilon} - H^{\varepsilon}.B^{\varepsilon} \to E^{0}.D^{0} - H^{0}.B^{0}$

As in example 2 there is also an equipartition of the energy trapped at a microscopic level between E.D and H.B

The quantities appearing in example 3 will look more natural if one understands that $(E^{\varepsilon}, B^{\varepsilon})$ are the coefficients of a 2-differential form ω^{ε} satisfying $d\omega^{\varepsilon} \to *$, and similarly for $(-H^{\varepsilon}, D^{\varepsilon})$. Then examples 1 and 3 are particular examples of

Example 4 : (differential forms)

$a^{\varepsilon} \to a^{0}$, $b^{\varepsilon} \to b^{0}$, $x \in R^{N}$, a^{ε} is a p-form, b^{ε} is a q-form
$da^{\varepsilon} \to *$;
$db^{\varepsilon} \to *$. Then $a^{\varepsilon} \wedge b^{\varepsilon} \to a^{0} \wedge b^{0}$. Of course one deduces $a^{\varepsilon} \wedge a^{\varepsilon} \to a^{0} \wedge a^{0}$ and $b^{\varepsilon} \wedge b^{\varepsilon} \to b^{0} \wedge b^{0}$

Sometimes some new equation $\sum_i \frac{\partial}{\partial x_i} f_i(u^{\varepsilon}) \to *$ is a consequence of H2 and it is then useful to add the $f_i(u^{\varepsilon})$ in the list H1 and the new equation in the list H2 and look what new information has been added. This is the case in the preceding example if one notices that $d(a^{\varepsilon} \wedge b^{\varepsilon}) = da^{\varepsilon} \wedge b^{\varepsilon} + (-1)^{p}\, a^{\varepsilon} \wedge db^{\varepsilon}$; then an iteration argument gives

Example 5 :

$U^{\varepsilon} = (u_{ij}^{\varepsilon}) \to U^{0} = (u_{ij}^{0})$ in $L^{q}(\Omega)$ weak, $q > N$, $\Omega \subset R^{N}$

$\frac{\partial u^{\varepsilon}_{ij}}{\partial x_k} - \frac{\partial u^{\varepsilon}_{ik}}{\partial x_j} \to *$. Then $\det U^{\varepsilon} \to \det U^0$

This result appears in the work of J. Ball in nonlinear elasticity and of Reshetnyak in pseudo conformal mapping at a time where only example 1 was known. This property of the determinant is linked with its use in defining topological degree : if $u^{\varepsilon}_{ij} = \frac{\partial v^{\varepsilon}_i}{\partial x_j}$ then quantities like $\int_{\Omega} \Psi(v^{\varepsilon}) \det U^{\varepsilon}\, dx$ do not change if one changes v^{ε} inside Ω. So, although v^{ε} has to be smooth in order to define it, this quantity is robust with respect to small wiggles in v^{ε} and this explains in a way why a limiting process can be used to define it for v continuous.

With this in mind it is then natural to think that if a topological invariant is obtained by integrating some quadratic form (or more generally some polynomial) it could be found using theorem 1 in a suitable way. I tried this for some invariant I had heard about : the first Chern class ; and it gave me the following.

<u>Example 6</u> : (Biancchi identity)

$R^{\varepsilon}_{ij} \to R^0_{ij}$ with $R^{\varepsilon}_{ij} \in E$ and $R^{\varepsilon}_{ij} + R^{\varepsilon}_{ji} = 0 \quad \forall i,j,\ x \in R^N,\ N \geq 4$

$\frac{\partial}{\partial x_k} R^{\varepsilon}_{ij} + \frac{\partial}{\partial x_j} R^{\varepsilon}_{ki} + \frac{\partial}{\partial x_i} R^{\varepsilon}_{jk} \to * \quad \forall i,j,k$

Then $\sum_{ijkl} \sigma(i,j,k,l)\, B(R^{\varepsilon}_{ij}, R^{\varepsilon}_{kl}) \to$ same quantity for R^0 if B is any symmetric bilinear form on $E \times E$ and $\sigma(i,j,k,l)$ is the signature of the permutation of the four distinct indices.

Here are two other examples of quadratic quantities adapted to a set of linear equations :

Example 7 :

$A_\mu^\varepsilon \rightharpoonup A_\mu^0$ with $A_\mu^\varepsilon \in E$, $x \in R^N$

$\frac{\partial A_\mu^\varepsilon}{\partial x_\nu} - \frac{\partial A_\nu^\varepsilon}{\partial x_\mu} \rightarrow *\ \forall \mu, \nu$. Then $B(A_\mu^\varepsilon, A_\nu^\varepsilon) \rightharpoonup B(A_\mu^0, A_\nu^0)$

if B is any antisymmetric bilinear form on ExE.

Example 8 :

$V_{ij}^\varepsilon \rightharpoonup V_{ij}^0$ with $\operatorname{tr} V^\varepsilon = 0$

$\frac{\partial V_{ij}^\varepsilon}{\partial x_k} - \frac{\partial V_{ik}^\varepsilon}{\partial x_j} \rightarrow *$. Then $(V^\varepsilon)^2 \rightharpoonup (V^0)^2$ (ie $\forall_{i,k} \sum_j V_{ij}^\varepsilon V_{jk}^\varepsilon \rightharpoonup$ same for V^0).

If example 7, suggested by Yang-Mills equation, has not been very useful, example 8, modeled on hydrodynamics (it first appeared in work by L. Mascarenhas) led me to some new remark for Euler equation. The basic idea I was following was that in a turbulent flow one expects that the quantities which are robust with respect to oscillations will be very useful and it is important to make them appear by algebraic manipulations of the equation.

If we write the Euler equation for $x \in R^N$

a) $\frac{Du_i}{Dt} + \frac{\partial p}{\partial x_i} = 0$ with $\frac{D}{Dt} = \frac{\partial}{\partial t} + \sum_j u_j \frac{\partial}{\partial x_j}$

$\operatorname{div} u = 0$

Then by taking derivative in x_j and noting V the matrix of entries $V_{ij} = \frac{\partial u_i}{\partial X_j}$ we obtain

$$\text{b)} \quad \frac{D}{Dt} V_{ij} + (V^2)_{ij} + \frac{\partial^2 p}{\partial x_i \partial x_j} = 0$$

and we see that V^2 appears. Actually the real difficulty for oscillating solutions lies more in understanding what happens of $\frac{D}{Dt}$ which contains u inside : there is no reason to believe that geometers are right by introducing covariant derivatives into the framework of this problem unless they can prove theorems for oscillating solutions or for a time interval large enough to contain (possible) apparition of singularities. Anticipating the next paragraph on homogenization one should think that the u_j explicitly written in equation a) are coefficient of a differential form and that weak topology is then natural for them but that the u hidden inside $\frac{D}{Dt}$ has a different interpretation and then a different weak-type topology associated to it : this fact has been known to specialists in numerical analysis who have introduced mixed eulerian-lagrangian coordinates to study this problem.

If we substract to b) the transposed equation we obtain $\frac{D}{Dt}(V-V^*) + V(V-V^*) + (V-V^*)V^*=0$ and then this gives

$$\text{c)} \quad \frac{D}{Dt} \det (V-V^*)=0$$

If this is useless in odd dimension it gives an invariant along characteristics in even dimension which generalizes the vorticity for N=2. As pointed out to me by D. Serre it is better to use the pfaffian theorem to describe this result ; if $\omega_1 = \sum_j u_j \, dx_j$ and $\omega_2 = d\omega_1$ (only x variables) is the vorticity we can take the exterior product of ω_2 with itself n times if $N=2n$

d) $\omega_2 \wedge \cdots \wedge \omega_2 = f\, dx$

and then e) becomes c)' $\frac{D}{Dt} f=0$.

In odd dimension, as a generalization of helicity u.curl u for $N=3$, D. Serre showed that the integral of $\omega_1 \wedge \omega_2 \wedge \cdots \wedge \omega_2$ is conserved, using some formalism of P. Olver. I know a more direct calculation which gives

e) $\frac{D}{Dt}\, \omega_1 \wedge \omega_2 \wedge \cdots \wedge \omega_2 + d(p - \frac{u^2}{2}) \wedge \omega_2 \wedge \cdots \wedge \omega_2 = 0$

which I will develop somewhere else with some other remarks (P. Constantin and S. Klainerman showed me during the seminar some of their results very similar to mine).

Another reason to look for special quadratic functions is that they also satisfy better estimates ; for example from b) we obtain $\Delta p + \mathrm{tr}\, V^2 = 0$ and for $N = 2$ we have the surprising fact that if $u \in (H^1(R^2))^2$ with div $u = 0$ then $p \in F\, L^1$ and $\frac{\partial p}{\partial x_j} \in L^2(R^2)$ (continuity of the pressure was noticed before in some work of L.E. Fraenkel). Such properties have been investigated by Hanouzet-Joly as a generalization of factorization property analog to $d\omega_1 \wedge d\omega_2 = d(\omega_1 \wedge d\omega_2)$ but the above result needs a more refined technique. In a way that needs to be studied more some of the long time existence theorems presented at the seminar by S. Klainerman and J. Shatah seem to rely partly on some particular property of these special quadratic forms.

To close this paragraph I will end by saying that it is not enough to apply corollary of theorem 1 to solve problems : thinking in terms of oscillations can lead to some application of it but then it only puts you on an interesting track ;

trying to understand why these special quadratic functions occur can point out the right framework for the problem at hand.

II. HOMOGENIZATION

I will first present a model equation which I first studied to test the methods I knew on turbulence ; my point of view was to try to understand oscillations of solutions of Navier Stokes equations for small viscosity, and of course with the actual tool I could not do it ; nevertheless, although far removed from a physical situation, something interesting can be learned from this model (there are some similarities with computation usually done in quantum mechanics but this matter requires more development). I do want to point out the important fact that theorems in homogenization have to be local in order to study separately the interaction with the boundary (functional analytic treatment of Navier Stokes equation in a bounded domain and asymptotic behaviour of solutions can only tell how turbulence survives after multiple reflection on the boundary ; turbulence has to be studied in terms of partial differential equations and not in terms of boundary value problems) ; this is the reason of the absence of boundary conditions here.

In a domain Ω of R^3 we consider the following equation :

$$(\alpha) \quad \begin{cases} -\nu \, \Delta \, u_\varepsilon + u_\varepsilon \times \text{curl} \, (v_0 + \lambda \, v_\varepsilon) = f - \text{grad} \, q_\varepsilon \\ \text{div} \, u_\varepsilon = 0 \end{cases}$$

v_0 is given in $(L^3(\Omega))^3$, f in $(H^{-1}(\Omega))^3$. $\lambda \geqslant 0$ is a parameter that measures the strength of the oscillations v_ε which is a sequence satisfying $v_\varepsilon \rightarrow 0$ in $(L^3(\Omega))^3$ weak. Assuming that u_ε

stays bounded in $(H^1(\Omega))^3$, for example because there are reasonable boundary conditions, we can extract a subsequence such that $u_\varepsilon \to u_0$ in $(H^1(\Omega))^3$ weak. We want to know what kind of equation satisfies u_0 : the problem is that the oscillations on v_ε induce wiggles in u_ε and that the weak limit of $u_\varepsilon \times \text{curl } v_\varepsilon$ is not 0. The answer is the following.

Theorem 2

There is a subsequence such that there exists a non negative symmetric matrix $M(x)$ (depending only on the sequence v_ε) with the property that

$$(\beta) \quad \begin{aligned} &-\nu\, \Delta u_0 + u_0 \times \text{curl } v_0 + \lambda^2 M u_0 = f - \text{grad } q. \\ &\text{div } u_0 = 0 \end{aligned}$$

$$(\gamma) \quad \nu\, |\text{grad } u_\varepsilon|^2 \to \nu\, |\text{grad } u_0|^2 + \lambda^2 (Mu_0, u_0) \text{ in } D'(\Omega)$$

M is obtained in the following way : take $k \in R^3$, solve

$$(\delta) \quad \begin{aligned} &-\nu\, \Delta w_\varepsilon + k \times \text{curl } v_\varepsilon = -\text{ grad } r_\varepsilon \\ &\text{div } w_\varepsilon = 0 \end{aligned}$$

with reasonable boundary conditions so that $w_\varepsilon \to 0$ in $(H^1)^3$ weak.

Then we have (for a subsequence)

$$(\gamma) \quad \begin{aligned} &w_\varepsilon \times \text{curl } v_\varepsilon \to Mk \text{ in } (H^{-1}\Omega))^3 \text{ weak} \\ &\nu\, |\text{grad } w_\varepsilon|^2 \to (Mk, k) \text{ in } D'(\Omega) \end{aligned}$$

The term $\lambda^2 M$ then corresponds to a correction, analogous to quantum effects calculations, obtained by solving a linear equation and then averaging quadratic quantities made out it (of course decomposition of v_ε on a basis of eigenvectors is only one way to make the computation).

Let us look now at a more technical situation where oscillating coefficients appear in higher order terms.

The basic example is

$$(1) \quad -\sum_{i,j} \frac{\partial}{\partial x_i}\left(a^\varepsilon_{ij}(x)\frac{\partial u^\varepsilon}{\partial x_j}\right) = f$$

If we note (2) $\begin{array}{l} E^\varepsilon = -\operatorname{grad} u^\varepsilon \\ D^\varepsilon = a^\varepsilon E^\varepsilon \quad \text{i.e. } D^\varepsilon_i = \sum_j a^\varepsilon_{ij} E^\varepsilon_j \end{array}$

then we have (3) $\begin{array}{l} \operatorname{curl} E^\varepsilon = 0 \\ \operatorname{div} D^\varepsilon = f \end{array}$

which is the framework of example 1 of the first paragraph. If $E^\varepsilon \to E^0$ and $D^\varepsilon \to D^0$ the main question is to understand the relation between D^0 and E^0. Assuming that a^ε is symmetric and satisfies (4) $\alpha I \leqslant a^\varepsilon(x) \leqslant \beta I$ a.e. with $0 < \alpha \leqslant \beta < +\infty$ we have the following.

Theorem 3

There is a subsequence and a^0 (depending only upon the a^ε) satisfying also (4) and such that $D^0 = a^0 E^0$

We will denote by $a^\varepsilon \overset{H}{\to} a^0$ this new type of weak convergence which means that $E^\varepsilon \to E^0$, $D^\varepsilon = a^\varepsilon E^\varepsilon \to D^0$, $\operatorname{curl} E^\varepsilon \to *$, $\operatorname{div} D^\varepsilon \to *$ implies $D^0 = a^0 E^0$.

In the case where $a_\varepsilon = \chi_\varepsilon \alpha I + (1-\chi_\varepsilon)\beta I$ where χ_ε is a characteristic function and $\chi_\varepsilon \to \Theta$ in $L^\infty(\Omega)$ weak *, we will say that we have a mixture of two distinct isotropic materials, $\Theta(x)$ denoting the local proportion of the material with parameter α (dielectric permittivity for example), and a^0 will represent the effective parameter for the mixture.

It is important to notice that, except in dimension 1, a^0 cannot be determined from Θ but has nevertheless to satisfy some constraints. It is a puzzling fact that, when dealing with mixtures, physicists are ready to imagine lots of different theories which give different values of a^0 in term of Θ and it is quite difficult to understand what information they use on the sequence a^ε ; in the above framework we have characterized all the couples (Θ, a^0) that could be obtained and this is given by the following :

Theorem 4 [3] the eigenvalues $\lambda_1, \dots \lambda_N$ of $a^0(x)$ satisfy

$$(5) \quad \mu_-(0) \leq \lambda_j \leq \mu_+(\Theta)$$

$$(6) \quad \begin{aligned} &\sum_j \frac{1}{\lambda_j - \alpha} \leq \frac{1}{\mu_-(\Theta)-\alpha} + \frac{N-1}{\mu_+(\Theta)-\alpha} \\ &\sum_j \frac{1}{\beta - \lambda_j} \leq \frac{1}{\beta - \mu_-(\Theta)} + \frac{N-1}{\beta - \mu_+(\Theta)} \end{aligned}$$

where $\mu_-(\Theta) = \left(\frac{\theta}{\alpha} + \frac{1-\Theta}{\beta} \right)^{-1}$ and $\mu_+(\Theta) = \Theta\alpha + (1-\Theta)\beta$

It does not means that if you mix two different materials with some given proportion in a "naive" way, you could obtain any of the a^0 described by (5) (6), even when restricting to isotropic a^0 ; it seems reasonable to think that a given "manufacturing process" can only create a smaller set of a^0, but a lot remains to be done in that direction.

In some cases the mixture is done by some evolution equation (like in the case of a mixture of two immiscible fluids) and it is difficult to understand if this creates only special configurations. We have studied a more simpler framework, where the mixing process consists in looking for the optimal configuration for some criterion : there are examples (which occur even in real engineering problems) where the optimal solution has to be a mixture (to be defined in some way). I will describe here two simple problems and give the result of the analysis, carried in [2] in a more general situation.

Let Ω be a bounded open set of R^N (N=2 or 3), and choose a set ω inside Ω of given measure m ; fill ω with a material of parameter α and the complement with a material of parameter β. If χ is the characteristic function of ω and $a = \chi\alpha I+(1-\chi)\beta I$ we then solve

$$\text{(a)} \quad \begin{aligned} &- \operatorname{div}(a \operatorname{grad} u) = 1 \\ &u|_{\partial\Omega} = 0 \end{aligned}$$

which we can interpret as giving the equilibrium temperature under 0 boundary conditions and uniform source of heat. We are interested in the two distinct problems.

<u>Problem 1</u> : find the best ω of measure m which maximizes $\int_\Omega u(x)dx$.

<u>Problem 2</u> : find the best ω of measure m which minimizes $\int_\Omega u(x)dx$.

So we want to find the extremal values of the average temperature in Ω. The difficulty is that, in most of the cases, no optimal configuration exists because a mixture of the two materials gives a better value to the criterion. Fortunately the analysis shows that only special a^0 of the set (5) (6) do appear in the optimal solutions and that they can be computed by solving the new (relaxed) following problems :

Problem 1' : find Θ such that $0 \leq \Theta(x) \leq 1$, $\int_\Omega \Theta(x)dx = m$ which maximizes $\int_\Omega u(x)\, dx$ but where u is solution of

$$(b) \quad \begin{cases} -\mathrm{div}(\mu_-(\Theta)\ \mathrm{grad}\ u) = 1 \\ u|_{\partial\Omega} = 0 \end{cases}$$

Problem 2' : find Θ such that $0 \leq \Theta \leq 1$, $\int_\Omega \Theta\, dx = m$ which minimizes $\int_\Omega u\, dx$ but where u is solution of

$$(c) \quad \begin{cases} -\mathrm{div}(\mu_+(\Theta)\ \mathrm{grad}\ u) = 1 \\ u|_{\partial\Omega} = 0 \end{cases}$$

If a solution (Θ,u) of problem 1' is found, the optimal mixture "consists" of slices perpendicular to grad u with the right proportion ; on the other hand if a solution of problem 2' is found, the optimal mixture "consists" of filaments parallel to grad u with the right proportion.

Problems 1' and 2' do have solutions and one can derive numerical algorithms for them, but if the solution does take values of Θ between 0 and 1 the initial problem is ill posed and numerical methods when applied to the initial problem will give a wrong answer in general.

Even if a classical solution exists, the conditions of optimality obtained by this method are stronger than the one obtained by classically moving the boundary of ω.

Although the idea has not been yet applied with success, it seems that this method will become able to handle instabilities of interfaces by first describing the case were the interface is so wild that a mixing region has to be considered and introducing the right effective parameter in this region.

III. OSCILLATIONS IN SEMILINEAR HYPERBOLIC SYSTEMS

We consider now a non linear evolution equation without oscillating coefficients, but we take an oscillating initial data and we want to describe the possible oscillations in the solution : do they persist and propagate, do they interact, are they created ?

We consider first a scalar hyperbolic equation

$$(1) \quad \frac{\partial u}{\partial t} + \frac{\partial}{\partial x} f(u) + g(u) = 0$$

and, of course, we only consider solutions satisfying the entropy conditions. The main result is that genuine non linearity of f prevents oscillations ; more precisely we have the following

Theorem 5 : Assume that there is no interval where f is affine and let u^{ε} be a sequence of solutions of (1) satisfying the entropy conditions and such that

$$(2) \quad u^{\varepsilon} \rightarrow u^{0} \text{ in } L^{\infty}(\omega) \text{ weak * } (\omega \text{ is an open set in } (x,t)$$

plane) then this implies

$$(3) \quad u^{\varepsilon} \rightarrow u^{0} \text{ in } L^{p}_{loc}(\omega) \text{ strong } \forall\, p < +\infty$$

This shows that if one takes an oscillating data, then the oscillations are killed in an arbitrary short time (interesting phenomena of formation and interaction of shocks occur in a boundary layer near time 0 but the study of this requires much more stronger assumptions on f).

The generalization to systems is not so easy : some analogous result for systems of two equations with genuine nonlinearity have been obtained by R. DiPerna ; in some linearly degenerate cases there are oscillating solutions as has been noticed by D. Serre.

We now switch to semilinear hyperbolic systems where oscillations do persist and propagate. The phenomena will be studied on two model equations : the Carleman model and the Broadwell model.

The Carleman model is

$$(4)\quad \begin{aligned} &\frac{\partial u}{\partial t} + \frac{\partial u}{\partial x} + u^2 - v^2 = 0 \\ &\frac{\partial v}{\partial t} - \frac{\partial v}{\partial x} - u^2 + v^2 = 0 \end{aligned}$$

We take initial data which are oscillating

$$(5)\quad \begin{aligned} &u(x, 0) = \psi_\varepsilon(x) \\ &v(x,0) = \Psi_\varepsilon(x) \end{aligned} \qquad \text{and } 0 \leqslant \psi_\varepsilon, \Psi_\varepsilon \leqslant M$$

Then we have $0 \leqslant u^\varepsilon, v^\varepsilon \leqslant M$ for $t \geqslant 0$ and we would like to describe the weak limits of functions of $(u^\varepsilon, v^\varepsilon)$. In order to do this, the knowledge of weak limits of ψ_ε and Ψ_ε is not enough and we need the complete knowledge of oscillations of the sequence ψ_ε and the same for Ψ_ε. More precisely if

$$(6)\quad \begin{aligned} &\psi_\varepsilon^m \to \phi_m \quad \text{in } L^\infty(R) \text{ weak } * \ \forall m \geqslant 1 \\ &\Psi_\varepsilon^m \to \Psi_m \quad \text{in } L^\infty(R) \text{ weak } * \ \forall m \geqslant 1 \end{aligned}$$

we can answer the question by the

Theorem 6 : If (5) (6) hold then the solution $(u^\varepsilon, v^\varepsilon)$ of (4) satisfies (7) $u_\varepsilon^m v_\varepsilon^p \to U_m V_p$ in $L^\infty(R\times]0,\infty[)$ weak * where U_m, V_m, $m \geqslant 1$, is the unique solution of the infinite system of equations

$$
(8)\quad
\begin{cases}
\left(\frac{\partial}{\partial t} + \frac{\partial}{\partial x}\right) U_m + m\, U_{m+1} - m\, U_{m-1}\, V_2 = 0 \quad \forall\, m \geq 1 \\
\left(\frac{\partial}{\partial t} - \frac{\partial}{\partial x}\right) V_m - m\, U_2\, V_{m-1} + m\, V_{m+1} = 0 \quad \forall\, m \geq 1 \\
U_m(x,0) = \Phi_m \;;\; V_m(x,0) = \Psi_m \qquad \forall\, m \geq 1
\end{cases}
$$

$$
0 \leq U_m, V_m \leq M^m \ \forall\, m \quad (U_0 = V_0 = 1)
$$

We should remark that no relation between oscillations of ψ_ε and Ψ_ε is necessary : oscillations of u_ε and v_ε immediately become locally independent, a fact expressed by (7) and which of course is a simple application of compensated compactness.

We should remark that there are constraints between the Φ_m or the U_m : for example $U_2 \geq U_1^2$ and $U_3 \geq \frac{U_2^2}{U_1}$. If we then introduce

$$
(9)\quad
\begin{cases}
\sigma_u(x,t) = (U_2 - U_1^2)^{1/2} \\
\sigma_v(x,t) = (V_2 - V_1^2)^{1/2}
\end{cases}
$$

which measure the strength of the oscillations in u^ε and v^ε; then we have after a simple calculation the

Corollary : σ_u and σ_v satisfy the following inequalities

$$
(10)\quad
\begin{cases}
\left(\frac{\partial}{\partial t} + \frac{\partial}{\partial x}\right) \sigma_u + \frac{U_2}{U_1}\, \sigma_u \leq 0 \\
\left(\frac{\partial}{\partial t} - \frac{\partial}{\partial x}\right) \sigma_v + \frac{V_2}{V_1}\, \sigma_v \leq 0
\end{cases}
$$

These inequalities show that the oscillations propagate along

characteristics, cannot be created and decay in strength.
The Broadwell model is

$$\frac{\partial u}{\partial t} + \frac{\partial u}{\partial x} + uv - w^2 = 0$$

$$(11) \quad \frac{\partial v}{\partial t} - \frac{\partial v}{\partial x} + uv - w^2 = 0$$

$$\frac{\partial w}{\partial t} \qquad - uv + w^2 = 0$$

and we will see that its properties are quite different. We take initial data which are oscillating

$$u(x,0) = \phi_\varepsilon$$

$$(12) \quad v(x,0) = \Psi_\varepsilon \qquad 0 \leq \phi_\varepsilon, \Psi_\varepsilon, \chi_\varepsilon \leq M$$

$$w(x,0) = \chi_\varepsilon$$

Then we have $0 \leq u_\varepsilon, v_\varepsilon, w_\varepsilon \leq F(M,t)$ for $t > 0$ (the best bound $F(M,t)$ is not known).
If we extract subsequences such that $u_\varepsilon^m \to U_m$, $v_\varepsilon^m \to V_m$, $w_\varepsilon^m \to W_m$ in L^∞ weak * for $m \geq 1$, we have

$$u_\varepsilon^m v_\varepsilon^p \to U_m V_p$$

$$(13) \quad v_\varepsilon^m w_\varepsilon^p \to V_m W_p \qquad \text{in } L^\infty \text{ weak } *$$

$$w_\varepsilon^m u_\varepsilon^p \to W_m U_p$$

but the first important difference is

$$(14) \quad u_\varepsilon v_\varepsilon w_\varepsilon \to X \text{ in } L^\infty \text{ weak } *$$

and in general $X \neq U_1 V_1 W_1$

Using (13) we obtain

$$(15)\quad \begin{aligned} &(\frac{\partial}{\partial t} + \frac{\partial}{\partial x})\, U_m + mV_1U_m - mU_{m-1}\, W_2 = 0 \quad \forall\, m \geq 1 \\ &(\frac{\partial}{\partial t} - \frac{\partial}{\partial x})\, V_m + mU_1V_m - mV_{m-1}\, W_2 = 0 \quad \forall\, m \geq 1 \end{aligned}$$

from which we have

$$(16)\quad \begin{aligned} &(\frac{\partial}{\partial t} + \frac{\partial}{\partial x})\, \sigma_u + V_1\, \sigma_u = 0 \\ &(\frac{\partial}{\partial t} - \frac{\partial}{\partial x})\, \sigma_v + U_1\, \sigma_v = 0 \end{aligned}$$

This shows that oscillations in u^ε and v^ε propagate along characteristics, are not created and decay in strength.
For describing the oscillations of w^ε we have

$$(17)\quad \begin{aligned} &\frac{\partial}{\partial t} W_1 - U_1V_1 + W_2 = 0 \\ &\frac{\partial}{\partial t} W_2 - 2\, X + 2W_3 = 0 \end{aligned}$$

and noting $\sigma_w = (W_2 - W_1^2)^{1/2}$ and using

$$(18)\quad \left| X - U_1\, V_1\, W_1 \right| \leq \sigma_u\, \sigma_v\, \sigma_w$$

we only obtain for σ_w the inequality

$$(19)\quad \frac{\partial}{\partial t} \sigma_w + \frac{W_2}{W_1} \sigma_w \leq \sigma_u\, \sigma_v$$

This shows that oscillations in w^ε propagate along characteristics but could be created by interaction of oscillations in u^ε and oscillations in v^ε. This is indeed the case.
As was pointed out by G. Papanicolaou a complete analysis can be done if the initial data have a periodic oscillating structure and we then proved the following

Theorem 7 : Assume that

$$(20)\quad u^{\varepsilon}(x,0) = a\left(x, \frac{x}{\varepsilon}\right) \; ; \; v^{\varepsilon}(x,0) = b\left(x,\frac{x}{\varepsilon}\right) \; ; \; w^{\varepsilon}(x,0) = c\left(x,\frac{x}{\varepsilon}\right)$$

where a,b,c are (smooth enough) functions of period 1 in the second variable. Then if we solve the system

$$(21)\quad \begin{aligned}
&\left(\frac{\partial}{\partial t} + \frac{\partial}{\partial x}\right)A(x,y,t)+A(x,y,t) \int_0^1 B(x,z,t)dz-\int_0^1 C^2(x,z,t)dz=0\\
&\left(\frac{\partial}{\partial t} - \frac{\partial}{\partial x}\right)B(x,y,t)+B(x,y,t) \int_0^1 A(x,z,t)dz-\int_0^1 C^2(x,z,t)dz=0\\
&\frac{\partial}{\partial t} C(x,y,t)+C^2(x,y,t)-\int_0^1 A(x,y-z,t)B(x,y+z,t)dz = 0\\
&A(x,y,0)=a(x,y) \; ; \; B(x,y,0)=b(x,y) \; ; \; C(x,y,0)=c(x,y)
\end{aligned}$$

we then have

$$(22)\quad \begin{aligned}
&u_{\varepsilon}(x,t) - A\left(x,\frac{x-t}{\varepsilon},t\right) \to 0 \text{ in } L^p_{loc} \text{ strong } \forall\, p<+\infty\\
&v_{\varepsilon}(x,t) - B\left(x,\frac{x+t}{\varepsilon},t\right) \to 0 \text{ in } L^p_{loc} \text{ strong } \forall\, p<+\infty\\
&w_{\varepsilon}(x,t) - C\left(x,\frac{x}{\varepsilon}, t\right) \to 0 \text{ in } L^p_{loc} \text{ strong } \forall\, p<+\infty
\end{aligned}$$

In this situation $U_m = \int_0^1 A^m(x,y,t)dy$ and similarly for V_m, W_m. One has a similar type of equations for the Carleman model. The difficulty of describing the oscillations for the Broadwell model lies in the fact that some kind of correlation between oscillations is needed and this is automatically taken into account in the periodically modulated case by the integral appearing in the equation for $\frac{\partial C}{\partial t}$ in (21).

It is clear that the method has to be improved in order to handle more general problems.

BIBLIOGRAPHIE

As I said in the introduction this is mainly an up to date english version of [1] whose bibliography should be added here ; [2] and [3] have been written since and contain some of the missing details.

As I quoted a few persons who talked at the Seminar, I refer to their lecture notes for further references.

[1] Tartar L.
Etude des oscillations dans les équations aux dérivées partielles non linéaires. Trends and Applications of Pure Mathematics to Mechanics, Ciarlet-Roseau ed. Lecture Notes in Physics Springer 195 (1984), p.384-412.

[2] Murat F.- Tartar L.
Calcul des variations et homogénéisation, to appear in Collection de la Direction des Etudes et Recherches d'Electricité de France, Eyrolles Paris (1984) (Cours de l'Ecole d'Eté CEA-EDF-INRIA sur l'homogénéisation).

[3] Tartar L.
Estimations fines de coefficients homogénéisés ; to appear in Research Notes in Mathematics Pitman (1984) (Colloque De Giorgi, Kree ed.).

Lectures in Applied Mathematics
Volume 23, 1986

NONLINEAR P.D.E. AND THE WEAK TOPOLOGY

Ronald J. DiPerna[1]

ABSTRACT. We shall discuss some new results dealing with oscillations in solutions to nonlinear systems of conservation laws.

We are concerned with solutions to nonlinear systems of conservation laws of hyperbolic and elliptic type

$$\sum_{j=0}^{m-1} \partial_j \, g_j(u) = 0 \tag{1.1}$$

$$u: \Omega \subset R^m \to R^n \qquad g_j: R^n \to R^p$$

where $\partial_j = \partial/\partial y_j$. We shall outline a general program and present some new results dealing with oscillations in weakly convergent solution sequences to (1.1). We shall also treat approximate solutions generated by finite difference schemes and diffusive regularizing operators. We refer the reader to [21] for details.

For systems of both hyperbolic type and elliptic type we shall present some new results that establish strong convergence of sequences of exact and approximate solutions without using uniform bounds on the derivatives. Weak convergence is transformed into strong convergence using the Tartar-Murat theory of compensated compactness and the Young measure. In this paper we shall be concerned exclusively with questions of local convergence, i.e. convergence on compact subsets of the domain of defi-

[1] 1980 Mathematics Subject Classification 35L65.
Supported by NSF Foundation under grant MCS-83-01135

nition Ω. The analysis of initial and/or boundary layers is a separate topic.

Several preliminary remarks are in order concerning the standard notions of weak and strong convergence and their relationship to nonlinear differential equations. Many situations arise in which one is presented with a sequence of exact or approximate solutions u^ε to (1.1) whose amplitude is bounded uniformly with respect to ε. In both the hyperbolic and elliptic settings the amplitude bound typically results from a maximum principle or energy argument and ensures local L^p control of the form

$$\int_K |u^\varepsilon|^p \, dy \leq \text{const}, \tag{1.2}$$

where the constant depends on the compact subset K in question. As a consequence of (1.2), one may at least assert the existence of a subsequence that converges in an appropriate weak topology:

$$u = w - \lim u^{\varepsilon_j} .$$

If $p = \infty$ the weak-star topology of L^∞ is relevant. If $1 < p < \infty$ the weak topology of L^p is relevant. If $p = 1$ the weak-star topology of the space BM of bounded measures is relevant. It is well known that the unit ball of the spaces L^∞, L^p, with $1 < p < \infty$ and BM are compact in the aforementioned topologies. Although these topologies are weak enough to ensure compactness, they are too weak to guarantee continuity of nonlinear state variables $g\colon R^n \to R^p$. Indeed, in general

$$g(w - \lim u^\varepsilon) \neq w - \lim g(u^\varepsilon) \tag{1.3}$$

if no uniform control on derivatives is assumed. The essential problem associated with this lack of continuity in the weak topology as expressed by (1.3) does not, of course, arise from the growth of the state variable g at infinity but rather from oscillations in u^ε.

For technical simplicity we shall henceforth assume that all sequences under consideration are bounded in L^∞

$$(1.4) \qquad |u^{\varepsilon}|_{\infty} \leq \text{const.},$$

uniformly with respect to ε. A generalization to L^p follows with appropriate growth restrictions on g. We remark that in the presence of a uniform L^{∞} bound (1.4) the basic notions of weak convergence coincide: convergence in the weak star topology of L^{∞}, convergence in the sense of distributions. In this situation we shall simply use the term weak convergence.

Weak convergence is classically contrasted with strong convergence. Weak convergence means convergence of local averages, i.e.

$$u = w - \lim u^{\varepsilon} \text{ if and only if}$$

$$\int_K u(y)dy = \lim \int_K u^{\varepsilon}(y)dy ,$$

for all compact K in Ω. Strong convergence means norm convergence, i.e.

$$u = s - \lim u^{\varepsilon} \text{ if and only if}$$

$$\lim \int_K |u(y) - u^{\varepsilon}(y)|dy = 0$$

for all compact K in Ω. Once again, the basic notions of strong convergence coincide if (1.4) is assumed: convergence in the strong topology of L^1_{loc} coincides with convergence in the strong topology of L^p_{loc}, $1 < p < \infty$. In this situation we shall simply use the term strong convergence.

A general problem in the theory of conservation laws (1.1) is the following. Given a weakly convergent sequence of exact or appropriate solutions u^{ε} to (1.1), determine whether or not u^{ε} contains a strongly convergent subsequence. If u^{ε}, or a subsequence theoreof, converges strongly one may pass to the limit in a general state variable g: if g is continuous from R^n to R^p then g is continuous in the strong topology, i.e.

$$g(s - \lim u^{\varepsilon}) = s - \lim g(u^{\varepsilon}).$$

As a corollary one may assert that the strong limit of sequence of solutions to (1.1) is again a solution to (1.1). In contrast one may not assert in general that the weak limit of sequence of solutions to (1.1) is again a solution to (1.1) without appealing to additional information.

A specific problem in this area is to determine the circumstances under which strong convergence may be deduced from weak convergence without using uniform derivative estimates. We remark in passing that if one either assumes or proves a uniform bound on derivatives in a pointwise or average sense then weak convergence is immediately transformed into strong convergence by classical functional analysis. If u^ε converges weakly to u and if, for example, the gradients are uniformly bounded in L^p_{loc},

$$\int_K |\nabla u^\varepsilon|^p \, dy \leq \text{const},$$

then u^ε converges strongly to u. In general, weak convergence plus compactness in the strong topology implies strong convergence. If apriori derivative estimates are available, no difficulty arises in proving convergence of approximation methods.

The general problem with which we are concerned may be stated as follows: describe the oscillations in a weakly convergent sequence of functions

$$z^\varepsilon : R^m \to R^N$$

subject to linear differential constraints of the form

$$\sum_{j=0}^{m-1} A_j \, \partial_j \, z^\varepsilon = \phi^\varepsilon \tag{1.5}$$

and nonlinear algebraic constraints of the form

$$\{z^\varepsilon(y)\} \subset M \tag{1.6}$$

for almost all y in R^m. Here A_j denotes a constant $s \times N$ matrix with s arbitrary and fixed. M is a subset of the ambient state space R^N and is usually a manifold. In the context of hyperbolic conservation laws (1.1), the distributions ϕ^ε typically vanish or maintain a distinguished sign reflecting the presence of an entropy condition. In the context of elliptic systems the distributions ϕ^ε typically vanish. Thus, for many purposes and, in particular, for the purpose of this introduction one may regard the object of study as a weakly convergent sequence of solutions or subsolutions to a first order system of differential equations with values in a submanifold of state space.

We remark in passing that the study of approximation methods for (1.1) involves distributions ϕ^ε that represent singular perturbations, namely higher order discrete or analytic operators multiplied by a small coefficient ε. In this situation it turns out that the natural condition on the sequence ϕ^ε is that it lies in a compact subset of the negative Sobolev space $W^{-1,2}$. The detailed structure of ϕ^ε does not play an essential role in the compactness theory for conservation laws; ϕ^ε is treated as a lower-order term. We refer the reader to the introductory section of [6] for a discussion of the $W^{-1,2}$ compactness condition and for applications of the compensated compactness theory to finite difference schemes.

A specific goal in this program is to describe the Young measure associated with a weakly convergent function sequence z^ε which satisfies linear differential constraints (1.5) and nonlinear algebraic constraints (1.6). We are particularly interested in determining how differential and algebraic constraints collaborate to suppress oscillations in z^ε. The problems are to be posed and answered in terms of the coefficient matrices A_j and the constitutive set M.

The general framework as presented by (1.5) and (1.6) is discussed in detail in section 2 of [21] along with the relevant background dealing with the Young measure. Section 2 of [21] includes a self-contained introduction to compensated compactness and the Young measure. The prototypical examples from mechanics associated with hyperbolic, elliptic and mixed-type systems are discussed in sections 2, 5, 7 and 10 of [21].

In the setting of mechanics the domain space R^m and the range space R^N of z^ε represent a physical space and a state space respectively. The differential constraints (1.5) express the basic conservation laws of mass, momentum and energy that govern general media. The algebraic constraints (1.6) embody the constitutive relations that characterize a specific medium. The equations of compressible fluid dynamics and of compressible elastostatics provide basic models for the general theory concerning hyperbolic and elliptic systems. The small disturbance equation of transonic flow provides the basic model for a system of mixed type.

One of the goals is to identify those features of the algebraic structure of the symbol of the differential system (1.5) and the constitutive manifold M that admit or exclude oscillations in the sequence z^ε. In this paper we are primarily concerned with structural relationships between (1.5) and (1.6) that exclude oscillations from the sequence z^ε, in particular with mechanisms shared by hyperbolic and elliptic systems.

In the context of hyperbolic conservation laws we shall be concerned mainly with nondegenerate systems in one space dimension. A standard structural hypothesis for this class of systems is the condition of genuine nonlinearlity in the sense of Lax [11], i.e. strict monotonicity of the wave speeds as a function of the wave amplitude. The prototypical examples of genuinely nonlinear systems arise in elasticity and consist of two equations $p = 2$, that express the conservation laws of mass and momentum for general media, i.e. gases, liquids and solids. In the special case of solution sequences to this class of genuinely nonlinear systems of two conservation laws, we shall prove a general conjecture of Tartar [16] and establish, as a corollary, strong convergence without derivative estimates.

Tartar's conjecture may be stated roughly as follows. If the differential system (1.5) does not allow any codimension-one oscillations with values in the constitutive set M then the sequence z^ε contains no strong oscillations, i.e. z^ε converges in the strong topology. This conjecture leads to the general problem of determining the circumstances under which the removal of the highest order oscillations, namely the codimension-one oscillations, guarantees the absence of all oscillations. The evidence accumulated so far indicates that the conjecture is true at least in the case of small oscillations, i.e. in the case where

$$|z^\varepsilon - \bar{z}|_\infty << 1$$

for some fixed state $\bar{z}$. In general one may anticipate the necessity of formulating a hierarchy of conditions associated with certain canonical oscillations in order to eliminate all possible strong oscillations for systems in several dimensions. At the current stage of development, attention is focused on the implications of the removal of codimension-one oscillations.

A precise formulation of Tartar's conjecture is given in section 2 of [21]. It turns out that a geometric condition determines whether or not it is possible for a general weakly convergent sequence z^ε satisfying differential and algebraic constraint to contain codimension-one oscillations. The condition is stated in the state space R^N in terms of the separation of the wave cone Λ of (1.5) and the constitutive set M. It is conjectured that if Λ and M are separated, i.e. if (1.5) and (1.6) jointly exclude codimension-one oscillations, then the Young measure associated with z^ε is a Dirac mass, i.e. z^ε converges strongly.

In sections 3, 4 and 10 of [21] we are concerned in part with the formulation of structural hypothesis on the flux functions g_j of (1.1) which guarantee the separation of the associated wave cone Λ from the constitutive manifold M. In this regard we first note that the separation property does not hold for hyperbolic systems in one (and several) space dimensions:

$$\partial_t u + \partial_x f(u) = \phi, \qquad u: R^p \to R^p, \tag{1.7}$$

cf. section 3. There is no loss of generality in assuming that flux function g_0 is the identity. We show in section 4, however, that if the system (1.7) is genuinely nonlinear in the sense of Lax then the following augmented system of $p + 1$ equations that includes a Lax entropy form does have the separation property:

$$\partial_t u_i + \partial_x f_i(u) = \phi_i, \qquad 1 \le i \le p$$

$$\partial_t \eta(u) + \partial_x q(u) = \phi_{p+1}.$$

Here (η,q) denotes any generalized entropy pair for (1.7) for which η is strictly convex. The separation property is an immediate corollary of a result of Lax [12] which implies that there exist no shock waves in a genuinely nonlinear system that conserves a strictly convex entropy field. For example, in the setting of genuinely nonlinear elasticity there exist no shock waves that simultaneously conserve mass, momentum and mechanical energy.

It is appropriate to remark at this point that, for hyperbolic systems of conservation laws with degenerate wave speeds, it is not possible in general to separate the wave cone from the constitutive manifold by augmenting the system with a finite number of entropy forms. In particular, if one or more of the wave speeds is linearly degenerate and thus allows contact discontinuities, then codimension-one oscillations may exist. The intersection of the wave cone and the manifold has a relatively simple structure and leads to an interesting problem of determining the nontrivial structure of the Young measure.

In sections 5 and 9 we consider a class of systems of two equations (1.7) with the form of the Lagrangian equations of elasticity, i.e.

$$\begin{aligned} \partial_t v - \partial_x u &= \phi_1 \\ \partial_t u - \partial_x \sigma(v) &= \phi_2 . \end{aligned} \tag{1.8}$$

We recall that (1.8) is strictly hyperbolic if $\sigma' > 0$ and genuinely nonlinear if $\sigma'' \neq 0$. We introduce a doubly augmented system of four equations involving two natural entropy forms,

$$
(1.9) \qquad \begin{aligned}
\partial_t v - \partial_x u &= \phi_1 \\
\partial_t u - \partial_x \quad &= \phi_2 \\
\partial_t \eta_j + \partial_x q_j &= \phi_{j+2}, \qquad j = 1,2,
\end{aligned}
$$

and we prove that the Young measure associated with any weakly convergent sequence $(u^\varepsilon, v^\varepsilon)$ having uniformly small oscillation is a Dirac mass, if $\sigma' > 0$ and $\sigma'' \neq 0$. As stated above, each of the sequences of distributions $\phi_j(u^\varepsilon, v^\varepsilon)$, $1 \leq j \leq 4$, is assumed to lie in a compact subset of $W^{-1,2}$. Thus, one may show that the Young measure reduces to a Dirac mass i.e. a monatomic measure with unit mass, if its support is sufficiently small and therefore that $(u^\varepsilon, v^\varepsilon)$ converges strongly. The reduction of the Young measure ν is established with the aid of a basic functional equation for ν introduced by Tartar [15] which will be discussed below.

We remark that the entropy forms in the augmented system (1.9) arise naturally from Noether's theorem through the temporal and spatial translation invariance of the integrand in the associated variational principle. The pair of state variables (η_1, q_1) denotes a generalized entropy pair associated with temporal translation invariance: η_1 is the mechanical energy and q_1 is the negative power supplied by the stress tensor,

$$
(1.10) \qquad \eta_1 = \frac{1}{2} u^2 + \Sigma(v) \qquad q_1 = -u\sigma(v)
$$

where the stored-energy function Σ is the primitive of σ. The dual entropy pair (η_2, q_2) is associated with spatial translation invariance and is given by

$$
(1.11) \qquad \eta_2 = uv \qquad q_2 = \frac{1}{2} u^2 + \tau(v)
$$

where τ is the Legendre transform of Σ. The duality is a reflection of the fact that the Lagrangian equations are invariant under the simultaneous interchange of space with time and stress with strain.

As a corollary of the aforementioned compactness theorem for elasticity, we may state the following result. Suppose $(u^\varepsilon, v^\varepsilon)$ is a weakly convergent sequence of solutions to the system

$$\partial_t v - \partial_x u = 0$$
$$\partial_t u - \partial_x \sigma(v) = 0.$$

If $(u^\varepsilon, v^\varepsilon)$ has uniformly small oscillation and if each of the following sequences of entropy fields

$$\partial_t \eta_j(u^\varepsilon, v^\varepsilon) + \partial_x q_j(u^\varepsilon, v^\varepsilon) \qquad j = 1,2, \tag{1.12}$$

given by (1.10) and (1.11) lies in a compact subset of $W^{-1,2}$, then $(u^\varepsilon, v^\varepsilon)$ converges strongly. In this connection we note that if the solutions $(u^\varepsilon, v^\varepsilon)$ satisfy the Lax entropy fields using just the fact that the L^∞ norm of $(u^\varepsilon, v^\varepsilon)$ is uniformly bounded. Thus, the hypothesis of compact entropy fields is natural from the viewpoint of both initial and boundary value problems.

We mention this corollary in connection with previous work on strong convergence of solutions to hyperbolic conservation laws using compensated compactness [6,7]. In [6] it was shown that, for a general class of systems of two genuinely nonlinear conservation laws in one space dimension, the $W^{-1,2}$-compactness of all entropy fields implies the L^1-strong compactness of solutions. This result applies, in particular, to solutions that satisfy the Lax entropy inequality: if u^ε is a uniformly bounded sequence of solutions satisfying the Lax entropy inequality [12] then all entropy fields are compact. This work leads to the question of whether or not knowledge of the compactness of just the physical entropy fields is sufficient to deduce the compactness of uniformly bounded solutions. The question can be answered in the affirmative at least in the setting of Lagrangian elasticity and small oscillations.

The compactness results of [6,7] for hyperbolic systems of two conservation laws were also established by showing that the Young measure ν associated with a weakly convergent solution sequence reduces to a Dirac mass. The proof of reduction of the Young measure ν is based on an analysis of a commutativity relation of the form

$$\nu \circ B = B \circ \nu \tag{1.13}$$

where B denotes the symplectic (antisymmetric, bilinear) form acting on all entropy pairs. Equation (1.13) is shorthand for the statement that

$$\langle\nu, \eta\tilde{q} - \tilde{\eta}q\rangle = \langle\nu,\eta\rangle\langle\nu,\tilde{q}\rangle - \langle\nu,\tilde{\eta}\rangle\langle\nu,q\rangle \tag{1.14}$$

for all entropy pairs (η,q) and $(\tilde{\eta},\tilde{q})$. The bracket in (1.14) denotes the expected value of the indicated variable, e.g.

$$\langle\nu,\eta\rangle = \int_{R^p} \eta(\lambda)d\nu(\lambda) .$$

The functional equation (1.14) was introduced by Tartar in [15]; if all entropy fields are compact, then the Young measure ν commutes with B acting on all entropy pairs. In [15], Tartar conjectured that any probability measure satisfying (1.14) for all entropy pairs of a nondegenerate hyperbolic system (1.7) is a Dirac mass and proved the conjecture in the case of a scalar equation. The conjecture was subsequently verified for nondegenerate systems of two equations using the Lax progressing entropy waves [6,7]. As a corollary, the first large data existence theorems for the Cauchy problem for elasticity and isentropic gas dynamics were obtained along with the first convergence results for first order accurate finite difference schemes, cf. [6,7]. We refer the reader to the introductory sections of [6] for additional background and a more precise formulation of results dealing with Tartar's functional equation (1.14) in setting of nondegenerate systems. More recently, the functional equation (1.14) has been analyzed for degenerate systems of two hyperbolic conservation laws by D. Serre [19].

In section 10 of [21], we show that the wave cone and constitutive manifold are separated for the small disturbance equations of transonic flow augmented by a single convex entropy form that is associated with spatial translation invariance in either of the two independent variables. The small disturbance equations provide the canonical example of a system of two conservation laws which changes type from hyperbolic to elliptic as the state variable corsses a hypersurface. Thus, the separation hypothesis of Tartar's conjecture is relevant.

Sections 6 and 7 are concerned with the compactness of solutions to general elliptic systems in two independent variables and to the special equations of hyperelasticity having a stored-energy function that is polyconvex in the sense of J. Ball [2]. In section 6 we establish the strong compactness of solutions with small oscillation to general elliptic systems (1.1) in two independent variables without using derivative estimates. It is shown that if the support of the associated Young measure is sufficiently small then it must reduce to a Dirac mass. Unlike the hyperbolic case, additional nonlinear entropy forms are not required to establish compactness for elliptic solution sequences, at least locally. We also obtain a similar result dealing with compactness of solutions with large oscillation for a special class of elliptic systems. The proofs are based upon a study of the coercive algebraic structure of the equations in state space and employ the basic div-curl lemma of Tartar and Murat.

In section 7 we consider the equations of hyperelasticity with a strictly polyconvex stored energy function and establish compactness of solution sequences with small oscillation: if u^ε is a sequence of hyperelastic deformations with small oscillation then there exists a subsequence which converges in the strong topology of L^1_{loc}. The proof is based upon a decomposition of the basic elliptic form associated with the system into a canonical quadratic part and a null-Lagrangian form. We refer the reader to J. Ball [2,3] for a general discussion of elasticity in the context of the calculus of variations and for an introduction to the notions of polyconvexity and null-Lagrangian forms.

In the setting of elliptic systems (1.1) in two independent variables, our proof of compactness of solution sequences with small oscillation is based on a study of Tartar's functional equation (1.14). In the context of elliptic equations the flux functions (g_{i0}, g_{i1}) act as "entropy pairs". If the Young measure commutes with the bilinear form B acting on just the flux functions of an elliptic system, then it must reduce to a point mass if its support is sufficiently small. Thus, the commutativity relation (1.14) provides a viewpoint for studying compactness to systems of both hyperbolic and elliptic type in two independent variables. One topic for future investigation deals with the structure of the Young measure for systems of mixed type in two independent variables.

These results on 2-d elliptic systems and on polyconvex elasticity provided motivation for a general theorem of Tartar [16] that establishes compactness of solution sequences with small oscillation to elliptic systems in an arbitrary number of state variables. This theorem follows as a corollary of a general result [16] that guarantees compactness provided that the wave cone is separated from the tangent space to the manifold. In other words, if the wave cone is separated from the constitutive manifold by a hyperplane, then the Young measure reduces to a point mass if its support is sufficiently small.

It turns out that this type of maximal separation of wave cone and manifold occurs if and only if the system is elliptic. In contrast, if the system is hyperbolic, then the wave cone intersects the tangent plane to the constitutive manifold in a maximal number of directions, although it may miss the manifold itself if appropriate entropy forms are introduced, cf. section 3 of [21]. These observations motivate the general problem of determining the geometric relationship between the wave cone and constitutive manifold for general systems of conservation laws in the context of the Young measure. The augmented system (1.9) of Lagrangian elasticity exhibits, roughly speaking, a fourth order separation between wave cone and manifold which can be utilized to prove that the Young measure is a Dirac mass with the aid of the basic div-curl lemma of Tartar and Murat from the compensated compactness theory. This mild form of separation provides an interesting contrast with the second order separation presented by elliptic systems.

The program discussed here deals with the structure of oscillations at a fixed point in space and time. We refer the reader to [8] and [17] for work on the dynamic behavior.

BIBLIOGRAPHY

1. J. M. Ball, On the calculus of variations and sequentially weakly continuous maps. Lecture Notes in Mathematics, Vol. 564 Springer-Verlag, 1976.

2. J. M. Ball, Convexity conditions and existence theorems in nonlinear elasticity, Arch. Rat. Mech. Anal. 63 (1977), 337-403.

3. J. M. Ball, J. C. Currie and P. J. Olver, Null Lagrangians, weak continuity and variational problems of arbitrary order, J. Functional Analysis 41 (1981), 135-175.

4. B. Dacorogna, Weak continuity and weak lower semicontinuity of nonlinear functionals, Lecture Notes in Mathematics, Springer-Verlag, 1982.

5. R. J. DiPerna, Uniqueness of solutions to hyperbolic conservation laws, Indiana Univ. Math. J. 28 (1979), 137-188.

6. R. J. DiPerna, Convergence of approximate solutions to conservation laws, Arch. Rat. Mech. Anal. 82 (1983), 27-70.

7. R. J. DiPerna, Convergence of the viscosity method for isentropic gas dynamics, Comm. in Math. Phys. 91 (1983), 1-30.

8. R. J. DiPerna, Measure-valued solutions to conservation laws, Arch. Rat. Mech. Anal. to appear (1985).

9. J. Glimm and P. D. Lax, Decay of solutions of systems of nonlinear hyperbolic conservation laws, AMS Memoir 101 (1970).

10. B. Keyfitz, Solutions with shocks, an example of an L^1-contractive semigroup, Comm. Pure Appl. Math. 24 (1971), 125-132.

11. P. D. Lax, Hyperbolic systems of conservation laws, II, Comm. Pure Appl. Math. 10 (1957), 537-566.

12. P. D. Lax, Shock waves and entropy, in Contributions to Nonlinear Functional Analysis, ed. E. A. Zarantonello, Academic Press (1971).

13. F. Murat, Compacité par compensation, Ann. Scuola Norm. Sup. Pisa 5 (1978), 489-507.

14. F. Murat, Compacité par compensation: Condition necessaire et suffisante de continuite' faible sous une hypotheses de rang constant, Ann. Scuola Norm. Sup. Pisa 8 (1981), 69-102.

15. L. Tartar, Compensated compactness and applications to partial differential equations, in Research Notes in Mathematics, Nonlinear Analysis and Mechanics: Heriot-Watt Symposium, Vol. 4, ed. R. J. Knops, Pitman Press, 1979.

16. L. Tartar, The compensated compactness method applied to systems of conservation laws, in Systems of Nonlinear Partial Differential Equations, ed. J. M. Ball, NATO ASI Series, C. Reidel Pub. Co. (1983).

17. L. Tartar, Solutions oscillantes des equations de Carleman, Seminaire Goulaouic-Meyer-Schwarz, Jan. 1983.

18. A. I. Vol'pert, The space BV and quasilinear equations, Math. USSR Sb. 2 (1967), 257-267.

19. D. Serre, private communication.

20. N. Kruzkov, First order quasilinear equations in several independent variables, Math. USSR Sb. 10 (1970), 127-243.

21. R. DiPerna, Compensated compactness and general systems of conservation laws, to appear.

Department of Mathematics
Duke University
Durham, NC 27706

Current Address:
Department of Mathematics
Princeton University
Princeton, NJ 08544

Lectures in Applied Mathematics
Volume 23, 1986

SHOCK WAVES AND DIFFUSION WAVES

Tai-Ping Liu

ABSTRACT. Shock waves exist in compressible materials and as surface waves on incompressible materials. In this article we describe recent progress on the study of the nonlinear stability and instability of shock waves. The importance of the role played by diffusion waves is explained.

KEY WORDS AND PHRASES: shock waves, viscous shock waves, linear and nonlinear diffusion waves, conservation laws.

There has been intense interest in the study of shock waves in recent decades. Shock waves are observed in many physical situations and can be easily produced experimentally to study the properties of the media. Thus they are carried by the compressible Euler equations, compressible Navier-Stokes equations, Boltzmann equations, elasticity models, shallow water equations and so forth [1], [12].

The system of partial differential equations of the simplest form which carries shock waves is the system of hyperbolic conservation laws.

$$(1) \qquad \frac{\partial u}{\partial t} + \frac{\partial f(u)}{\partial x} = 0, \quad -\infty < x < \infty,\ t \geq 0,$$

1980 Mathematics Subject Classification: Primary 34K55,76N10
Secondary 35B40,35L65.
Partially supported by NSF Grant No. 01523782

where $u = u(x,t)$, the density, and $f(u)$, the flux are n-vectors. The system is assumed to be hyperbolic, that is, $\partial f(u)/\partial u$ has real eigenvalues $\lambda_1(u) \le \lambda_2(u) \le \ldots \le \lambda_n(u)$. For linear system. $\partial f(u)/\partial u$ a constant matrix, (1) can be diagonalized and a general solution is decomposed into i-modes, $i = 1,2,\ldots,n$ propagating along i-characteristic curves $dx/dt = \lambda_i$. Much richer wave phenomena result when the system is nonlinear, that is, the characteristic speed $\lambda_i = \lambda_i(u)$, $i = 1,2,\ldots,n$, depend on the basic dependence variables u. It turns out that the sign of $\nabla\lambda_i(u) \cdot r_i(u)$ is important. When $\nabla\lambda_i(u) \cdot r_i(u) \neq 0$ for all u, the i-field is convex and when $\nabla\lambda_i(u) \cdot r_i(u) \equiv 0$ for all u, the i-field is linear (called genuinely nonlinear and linearly degenerate in [5]). For compressible Euler equations, $n = 3$, two fields are convex and one linear. For models for elasticity and multiphase flows $\nabla\lambda_i(u) \cdot r_i(u)$ may change signs and wave structure becomes more complicated. To understand this, consider scalar equation $n = 1$. It is convex if $f''(u) \neq 0$ and linear if $f''(u) \equiv 0$. Suppose that $f''(u) > 0$, e.g. $f(u) = u^2/2$. When the initial data $u(x,0)$ is a monotonically increasing function the characteristic lines $dx/dt = f'(n)$ diverge and give rise to an expansion wave. Conversely, a monotonically decreasing data give rise to compression waves which eventually break into discontinuity waves, the shock waves. A shock wave is produced out of the collision of characteristic waves and so characteristic curves impinge on the shock wave from both sides. This helps to stabilize the shock wave. It also is responsible for the dissipation of the solution. Consider, for instance, a solution with compact support and consisting of an expansion wave sandwiched by two shock waves. The expansion wave propagates along characteristic curves which eventually run into the shock waves and is cancelled. In fact a solution with compact support to convex scalar conservation law decays at the rate $t^{-1/2}$ and converges to N-wave

$$u(x,t) \to \begin{cases} 0 & \text{for } x > (2q)^{-1/2}\, t^{1/2}, \\ x/t & \text{for } -(2p)^{-1/2}\, t^{1/2} < x < (2q)^{-1/2}\, t^{1/2}, \\ 0 & \text{for } x < -2(2p)^{-1/2}\, t^{1/2}, \end{cases}$$

in $L_1(x)$, where the constant $p, q, p \le 0 \le q$, are time invariants of the solution

$$p = \min_x \int_{-\infty}^{x} u(y,t)dy,$$

$$q \equiv \max_x \int_{x}^{\infty} u(y,t)dy, \quad t \ge 0,$$

[5], Figure 1.

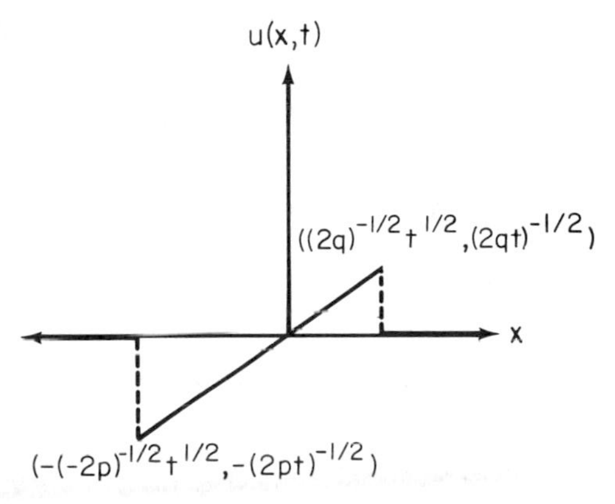

Figure 1

N-wave

These are nonlinear hyperbolic waves. Linear hyperbolic waves are traveling waves and are carried by $u_t + cu_x = 0$, c constant. For general system (1), a perturbation of a shock wave gives rise to nonlinear hyperbolic waves for convex fields and linear hyperbolic waves for linear fields, [6].

For actual physical situations, other effects are present. We start with the viscous effects and consider viscous conservation laws :

$$\frac{\partial u}{\partial t} + \frac{\partial f(u)}{\partial x} = \frac{\partial}{\partial x}\left(B(u)\right)\frac{\partial u}{\partial x}), \quad u \in \mathbb{R}^n, \tag{2}$$

where the viscosity matrix $B(u)$ is a smooth function of u. One important physical example is the compressible Navier-Stokes equations. Stokes pointed out that viscous effects tend to smooth out the shock waves and one obtains viscous shock waves $u(x,t) = \phi(x-\sigma t)$ which becomes the discontinuous shock waves for (1) as $B(u)$ vanishes, Figures 2 and 3. In spite of its obvious importance,

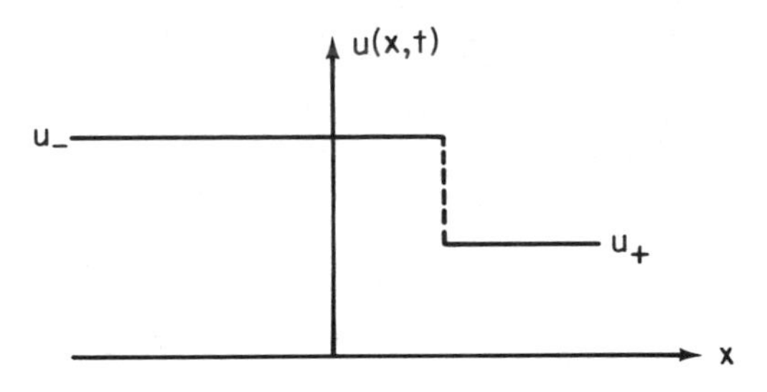

Figure 2
Inviscid shock wave

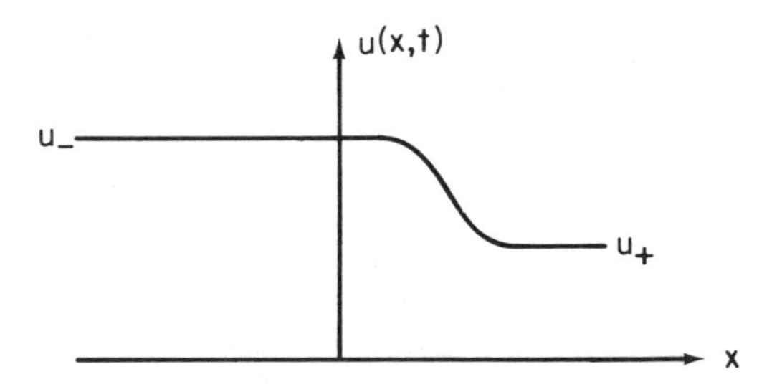

Figure 3
Viscous shock wave

there has been no satisfactory study on the nonlinear stability of viscous shock waves until recently, when it is noticed that a perturbation of viscous shock waves, in addition to translating the shock waves, also gives rise to diffusion waves, [9]. For linear fields, the diffusion waves are constructed out of solutions of heat equations and, for convex fields out of solutions of the Burgers equation

$$\frac{\partial u}{\partial t} + u\frac{\partial u}{\partial x} = \alpha\frac{\partial^2 u}{\partial x^2}, \quad u \in \mathbb{R}^1, \quad \alpha > 0. \tag{3}$$

It is worth noting that the diffusion waves for (3) are qualitatively different from the N waves even $\alpha \to 0_+$, Figures 4.1 and 4.2.

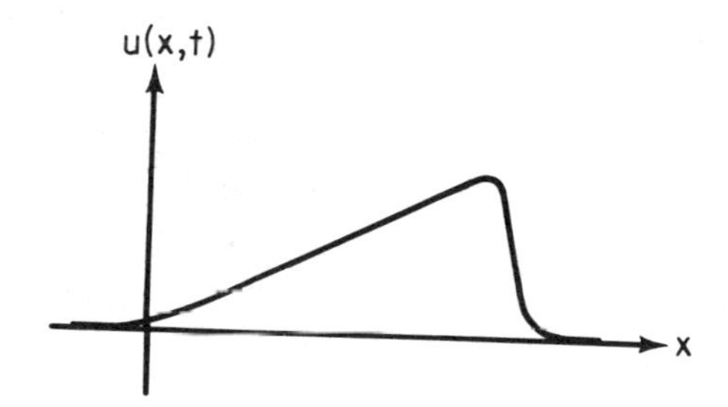

Figure 4.1
Nonlinear diffusion wave (positive mass)

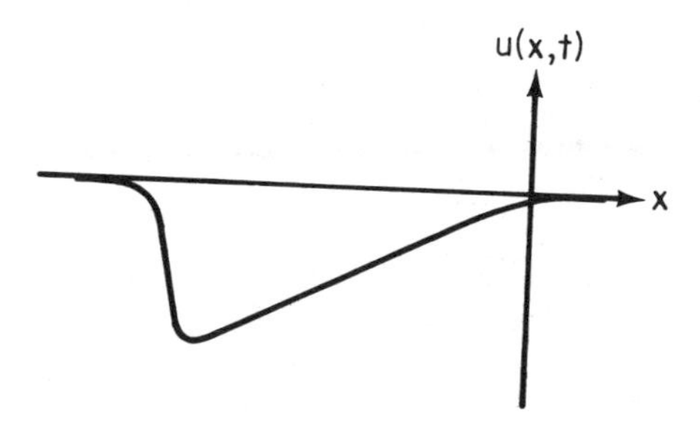

Figure 4.2
Nonlinear diffusion wave (negative mass)

The linear diffusion waves for (2) are governed by heat equation and are completely different from the linear hyperbolic waves for the linear fields of (1). Thus there exists essential differences between (1) and (2) on the level of diffusion waves. Moreover the differences do not go away as the viscosity matrix $B(u)$ tends to zero. In particular, the compressible Euler equations approximate the compressible Navier-Stokes equations for small viscosity only for local and intermediate times and not for large time.

For the study of nonlinear waves for system (1) in [6] and [8], one uses characteristic method and the technique of nonlinear superpositions introduced by [2], [3] and [7]. In studying system (2), [9] introduces a combination of characteristic method and energy method for the stability analysis. The analysis demands a new understanding of diffusion waves in terms of their compression and weak expansion. [9] studies systems which are uniformly parabolic. The ideas are generalized and applied to the compressible Navier-Stokes equations in [10]. For works preceding [9], see [4] and [11].

Diffusions waves are present not only for parabolic systems. Consider, for example, the following 2×2 conservation laws with damping:

$$
(4) \qquad \begin{cases} \dfrac{\partial v}{\partial t} - \dfrac{\partial w}{\partial x} = 0 \\[2ex] \dfrac{w}{\partial t} + \dfrac{\partial p(v)}{\partial x} = -\alpha w, \ \alpha > 0, \ p'(v) < 0. \end{cases}
$$

Suppose that the initial data have compact support. Let $\phi(x,t)$ be a solution ϕ of the the heat equation:

$$
(5) \qquad \int_{-\infty}^{\infty} \phi(x,t)dx = \int_{-\infty}^{\infty} v(x,0)dx,
$$

$$
\frac{\partial \phi}{\partial t} = \frac{|p'(0)|}{\alpha} \frac{\partial^2 p}{\partial x^2},
$$

and set

$$\psi(x,t) \equiv \frac{|p'(0)|}{\alpha} \frac{\partial\phi}{\partial x} .$$

Write

$$v(x,t) \equiv \phi(x,t) + \bar{v}(x,t),$$

$$w(x,t) = \psi(x,t) + \bar{w}(x,t).$$

It can be shown that $(\bar{v},\bar{w})$ satisfies a linear system with fast decaying errors. In other words, $(\bar{v},\bar{w})$ decays faster than (ϕ,ψ). Thus the large-time behavior of the solution (v,w) are governed by the heat equation and waves eventually do not propagate along the characteristic directions $dx/dt = \pm\sqrt{p'(v)}$. When $v(+\infty,0) \neq v(-\infty,0)$, the large-time behavior is governed by a nonlinear parabolic equation. This is reported in an upcoming paper by the author.

Studies are also underway for conservation laws with delay effects. Again, though such systems are hyperbolic, diffusion waves are present. Diffusion waves of the general description presented in [9] also exist in other physical models such as combustions and kinetic theory of gases.

As we have seen for system (4), the number of waves in the asymptotic state may be less than the number of modes for the associated hyperbolic conservation laws. For system (4), only one density is conserved, (5), and there is only one wave in the asymptotic state. This is the general rule applicable to other systems. This explains why a perturbation of a shock wave gives rise to diffusion waves if there are more than one conserved quantities. Another important point we have learned is that the classification of a system by linear criterion such as hyperbolicity and parabolicity is useful only for the study of local behavior of solutions. For the physically important large-time behavior, a hyperbolic system may carry waves governed by a parabolic system and, conversely, waves for nonlinear parabolic systems often behave hyperbolically. The usual justification of the equivalence of two physical systems of letting parameters such as viscosity, heat conductivity and mean free path to go to zero can be replaced by the study of the similarities and differences of the large-time behavior of the solutions for the two systems. This viewpoint has physical significance as these physical parameters, though small, never vanish and time does become large. Moreover, the usual approximation involves essentially Taylor expansion of a given function which makes sense for diffusion waves as time becomes large.

BIBLIOGRAPHY

1. Courant, R. and Friedrich, K. O., Supersonic Flow and Shock Waves, Intersciences, N.Y., 1948.

2. Glimm, J., Solutions in the large for nonlinear hyperbolic systems of conservation laws, Comm. Pure Appl. Math. 18 (1965), 697-715.

3. Glimm, J. and Lax, P. D., Decay of solutions of nonlinear hyperbolic conservation laws, Memoirs, Amer. Math. Soc. 101 (1970).

4. Goodman, J., Nonlinear asymptotic stability of viscous shock profiles for conservation laws, preprint.

5. Lax, P. D., Hyperbolic system of conservation laws, II, Comm. Pure Appl. Math., 10 (1957), 537-566.

6. Liu, T.-P., Linear and nonlinear large-time behavior of solutions of general systems of hyperbolic conservation laws, Comm. Pure Appl. Math. 30 (1977), 767-796.

7. __________, Deterministic of Glimm scheme, Comm. Math. Phys. 57 (1977), 135-148.

8. __________, Admissible solutions of hyperbolic conservation laws, Memoirs, Amer. Math. Soc. 240 (1981).

9. __________, Nonlinear stability of shock waves for viscous conservation laws, Memoirs, Amer. Math. Soc. (1985).

10. __________, Shock waves for the compressible Navier-Stokes equations are nonlinear stable, to appear.

11. Natzumura, A. and Nishihara, K., On a stability of traveling wave solutions of a one-dimensional model system for compressible viscous fluid, preprint.

12. Whitham, G., Linear and Nonlinear Waves, John Wiley and Sons, 1974

DEPARTMENT OF MATHEMATICS
UNIVERSITY OF MARYLAND
COLLEGE PARK, MD 20742

Lectures in Applied Mathematics
Volume 23, 1986

THE NULL CONDITION AND GLOBAL EXISTENCE TO NONLINEAR WAVE EQUATIONS

S. Klainerman *

Let $\mathbb{R}^{(4)}$ denote the four dimensional space-time Minkowski space with metric $\eta_{ab} = \mathrm{diag}(-1,1,1,1)$ and coordinates (x^a), $a = 0,1,2,3$ where $x^0 = t$, $x = (x^1,x^2,x^3)$ are the time and space variables. Throughout this paper we will use the usual geometric conventions of raising and lowering indices with regard to η and the summation convention of repeated upper and lower indices. In particular we will write $x_a = \eta_{ab}x^b$ and denote by $\Box$ the D'Alembertian of $\mathbb{R}^{(4)}$, $\Box = -\eta^{ab}\partial_a\partial_b$ with ∂_a denoting the partial derivatives with regard to x^a. We will consider systems of nonlinear wave equations of the type

$$\Box u = F(u,u',u'') \tag{1}$$

where u is a p-vector, $u = (u^I)_{I=1,2,\ldots,p}$ defined on $\mathbb{R}^{(4)}$ with u', u'' denoting its first and second derivatives with respect to (x^a). In other words $u' = (u^I_a)$, $u'' = (u^I_{ab})$, $I = 1,2,\ldots,p$ and $a,b = 0,1,2,3$ where $u^I_a = \partial_a u^I$, $u^I_{ab} = \partial_a\partial_b u^I$. Here $F = (F^I)_{I=1,\ldots,p}$ is a smooth function of u,u',u'' defined in a neighborhood of $(u,u',u'') = 0$ in $\mathbb{R}^{21p}$ and satisfying certain

---

1980 Mathematics Subject Classification. 35L70

[1]Grant No. MCS-82-01599, and a Sloan Foundation Fellowship.

assumptions. The first two assure that (1) is hyperbolic around the trivial solution $u \equiv 0$

(H_1) $$F = 0(|u| + |u'| + |u''|)^2$$

for small u,u',u''

(H_2) $$\frac{\partial F^I}{\partial u^J_{ab}} = 0 \quad \text{for all} \quad I \neq J$$

and all $a,b = 0,1,2,3$. Without loss of generality we can also assume here that $\frac{\partial F^I}{\partial u^J_{00}} = 0$.

The assumption (H_2) implies that the system (1) is diagonal with respect to the second derivatives. Though it is not a necessary condition for well-posedness it is verified by most examples of interest. For simplicity of the exposition we will also assume that the system is quasilinear i.e. F^A is linear with respect to (u^I_{ab}).

As it has been shown in ([1], [2]), for scalar equations, the generic nonlinearity condition (H_1) is not sufficient to assure global existence of smooth solutions to (1) (see also [3], [4] for results on almost global existence). On the other hand it is known ([5], [6], [7]) that if F does not contain any quadratic term, i.e. $F = 0(|u| + |u'| + |u''|)^3$, then (1) admits globally smooth solutions for all sufficiently small initial data. The aim of this paper is to show that this sufficient condition for global existence can be substantially relaxed by allowing certain degenerate types of quadratic nonlinearities.

Definition 1 ([8]). Let $F = F(u,v,w)$ a real valued function in the variables $u = (u^I)$, $v = (v^I_a)$, $w = (w^I_{ab})$ with $I = 1,\dots,p$ and $a < b$ running from 0 to 3, smoothly defined in a neighborhood of the origin in $\mathbb{R}^p \times \mathbb{R}^{4p} \times \mathbb{R}^{10p}$. We say that $F(u,u',u'')$ (where u', u'' denote the first and second partial derivatives of u) satisfies the "Null Condition" if, for any vector (X^a), null with respect to η, (i.e. $\eta_{ab}X^aX^b = 0$) the following identities hold for all $I,J = 1,\ldots,p$ and all u,v,w,

$$N_{(i)} \qquad \frac{\partial^2 F}{\partial v_a^I \partial v_b^J} X^a X^b = 0$$

$$N_{(ii)} \qquad \frac{\partial^2 F}{\partial v_a^I \partial v_{bc}^J} X^a X^b X^c = 0$$

$$N_{(iii)} \qquad \frac{\partial^2 F}{\partial v_{ab}^I \partial v_{cd}^J} X^a X^b X^c X^d = 0$$

Equivalently(*), see [9],

$$(N) \qquad F(u^I, X^a v_a^I, X^{ab} w_{ab}^I) = 0$$

for all u,v,w and all null vectors (X^a).

Given a system of type (1) with F verifying (H_1) we say that it satisfise the "Null Condition" if the quadratic part of F satisfies the Null Condition. More precisely,

$$(H_3) \qquad F = N(u',u'') + C(u,u',u'')$$

where N is a quadratic function in u', u'' which satisfies the null condition and,

$$C(u,u',u'') = O(|u| + |u'| + |u''|)^3$$

As in [3], [4] we will study the initial value problem

$$(I.V.P.) \qquad u(0,x) = \varepsilon f(x)\ , \quad u_t(0,x) = \varepsilon g(x)$$

with f,g given smooth functions with compact support and ε a small parameter. We remark, however, that less restrictive conditions can be assumed.

Theorem. Let the system (1) satisfy the hypotheses (H_1), (H_2), (H_3). Given $f,g \in C_0^\infty(\underline{R}^3)$ there exists an ε_0 sufficiently small, depending on only finitely many derivatives of F,f,g such

---

(*)The definitions are equivalent if F is C^2.

that, for any $0 < \varepsilon < \varepsilon_0$, (1) admits a unique solution $u \in C^\infty((0,\infty) \times \mathbb{R}^3)$ verifying (I.V.P.).

Theorem 1 was announced, in the scalar case, in [8]. A different proof of it was recently given by D. Christodoulou [9] based on the conformal method (previously introduced by him in [10] and [11]).

The simplest examples of functions which satisfies the Null Condition of Definition 1 are provided by

$$(Q) \qquad Q(f,g) = \eta^{ab}\partial_a f \cdot \partial_b g$$

$$(Q_{ab}) \qquad Q_{ab}(fg) = \partial_a f\ \partial_b g - \partial_b f\ \partial_a g\ , \quad a,b = 0,1,2,3\ .$$

These bilinear forms are the building blocks of any quadratic function $N = N(u',u'')$ verifying the Null Condition. Indeed the general form of such an expression, which is also linear in u'' and verifies H_2, is given by

$$(N') \qquad N^I = T_J^{Iabc}Q_{ab}(\partial_c u^I, u^J) + T_J^{Ia}Q(\partial_a u^I, u^J)$$

$$+ T_{JK}^{Iab}Q_{ab}(u^J, u^K) + T_{JK}^{I}Q(u^J, u^K)$$

for given, real, coefficients T_J^{Iabc}, T_J^{Ia}, T_{JK}^{Iab} and T_{JK}^{I}. Accordingly we will assume that the quadratic term N in (H_3) has the form given by (N′). The cubic term C has the form

$$(G) \qquad C^I = C^{Iab}(u,u')\ \partial^2_{ab}u^I + D^I(u,u')$$

where C^{Iab}, D^I are smooth in u,u' for all $I = 1,\dots,p$, $a,b = 0,1,2,3$, and

$$C^{Iab} = O(|u| + |u'|)^2$$
$$D^I = O(|u| + |u'|)^3$$

for all sufficiently small u,u'.

In the end of this introduction we would like to draw attention to a recent paper of Hörmander [15] which generalizes and sharpens the decay estimate of [4] (and section 2 here). Moreover he derives sharp interpolation inequalities which allows one to give a precise smallness assumption in the hypotheses of our theorem.

1. The Null Condition and the Poincaré and Scaling Invariance of $\Box$.

Let $\Omega_{ab} = x_a\partial_b - x_b\partial_a$ denote the generators of the Lorentz group of $\underset{\sim}{R}^{(4)}$ with $\Omega_{ij} = x_i\partial_j - x_j\partial_i$, $i,j = 1,2,3$ and $\Omega_{i0} = x_i\partial_t + t\,\partial_i = L_i$ for $i = 1,2,3$. Also let $L_0 = x^a\partial_a = t\,\partial_t + x^i\partial_i$. Togehter, the operators ∂_a , Ω_{ab} form the generators of the Poincaré group of $\underset{\sim}{R}^{(4)}$. If we add L_0 the corresponding operators are generators of a subgroup of the Conformal Subgroup of $\underset{\sim}{R}^{(4)}$. For uniformity of notation we will denote them by $(\Gamma_\sigma), \sigma = 0,1,\ldots,10$ whenever we don´t need to distinguish between them. For convenience we may pick Γ_0 to be L_0. The invariance properties of $\Box$ with respect to the Poincaré group are simply expressed in the following

$$[\Gamma_\sigma , \Box] = 0 , \quad \sigma = 1,\ldots,10 \tag{1.1}$$

while the scaling invariance of implies

$$[\Gamma_0 , \Box] = -2\Box . \tag{1.1´}$$

Here [,] denotes the usual commutator of linear operators. Also notice that the commutator of any two Γ´s is a linear combination of the Γ´s and the commutator of any Γ with a partial derivative D is a linear combination of the D´s. We express it schematically in the form

$$[\Gamma,\Gamma] = \Gamma , \quad [\Gamma,\partial] = \partial \tag{1.2}$$

The precise commutator relations are

$$[\Omega_{ab},\partial_c] = \eta_{ac}\partial_b - \eta_{bc}\partial_a \tag{1.3$_i$}$$

$$[\Omega_{ab},\Omega_{cd}] = \eta_{ac}\Omega_{bd} - \eta_{bc}\Omega_{ad} + \eta_{ad}\Omega_{bc} - \eta_{bd}\Omega_{ac} \tag{1.3$_{ii}$}$$

$$[\Omega_{ab},L_0] = 0 , \quad [L_0,\partial_a] = -\partial_a . \tag{1.3$_{iii}$}$$

To distinguish between different types of derivatives we will use the following notation.

Notation. Given a vector function $u = (u^I)$ defined in $\underline{R}^{(4)}$ we write

$$\Gamma u = (\Gamma_\sigma u^I)\ , \qquad I = 1,\dots,p\ ; \qquad \sigma = 0,1,\dots,10$$

$$u' = Du = (\partial_a u^I)\ , \qquad I = 1,\dots,p\ ; \qquad a = 0,1,2,3\ .$$

$$Lu = (L_a u^I)\ , \qquad I = 1,\dots,p\ ; \qquad a = 0,1,2,3$$

$$\Omega u = (\Omega_{ab} u^I)\ , \qquad I = 1,\dots,p\ ; \qquad a,b = 0,1,2,3$$

$$\overline{\Omega}u = (L_0 u^I, \Omega u^I)\ , \qquad I = 1,\dots,p\ .$$

The absolute values $|\Gamma u|$, $|Du|$, $|Lu|$, $|\Omega u|$, $|\overline{\Omega}u|$, $|u|$ denote the sum of the absolute values of all components.

The crucial property of quadratic nonlinearities which satisfy the Null Condition is included in the following

Lemma 1.1. Given two C^1 functions f,g in $\underline{R}^{(4)}$ we have, at every point $(t,x) \in \underline{R}^4$,

(i) $$|Q_{ab}(f,g)| \leq c\,\frac{1}{M(t,|x|)}\,(|Df|\ |\Omega g| + |\Omega f|\cdot|Dg|)$$

(ii) $$|Q(f,g)| \leq c\,\frac{1}{M(t,|x|)}\,(|\Omega f|\cdot|Dg| + |Df|\cdot|\overline{\Omega}g|)$$

(iii) $$|Q(f,g)| \leq c\,\frac{1}{M(t,|x|)}\,(|\overline{\Omega}f|\cdot|Dg| + |Df|\cdot|\Omega g|)$$

where Q_{ab} , Q are the quadratic expressions (Q_{ab}), (Q) defined in the introduction and $M(t,|x|) = \max(t,|x|)$.

Proof: According to the definition of L_i we have the following formulae

(1.4) $$\partial_i = -\frac{x_i}{t}\,\partial_t + \frac{1}{t}\,L_i\ , \qquad i = 1,2,3$$

$$\partial_i = \frac{x_i}{|x|}\,\partial_r + \frac{1}{|x|}\,A_i\ , \qquad i = 1,2,3$$

Also

$$\partial_t = -\frac{1}{|x|}\partial_r + \frac{1}{|x|}\left(\frac{x^i}{|x|}L_i\right)$$

$$\partial_t = -\frac{|x|}{t}\partial_r + \frac{1}{t}L_0$$

with $\partial_r = \frac{x^i}{|x|}\partial_i$ the radial derivative and A^i a linear combination of the angular derivatives Ω_{jk} , $j,k = 1,2,3$, so that,

$$|A^i f| \leq \sum_{j,k}^{3} |\Omega_{jk} f| , \qquad i = 1,2,3.$$

Thus, for $a = i$, $b = j$ we have decomposing $\partial_i f$, $\partial_j f$,

$$Q_{ij}(f,g) = \frac{1}{t}[-\partial_t f \cdot \Omega_{ij} g + (L_i f\ \partial_j g - L_j f\ \partial_i g)]$$

$$Q_{ij}(f,g) = \frac{1}{|x|}[\partial_r f\ \Omega_{ij} g + (A_i f\ \partial_j g - A_j f\ \partial_i g)]$$

For $a = 0$, $b = j$, we decompose $\partial_j f$, $\partial_j g$ to find,

$$Q_{0j}(f,g) = \frac{1}{t}(\partial_t f\ L_j g - L_j f\ \partial_t g)$$

$$Q_{0j}(f,g) = \frac{x^j}{|x|}(\partial_t f\ \partial_r g - \partial_r f\ \partial_t g) + \frac{1}{|x|}(\partial_t f\ A_j g - A_j f\ \partial_t g)$$

$$= \frac{1}{|x|}\left[\frac{x^j}{|x|}((L_r f)\partial_r g - (\partial_r f)L_r g) + (\partial_t f\ A_j g - A_j f\ \partial_t g)\right]$$

with $L_r = \frac{x^i}{|x|}L_i$. Therefore, in all cases,

$$|Q_{ab}(f,g)| \leq \frac{1}{M(t,|x|)}(|Df|\ |\Omega g| + |\Omega f|\ |Dg|)$$

To prove (ii) we write, with the help of formula (1.4),

$$Q(f,g) = \partial_t f\ \partial_t g + \frac{|x|}{t}\ \partial_t f\ \partial_r g - \frac{1}{t}\sum_{i=1}^{3}(L_i f)(\partial_i g)$$

$$= \frac{1}{t}[(\partial_t f)(L_0 g) - \sum_i (L_i f)(\partial_i g)] .$$

Also,

$$Q(f,g) = \partial_t f\, \partial_t g - \partial_r f\, \partial_r g - \frac{1}{|x|} \sum_{i=1}^{3} (\partial_i f)(A_i g)$$

$$= \frac{1}{|x|} [((L_r f)\partial_t g - (\partial_r f)L_0 g) - \sum_{i=1}^{3} (\partial_i f)(A_i g)]$$

and consequently

$$|Q(f,g)| \leq \frac{1}{M(t,|x|)} (|Df|\ |\bar{\Omega} g| + |\Omega f|\ |Dg|)$$

The proof of (ii´) follows by symmetry.

We will next prove a lemma which shows that the form of Q and Q_{ab} does not change under differentiation by the operators Γ. For convenience we introduce the commutator between a linear operator L and quadratic nonlinear operator $Q = Q(f,g)$, to be the quadratic operator defined by

(1.5) $$[\Gamma,Q](f,g) = \Gamma Q(f,g) - Q(\Gamma f,g) - Q(f,\Gamma g)$$

with this definition we have, for Q and Q_{ab} the quadratic operators (Q), (Q_{ab}),

<u>Lemma 1.2.</u>

(i) $$[\Omega_{cd},Q_{ab}] = -\eta_{ac}Q_{bd} + \eta_{ad}Q_{bc} + \eta_{bc}Q_{ad} - \eta_{bd}Q_{ac}$$

(ii) $$[L_0,Q_{ab}] = 0$$

(iii) $$[\Omega_{ab},Q] = 0$$

(iv) $$[L_0,Q] = -2Q\ .$$

The proof is an immediate consequence of the commutator relations (1.3). Since we also have, trivially, $[\partial_c,Q_{ab}] = 0$ and $[\partial_c,Q] = 0$ for all $a,b,c = 0,1,2,3$ we will write, schematically

(1.6) $$[\Gamma,Q] = Q$$

meaning that the comutation of any Γ with any Q is a linear combination of the Q´s.

We will apply Lemmas 1.1 and 1.2 to derive estimates for generalized derivatives of $N(u',u'')$. We will denote by Γ^k any given product of k of the operators $\Gamma_0, \Gamma_1, \ldots, \Gamma_{10}$. For convenience we will denote by $|\Gamma^k u|$ the sum of the absolute values over all components $\Gamma^k u$ for all k-products Γ^k

$$(1.7) \qquad |\Gamma^k u| = \sum_{0 \le \alpha_1, \ldots, \alpha_k \le 10} |\Gamma_{\alpha_i} \cdot \ldots \cdot \Gamma_{\alpha_k} u|$$

Using repeatedly (1.6), Lemma 1.1 and the notation (1.7) we derive

Proposition 1.1. Let f,g be two C^{k+1} functions in $\mathbb{R}^{(4)}$. We have,

$$\text{(i)} \quad |\Gamma^k Q_{ab}(f,g)| \le M(t,|x|)^{-1} \sum_{i+j=k} (|D\Gamma^i f||\Gamma^{j+1} g| + |\Gamma^{i+1} f||D\Gamma^j g|)$$

$$\text{(ii)} \quad |\Gamma^k Q(f,g)| \le M(t,|x|)^{-1} \sum_{i+j=k} (|D\Gamma^i f||\Gamma^{j+1} g| + |\Gamma^{i+1} f||D\Gamma^{j+1} g|)$$

2. A Priori Uniform Decay Estimates

In what follows we derive uniform decay estimates for solutions of (1). Our method consists in deriving generalized Sobolev estimates similar to those previously presented in [4]. The following notation will be used:

Notation 2.1. Given $u = u(t,x)$, a smooth function in $\mathbb{R}^{(4)}$ we write

$$\|u(t)\|_k^2 = \sum_{|\alpha| \le k} \|\Gamma^\alpha u(t)\|_{L^2(\mathbb{R}^n)} = \|u(t)\|_{\Gamma,k}^2$$

$$|||u(t)|||_k = \sum_{|\alpha| \le k} \left\| \frac{1}{|x|} \Gamma^\alpha u(t) \right\|_{L^1(\mathbb{R}^n)} = |||u(t)|||_{\Gamma,k}$$

with Γ^α any product $\Gamma_0^{\alpha_0} \Gamma_1^{\alpha_1} \ldots \Gamma_{10}^{\alpha_{10}}$ of the operators Γ defined in section 1, $|\alpha| = \alpha_0 + \alpha_1 + \ldots + \alpha_{10}$.

Proposition 2.1. Let $u = u(t,x)$ be a smooth function in $\mathbb{R}^{(4)}$ decaying sufficiently fast at $|x| \to \infty$ for any fixed positive t. Let $M(t,|x|) = \max(t,|x|)$. We have, for all $r = |x| \neq 0,t$,

(i) $$|u(t,x)| \leq c(1+M(t,|x|))^{-1/2} \cdot \|Du(t)\|_4$$

(ii) $$|u(t,x)| \leq c(1+M(t,|x|))^{-1}(1+|t - |x||)^{-1/2} \cdot \|u(t)\|_4$$

and also

(iii) $$|u(t,x)| \leq (1+M(t,|x|))^{-1} \; |||Du(t)|||_4$$

(iv) $$|u(t,x)| \leq (1+M(t,|x|))^{-1}(1+|t-|x||)^{-1}|||u(t)|||_4$$

Proof: The following form of the Sobolev inequality on the unit sphere S^2 in $\mathbb{R}^3$ will be used repeatedly (see [4]).

Lemma 2.2. Let $v = v(x)$ a smooth function in $\mathbb{R}^3$. We have, with $r = |x|$,

(i) $$|v(x)| \leq c\Big(\sum_{|\alpha|\leq 2} \int_{S^2} |\Omega^\alpha v(r\xi)|^2 \, dS_\xi\Big)^{1/2}$$

(ii) $$|v(x)| \leq c\Big(\sum_{|\alpha|\leq 2} \int_{S^2} |\Omega^\alpha v(r\xi)| \, dS_\xi\Big)$$

where Ω^α are products of the angular momentum operators $\Omega_{ij} = x_i\partial_j - x_j\partial_i$, $i,j = 1,2,3$.

The first step in the proof of the Proposition will be to deduce the following

(2.1) (i) $$|u(t,x)| \leq c|x|^{-1/2} \cdot \|Du(t)\|_2$$

(ii) $$|u(t,x)| \leq c|x|^{-1} \|u(t)\|_2^{1/2} \cdot \|Du(t)\|_2^{1/2}$$

(iii) $$|u(t,x)| \leq c|x|^{-1} \; |||Du(t)|||_2$$

this is an immediate consequence of Lemma 2.2 and the following

Lemma 2.3. Given $v = v(x)$ a smooth function in $\mathbb{R}^3$, decaying sufficiently fast for $|x| \to \infty$ we have the following,

(i) $$\Big(\int_{|\xi|=1} |v(r\xi)|^2 \, dS_\xi \Big)^{1/2} \le r^{-1/2} \Big(\int_{|y|\ge r} |Dv(y)|^2 \, dy \Big)^{1/2}$$

(ii) $$\Big(\int_{|\xi|=1} |v(r\xi)|^2 \, dS_\xi \Big)^{1/2}$$
$$\le 2r^{-1} \Big(\int_{|y|\ge r} |v(y)|^2 \, dy \Big)^{1/4} \Big(\int_{|y|\ge r} |Dv(y)|^2 \, dy \Big)^{1/4}$$

(iii) $$\int_{|\xi|=1} |v(r\xi)| \, dS_\xi \le r^{-1} \Big(\int_{|y|\ge r} |Dv(y)| \, dy \Big)$$

To prove these we first write

$$v(r\xi) = - \int_r^\infty \frac{d}{d\lambda} v(\lambda\xi) \, d\lambda$$

and deduce, by Cauchy-Schwartz,

$$|v(r\xi)| \le r^{-1/2} \Big(\int_r^\infty \lambda^2 \Big|\frac{d}{d\lambda} v(\lambda\xi)\Big|^2 \Big)^{1/2}$$

and

$$|v(r\xi)| \le r^{-1} \int_r^\infty \lambda \, \Big| \frac{d}{d\lambda} v(\lambda\xi)\Big| \, d\lambda$$

which integrated over S^2 prove the estimates (i), (iii) of the lemma. To prove (ii), we start instead with v^2 and write,

$$v^2(r\xi) = - 2 \int_r^\infty v \frac{d}{d\lambda} v(\lambda\xi) \, d\lambda$$

By Cauchy-Schwartz,

$$|v(r,\xi)|^2 \le 2 \frac{1}{r^2} \Big(\int_r^\infty \lambda^2 |v(\lambda\xi)|^2 \, d\lambda\Big)^{1/2} \Big(\int_r^\infty \lambda^2 |Dv(\lambda\xi)|^2 \, d\lambda\Big)^{1/2}$$

Integrating over S^2 and using Cauchy Schwartz once more we obtain the desired estimate and end the proof of the lemma.

Clearly, the inequalities (2.1) prove (i), (ii), (iii) of

the proposition for t close to r, i.e. $|t-r| < 1$. For $t \neq r$ we will prove instead the following,

$$\text{(2.2)}\quad \text{(i)}\qquad |u(t,x)| \leq c|t-r|^{-1/2} \cdot \|Du(t)\|_4$$

$$\text{(ii)}\qquad |u(t,x)| \leq c|t-r|^{-3/2} \|u(t)\|_4$$

and

$$\text{(iii)}\qquad |u(t,x)| \leq c|t-r|^{-1} \cdot |||Du(t)|||_4$$

$$\text{(iv)}\qquad |u(t,x)| \leq c|t-r|^{-2} |||u(t)|||_4$$

for all $t \neq |x| = r$.

To prove these we will use the following formula (see [4], 7a) for the radial derivative $\partial_r = \frac{x^i}{|x|} \partial_i$.

$$\text{(2.3)}\qquad \partial_r = \frac{1}{t^2-r^2} (tL_r - rL_0)$$

where $L_r = \frac{x^i}{|x|} L_i$.

In particular we deduce from (2.2)

$$\text{(2.4)}\qquad |\partial_r u(t,x)| \leq \frac{1}{|t-|x||} |Lu(tx)|$$

$$|\partial_r^2 u(t,x)| \leq \frac{1}{|t-|x||^2} (|Lu(tx)| + |L^2 u(t,x)|)$$

Now consider $w(t,r\xi) = (t-r)^2 u(t,r\xi)$ and observe that,

$$w = \partial_r w = 0 \qquad \text{for} \quad t = r$$

$$|\partial_r^2 w| \leq c(|u| + |Lu| + |L^2 u|)$$

Consequently, for any $t > r$

$$w(t,r\xi) = -\int_r^t (\lambda - r) \frac{d^2}{d\lambda^2} w(t,\lambda\xi)\, d\lambda$$

whence,

$$\text{(2.5)}\qquad |w(t,r\xi)| \leq c \int_r^t |\lambda-r| (|u|+|Lu|+|L^2 u|)\, d\lambda$$

Using Cauchy-Schwartz we deduce

$$|w(t,r\xi)| \leq c(t-r) \left(\int_r^t |\lambda-r| \, (|u| + |Lu| + |L^2u|)^2 \, d\lambda \right)^{1/2}$$

Integrating these last two inequalities over S^2 we find,

$$\int |w(t,r\xi)| \, dS_\xi \leq c \int_r^t \lambda \, d\lambda \int_{|\xi|=1} (|u|+|Lu|+|L^2u|) \, dS_\xi$$

$$\int |w(t,r\xi)|^2 \, dS_\xi \leq c(t-r)^2 \int_r^t \lambda \, d\lambda \int_{|\xi|=1} (|u|+|Lu|+|L^2u|)^2 \, dS_\xi$$

We now apply Lemma 2.3 on the right hand side of the last two inequalities,

$$\int_{|\xi|=1} |w(t,r\xi)| \, dS_\xi \leq c(t-r) \, |||Du(t)|||_2$$

$$\int_{|\xi|=1} |w(t,r\xi)|^2 \, dS_\xi \leq c(t-r)^3 \, \|Du(t)\|_2$$

Finally, applying Lemma 2.2 and recalling the definition of w, we find,

$$|u(t,x)| < c(t-|x|)^{-1/2} \, \|Du(t)\|_2$$

$$(|u(t,x)|) \leq c(t-|x|)^{-1} \, |||Du(t)|||_2$$

which prove (2.2_i) and (2.2_{iii}) for $t > |x|$.

To prove (ii) and (iii) we go back to (2.5) and, integrating over S^2 we find,

$$\int_{|\xi|=1} |w(t,r\xi)| \, dS_\xi \leq c|||u(t)|||_2$$

$$\left(\int_{|\xi|=1} |w(t,r\xi)|^2 \, dS_\xi \right) \leq c(t-r) \, \|u(t)\|_2$$

Therefore, by Lemma 2.2,

$$|u(t,x)| \le c(t-r)^{-3/2} \|u(t)\|_4$$

$$|u(t,x)| \le c(t-r)^{-2} |||u(t)|||_4$$

which ends the proof of (2.2) for $t > r$. For $r > t$ we write,

$$w(t,r\xi) = - \int_t^r \frac{d}{d\lambda} w(t,\lambda\xi)\, d\lambda$$

According to (2.4),

$$\left|\frac{d}{d\lambda} w(t,\lambda\xi)\right| \le |\lambda-t| \; (|u| + |Lu|)$$

and henceforth,

$$|w(t,r\xi)| \le \int_t^r (\lambda-t) \; (|u| + |Lu|)\, d\lambda \tag{2.6}$$

We now continue the proof precisely as before. Notice that we need in fact one less derivative i.e.,

(2.2′) (i) $|u(t,x)| \le c(r-t)^{-1/2} \|Du(t)\|_3$

(ii) $|u(t,x)| \le c(r-t)^{-3/2} \|u(t)\|_3$

(iii) $|u(t,x)| \le c(r-t)^{-1} |||Du(t)|||_3$

(iv) $|u(t,x)| \le c(r-t)^{-2} |||u(t)|||_3$

for all $r = |x| > t$.

Finally, the proof of the proposition follows immediately from (2.1) and (2.2) for $\max(t,r) \ge 1$ and from the classical Sobolev inequalities for $t \le 1$, $r \le 1$.

Remark. Each derivative improves the estimates of Proposition 2.1 by the factor $(1+|t-r|)^{-1}$. More precisely,

(i) $$|D^k u(t,x)| \le c(1+M(t,|x|)^{-1/2}(1+|t-|x||)^{-k} \|Du(t)\|_{4+k}$$

(ii) $$|D^k u(t,x)| \le c(1+M(t,|x|))^{-1}(1+|t-|x||)^{-k-1/2} \|u(t)\|_{4+k}$$

(iii) $\quad |D^k u(t,x)| \leq (1+M(t,|x|))^{-1}(1+|t-|x||)^{-k}|||Du(t)|||_{4+k}$

(iv) $\quad |D^k u(t,x)| \leq (1+M(t,|x|))^{-1}(1+\big|t-|x|\big|)^{-k-1}|||u(t)|||_{4+k}$

In the end of this section we appeal to an estimate (proved in [12], Theorem 1) of the $|||\ |||$ norm of solutions to the linear wave equation.

Proposition 2.2. Let $u = u(t,x)$ be a solution to the standard initial value problem,

$$\square u = 0 \ ; \quad u = 0 \ , \quad u_t = g \quad \text{at} \quad t = 0$$

with g a given smooth function, decaying sufficiently fast for $|x| \to 0$. Then,

(i) $$\int \frac{1}{|x|} |u(t,x)| \, dx \leq c \int |g(x)| \, dx$$

(ii) $$\int \frac{1}{|x|} |u'(t,x)| \, dx$$

$$\leq c \int \frac{1}{|x|} \left(1 + \log \frac{t+|x|}{|t-|x||}\right) (|\Omega g(x)| + |g(x)|) \, dx$$

Here, $\Omega g = (\Omega_{ij} g)$ with $i,j = 1,2,3$.

The proposition will be applied to the inhomogeneous problem, $u = F$.

Corollary. Let $u = u(t,x)$ be a solution of the inhomogeneous problem

$$\square u = F(t,x) \ ; \quad u = u_t = 0 \quad \text{at} \quad t = 0$$

with F a smooth function in $\mathbb{R}_+ \times \mathbb{R}^3$ vanishing sufficiently fast as $|x| \to \infty$ for each fixed t. Then,

$$\int \frac{1}{|x|} |u'(t,x)| \, dx$$

$$\leq c \int_0^t ds \int \frac{1}{|x|} \left(1 + \log \frac{t-s+|x|}{\big|t-s-|x|\big|}\right) (|F(s,x)| + |\Omega F(s,x)|) \, dx$$

3. Energy Estimates for Linear Equations

We start with linear scalar equations of the type,

$$- A^{ab}\,\partial^2_{ab} v(t,x) = f(t,x) \tag{3.1}$$

where A^{ab}, f are smooth functions of $(t,x) \in \underset{=}{R}^{(4)}$ and $(A^{ab})_{a,b=0,1,2,3}$ is a nondegenerate symmetric matrix of signature $(-1,1,1,1)$. We want to multiply (2.1) by Mv where M is the vector field $m^c\partial_c$ with $m^a(t,x)$ smooth functions in $\underset{=}{R}^{(4)}$. To do this we observe that,

$$\partial_a(A^{ab}v_bv_c) = A^{ab}v_{ab}v_c + A^{ab}v_bv_{ac} + (\partial_aA)^{ab}v_bv_c$$

$$\partial_b(A^{ab}v_av_c) = A^{ab}v_{ab}v_c + A^{ab}v_av_{bc}\ (\partial_bA^{ab})v_av_c$$

$$\partial_c(A^{ab}v_av_b) = A^{ab}v_{ac}v_b + A^{ab}v_av_{bc} + (\partial_cA^{ab})v_av_b$$

Adding the first two expressions and subtracting the third we find,

$$2A^{ab}v_{ab}v_c = \partial_a(A^{ab}v_bv_c) + \partial_b(a^{ab}v_av_c) - \partial_c(A^{ab}v_av_b)$$
$$- (\partial_aA^{ab})v_bv_c - (\partial_bA^{ab})v_av_c + (\partial_cA^{ab})v_av_b$$

Multiplying by m^c and adding we derive the formula

$$\partial_a(-A^{ab}v_b\cdot Mv + \frac{1}{2}\,m^aA^{bc}v_bv_c) \tag{3.2}$$
$$= Mv\cdot f - (\partial_aA^{ab})v_bMv + \frac{1}{2}\,(\partial_cA^{ab})m^cv_av_b$$
$$- (\partial_am^c)A^{ab}v_bv_c + \frac{1}{2}\,(\partial_cm^c)A^{ab}v_av_b$$

where $M = m^c\partial_c$. On the other hand, multiplying (3.1) by qv we derive,

$$\partial_a(-A^{ab}v_bqv) = qvf - \partial_a(A^{ab}q)v_bv - qA^{ab}v_av_b \tag{3.3}$$

Also,

$$(3.3') \quad \partial_a(-A^{ab}v_b qv + \frac{1}{2} A^{ab}q_b v^2) = qvf + \frac{1}{2} \partial_a(A^{ab}q_b)v^2 - qA^{ab}_a vv_b - qA^{ab}v_a v_b$$

Adding (3.2) and (3.3′) together we infer the following

$$(3.4) \quad \partial_a(-A^{ab}v_b Lv + \frac{1}{2} m^a A^{bc}v_b v_c + \frac{1}{2} A^{ab}q_b v^2)$$

$$= (Lv)f - (\partial_a A^{ab})v_b Lv + \tilde{A}^{ab}v_a v_b + \frac{1}{2} \partial_a(A^{ab}q_b)v^2$$

where $L = M+q = m^c\partial_c+q$ and,

$$\tilde{A}^{ab} = \frac{1}{2} M(A^{ab}) + \frac{1}{2} (\mathrm{Div}(M)-2q)A^{ab}-A^{bc}\partial_c m^a,$$

with $\mathrm{Div}(M) = \partial_c m^c$.

In particular if A^{ab} is the Minkowski metric η^{ab} we infer, the following

$$(3.5) \quad \partial_a(e^a_L) = (Lv)f + R_L$$

where,

$$e^a_L(v) = -\eta^{ab}v_b Lv + \frac{1}{2} m^a\eta^{bc}v_b v_c + \frac{1}{2} \eta^{ab}q_b v^2$$

$$R_L(v) = \frac{1}{2} (\mathrm{Div}(M)-2q)\eta^{ab} - \eta^{bc}\partial_c m^a + \frac{1}{2} (\eta^{ab}q_{ab})v^2$$

Adding (3.2) and (3.3) together, we obtain instead of (3.4),

$$(3.4') \quad \partial_a(-A^{ab}v_b Lv + \frac{1}{2} m^a A^{bc}v_b v_c) = (Lv)f$$

$$- (\partial_a A^{ab})v_b Lv + \frac{1}{2} [M(A^{ab})+(\mathrm{Div}(M)-2q)A^{ab}]v_a v_b - A^{ab}v_b[\partial_a,L]v$$

with $[\partial_a,L]$ the commutator between ∂_a and $L = m^c\partial_c+q$.

If A^{ab} is a perturbation of η^{ab} i.e. $A^{ab} = \eta^{ab} + p^{ab}$, for sufficiently small, symmetric P^{ab}, we can combine the formulas (3.5), (3.4′) in the following,

$$\partial_a(e^a_{A,L}) = (Lv)f + R_{A,L} \tag{3.4''}$$

where

$$e^a_{A,L}(v) = e_L(v) - P^{ab}v_bLv + \frac{1}{2} m^a P^{bc} v_b v_c$$

$$R_{A,L}(v) = R_L(v) - (\partial_a P^{ab})v_b Lv$$

$$+ \frac{1}{2} [M(P^{ab})+(Div(M)-2q)P^{ab}]v_a v_b - P^{ab}v_b[\partial_a, L]v$$

To summarize, any smooth solution of (3.1) with $A^{ab} = \eta^{ab}+P^{ab}$ and P^{ab} a "small" symmetric matrix (such that A^{ab} remains Larentzian), verifies the identity (3.4").

It is a well known fact that R_L vanishes identically if L is one of the operators ∂_a ,Ω_{ab} , L_0+1 or K_a-2x_a with K_a the conformal operators

$$K_a = - 2 x_a L_0 + \eta_{bc} x^b x^c \partial_a$$

and $x_a = \eta_{ab}x^b$. Indeed this can be easily verified from (3.5). Moreover, see [13], they are the only exact, first order multipliers. Among these, the only "timelike" multipliers are ∂_t and K_0+2t i.e. they are the only ones for which $e^0_L(v)$ is everywhere positive. The multiplier L_0+1 is only time-like in the interior of the light cone $\eta_{ab}x^a x^b \leq 0$. The multiplier K_0+2t has been used by C. Morawetz [14] in connection with local decay estimates for the wave equation outside an obstacle.

In what follows we will take L to be the first order operator

$$K = (1 + t^2 + |x|^2)\partial_t + 2tx^i\partial_i + 2t$$

We find $R_K(v) \equiv 0$ and

$$e^0_K = \frac{1}{2} (1+t^2+|x|^2)(v_t^2+ \sum_{i=1}^{3} v_i^2) + 2tx^i v_t v_i + 2tvv_t-v^2$$

Notice that the following formulas hold,

$$\sum_{i=1}^{3} v_i^2 = (\partial_r v)^2 + \frac{1}{r^2} |\Omega' v|^2$$

$$\sum_{i=1}^{3} (L_i v)^2 = (L_r v)^2 + \frac{t^2}{r^2} |\Omega' v|^2$$

with

$$|\Omega' v|^2 = \frac{1}{2} \sum_{i \neq j} |\Omega_{ij} v|^2$$

$\partial_r = \frac{x^i}{|x|} \partial_i$, the radial derivative, $r = |x|$, and

$$L_r = \frac{|x^i}{|x|} L_i = \partial_r + |x| \partial_t .$$

Therefore,

$$(3.6) \quad e_K^0 = \frac{1}{2} (v_t^2 + \sum_{i=1}^{3} v_i^2) + \frac{1}{2} [(L_0 v)^2 + (L_r v)^2] + \frac{t^2 + |x|^2}{2|x|^2} |\Omega' v|^2 + 2t v v_t - v^2$$

Integrating (3.5) over $\mathbb{R}^3$ we find (provided that v vanishes sufficiently fast at infinity),

$$\int e_K^0(v(t))\, dx = \int (Kv) f\, dx$$

Indeed the integral of $R_K(v)$ vanishes as K is an exact multiplier. In what follows we will show that $\int e_K^0(v)\, dx$ can be dominated from below by the following integral,

$$\int (|L_0 v|^2 + |\Omega v|^2 + |Dv|^2 + v^2)\, dx$$

where $|1\Omega v|^2 = \frac{1}{2} \sum_{a \neq 0} |\Omega_{ab} v|^2$, $|Dv|^2 = \sum_a |\partial_a v|^2$. To do this we note the following,

$$\int t v v_t = \int v L_0 v - \int x^i v\, \partial_i v$$

$$= \int v L_0 v + \frac{3}{2} \int v^2$$

Similarly,

$$\int tvv_t = \int \frac{t}{|x|} v\, L_r v - \int \frac{t^2}{|x|} v\, \partial_r v$$

$$= \int \frac{t}{|x|} v\, L_r v + \frac{1}{2} \int \frac{t^2}{|x|^2} v^2\, dx$$

Henceforth, writing,

$$\int (2vv_t - v^2)\, dx = \int \frac{3}{2} vL_0 v + \frac{5}{4} \int v^2 + \frac{1}{2} \int \frac{t}{|x|} v\, L_r v + \frac{1}{4} \int \frac{t^2}{|x|^2} v^2$$

and integrating (3.6) we derive,

$$\int e_K^0(v)\, dx = \frac{1}{2} \int |Dv|^2 + \frac{1}{2} \int \frac{t^2 + |x|^2}{|x|^2} |\Omega' v|^2$$

$$+ \frac{1}{2} \int [(L_0 v)^2 + 3vL_0 v + \frac{5}{2} v^2]$$

$$+ \frac{1}{2} \int [(L_r v)^2 + \frac{t}{|x|} vL_r v + \frac{1}{2} \frac{t^2}{|x|^2} v^2]$$

Clearly, the last two integrals are both positive. Moreover,

$$\frac{1}{2} \int ((L_0 v)^2 + 3vL_0 v + \frac{5}{2} v^2) \geq \frac{1}{20} \int ((L_0 v)^2 + v^2)$$

$$\frac{1}{2} \int ((L_r v)^2 + \frac{t}{|x|} vL_r v + \frac{1}{2} \frac{t^2}{|x|^2} v^2) \geq \frac{1}{20} \int (L_r v)^2$$

Consequently,

$$\int e_K^0(v)\, dx \geq \frac{1}{20} \int (|Dv|^2 + |L_0 v|^2 + |L_r v|^2 + \frac{t^2 + |x|^2}{|x|^2} |\Omega' v|^2 + v^2)$$

Finally, using the second formula above (3.6),

$$\int e_K^0(v)\, dx \geq \frac{1}{20} \int |\Gamma v|^2\, dx$$

with

$$|\Gamma v|^2 = (L_0 v)^2 + |\Omega v|^2 + |Dv|^2 + v^2$$

and

$$|\Omega v|^2 = \frac{1}{2} \Sigma \, |\Omega_{ab} v|^2$$

We have thus proved the following

Lemma 3.1. Let $v = v(t,x)$ be a smooth solution of (3.1) with $A^{ab} = \eta^{ab}$, which decays like $O(\frac{1}{|x|^2})$ as $|x| \to \infty$. Let K be the first order operator

$$K = \partial_0 + K_0 + 2t = (1+t^2+|x|^2)\partial_t + 2tx\, \partial_i + 2t$$

and

$$\|v(t)\|_E^2 = \int e_K^0(v)$$

with e_K^0 defined by (3.4). Then,

$$\frac{d}{dt} \|v(t)\|_E^2 = \int (Kv)f$$

and

$$4\|v(t)\|_{\Gamma,1}^2 \geq \|v(t)\|_E^2 \geq \frac{1}{20} \|v(t)\|_{\Gamma,1}$$

(see Notation 1.2).

We now assume that A^{ab} has the form

$$A^{ab} = \eta^{ab} + B^{[ab]} + C^{ab} \,, \quad B^{[ab]} = \frac{1}{2}(B^{ab} + B^{ba}) \tag{3.9}$$

where $B^{ab} = (T_J^{(ac)b} + T_J^a \eta^{bc})\partial_c w^J$, with $a,b,c = 0,1,2,3$, $J = 1,2,\ldots,p$, T_J^{acb}, T_J^a constants, $T_J^{(ac)b} = (T_J^{acb} - T_J^{cab})$ and $w^J = w^J(t,x)$, $C^{ab} = C^{ab}(t,x)$ smooth functions of $(t,x) \in \underline{R}^{(4)}$. Introduce the following notation

$$[w(t)] = \sum_J \max_x (1 + M(t,|x|))\, |w^J(t,x)| \tag{3.10}$$

$$[[C(t)]] = \sum_{a,b} \max_x (1 + M(t,|x|))^2 |C^{ab}(t,x)|$$

and

$$[w(t)]_s = \sum_{i=0}^{s} [\Gamma^i w(t)]$$

$$[[C(t)]]_s = \sum_{i=0}^{s} [\Gamma^i C(t)]$$

We will prove the following

Lemma 3.2. Let A^{ab} be a Lorentz metric defined by (3.9) with $w = (w^J)$ and C^{ab} smooth and verifying the condition

$$[w(t)]_2 + [[C(t)]]_2 < \delta$$

for some $0 < \delta < 1$ and $t \in [0,T]$. Let $v = v(t,x)$ be a smooth solution of (2.1) decaying like $O(1/|x|^2)$ as $|x| \to \infty$ and let $\bar{L}$ be the first order operator of Lemma 3.1, then, if δ is sufficiently small, there exists a sufficiently small positive constant m_0 such that for all $t \in [0,T]$,

$$\text{(i)} \qquad m_0^{-1} \|v(t)\|^2_{\Gamma,1} + \|v(t)\|^2_{E,A} = \int e^0_{A,K}(v(t))\, dx \geq m_0 \|v(t)\|^2_{\Gamma,1}$$

and, for some constant $c > 0$,

$$\text{(ii)} \qquad \int R_{A,K}(v(t))\, dx \leq c\, \delta \frac{1}{t+1} \|v(t)\|^2_{\Gamma,1}$$

Moreover the following energy inequality holds

$$\text{(iii)} \qquad \frac{d}{dt} \|v(t)\|_{E,A}$$

$$\leq \int (1+M(t,|x|))^2\, |f(t,x)|^2 dx\Big)^{1/2} + c\, \delta \frac{1}{(t+1)} \|v(t)\|_{E,A}$$

Proof: Integrate (3.4") for $L = K$, $A^{ab} = \eta^{ab} + P^{ab}$ and $P^{ab} = B^{[ab]} + C^{ab}$. We obtain,

$$\text{(3.11)} \qquad \int e^0_{A,K}(v)\, dx = \int Kv \cdot f\, dx + \int R_{A,K}(v)\, dx$$

where

$$\text{(3.11')} \qquad e^0_{A,K} = e^0_K + e_{B,K} + e_{C,K}$$

and,

$$e^{B,K} = -B^{(ab)} v_b Kv + \frac{1}{2} m^a B^{(bc)} v_b v_c$$

$$e^{C,K} = -C^{ab} v_b Kv + \frac{1}{2} m^a C^{bc} v_b v_c$$

Also,

(3.12) $$R^0_{A,K} = R_{B,K} + R_{C,K}$$

where, with $K = \partial_t + K_0 + 2t = (1+t^2+|x|^2)\partial_t + 2tx^i\partial_i + 2t$,

$$R_{B,K} = -(\partial_a B^{(ab)} v_b Kv) - B^{(ab)} v_b [\partial_a, K] v$$

$$+ \frac{1}{2}(\partial_t + K)^0 (B^{(ab)}) + 2tB^{(ab)}\ v_a v_b$$

and similarly for $R_{C,K}$.

We now write,

$$e^0_{B,\bar{L}} = -B^{ab} v_b Kv + \frac{1}{2}(1+t^2+|x|^2)\ B^{bc} v_b v_c$$

$$= -(T^{acb}_J + T^0_J \eta^{bc})\ w^J_c v_b Kv$$

$$+ \frac{1}{2}(1+t^2+|x|^2)\ (T^{(bd)c}_J + T^b_J \eta^{cd})\ w^J_d v_b v_c$$

According to the definition of Q, Q_{ac} given in the introduction we have,

$$T^{(ac)b}_J \partial_a f\ \partial_c g = T^{acb}_J\ Q_{ac}(f,g)$$

$$T^a_J \eta^{bc}\ \partial_b f\ \partial_c g = T^a_J\ Q(f,g)$$

Therefore,

$$e^0_{B,K}(v) = -T^{0cb}_J w^J_c v_b Kv - T^0_J\ Q(w,v)\cdot Kv$$

$$+ \frac{1}{2}(1+t^2+|x|^2)[T^{bdc}_J Q_{bd}(w^J,v)v_c + T^b_J Q(w^J,v)v_b]$$

Now using Lemma 1.1 to estimate Q_{ab}, Q and the remark that, since $K = \partial_t + x^a L_a + 2t = \partial_t + t(L_0+2) + x^i L_i$,

$$|Kv| \leq M(t,|x|)\ (|\Gamma v| + |v|)$$

we obtain, with some constant c, and any $t \in [0,T]$,

$$(3.13) \qquad |e^0_{B,K}(v)| \leq c[w]_1(|v|^2 + |\Gamma v|^2)$$

Similarly we find

$$(3.14) \qquad |e^0_{C,\tilde{L}}(v)| \leq c[[C]]_1(|v|^2 + |\Gamma v|^2)$$

The estimate (i) of the Lemma follows now from (3.13),(3.14) and Lemma 3.1, provided that δ is sufficiently small.

To estimate $R_{A,K}$ we use (3.12) and the definition of B. We have,

$$\partial_a B^{ab} v_b Kv = (T_J^{(ac)b} + T_J^a \eta^{bc})\, w^J_{ac} v_b Kv$$

$$= T_J^a\, Q(w_a^J, v)Kv$$

and, according to Lemma 1.1 and the definition of $K = \partial_t + x^a L_a + 2t$, we find, for any $t \in [0,T]$,

$$|\partial_a B^{ab} v_b Kv| \leq \frac{1}{t+1}\, [w]_1(|v|^2 + |\Gamma v|^2)$$

Similarly, since $[\partial_a, K] = 2L_a + 2\delta_{a0}$, we find

$$|B^{ab} v_b [\partial_a, K]v| \leq c\, \frac{1}{t+1}\, [w]_1(|v|^2 + |\Gamma v|^2)$$

Also,

$$2t\ B^{ab} v_a v_b = 2t(T_J^{(ac)b} + T_J^a \eta^{bc}) w^J_c v_a v_b$$

$$= 2t\ T_J^{acb}\, Q_{ac}(w^J, v) v_b + 2t T_J^a Q(w^J, v) v_a$$

and therefore according to Lemma 1.1,

$$|2t\ B^{ab}v_a v_b| \le c\ \frac{1}{(t+1)}\ [w]_1\ |\Omega v|\cdot|Dv| \le c\ \frac{1}{t+1}\ [w]\ |\Gamma v|^2$$

Finally, with $M = (1 + t^2 + |x|^2)\partial_t + 2tx^i\partial_i$ and $m^0 = (1+t^2+|x|^2)$, $m^i = 2tx^i$ we have

$$\frac{1}{2}\ (\partial_t+K^0)B^{ab} = \frac{1}{2}\ m^c\partial_c B^{ab}v_a v_b =$$

$$= \frac{1}{2}\ v_a v_b(T_J^{(ad)b} + T_J^a\eta^{bd})\ [\partial_d(m^c\partial_c w) - (\partial_d m^c)w_c^J]$$

$$= \frac{1}{2}\ T_J^{adb}Q_{ad}(Mw^J,v)v_b + \frac{1}{2}\ T_J^a\ Q(Mw,v)v_a$$

$$- T_J^{(ad)b}([\partial_d,M]w)v_a v_b - \frac{1}{2}\ T_J^a([\partial_d,M]w)\eta^{bd}v_a v_d$$

and find again, estimating term by term with the help of Lemma 1.1,

$$|\frac{1}{2}\ m^c\partial_c B^{ab}v_a v_b| \le c\ \frac{1}{(t+1)}\ [w]_2(|Dv|+|\Omega v|)(|Dv|)$$

$$\le c\ \frac{1}{(t+1)}\ [w]_2\ |\Gamma v|^2 + |v|^2)$$

Henceforth,

$$|R_{B,K}(v)| \le c\ \frac{1}{t+1}\ [w]_2(|v|^2 + |\Gamma v|^2) \tag{3.16}$$

Similarly, but easier

$$|R_{C,K}(v)| \le c\ \frac{1}{t+1}\ [[C]]_2(|v|^2 + |\Gamma v|^2) \tag{3.17}$$

and the estimate (3.12) follows by integration.

4. Proof of the Theorem

We will now apply our previous considerations to the nonlinear problem

(4.1) $$\Box u^I = F^I(u,u',u'')$$

where $F = (F^I)_{I=1,\ldots,p}$ satisfies the Null condition given in the introduction. Thus F^I has the form

(4.2) $$F^I = T_J^{Iabc}Q_{ac}(\partial_b u^I,u^J) + T_J^{Ia}\, Q(\partial_a u^I u^J)$$
$$+ T_{JK}^{Iab}\, Q_{ab}(u^J,u^K) + T_{JK}^{I}\, Q(u^J,u^K)$$
$$+ C^{Iab}(u,u')\, \partial_{ab}^2 u^I + D^I(u,u')$$

with Q, Q_{ab} defined by (Q), (Q_{ab}) in the introduction, T_J^{Iacb}, T_J^{Ia}, T_{JK}^{Iab}, T_{JK}^{I} given constants and C^{Iab}, D^I smooth functions of u,u' vanishing, at least, of order two, resp. three, at $u,u' \equiv 0$. We want to derive generalized energy estimates for (4.1). Thus consider Γ^α to be any particular product of the vector fields $\Gamma_0,\Gamma_1,\ldots\Gamma_{10}$ defined in section 1. We write, from (3.1),

(4.3) $$v^I = T_J^{Iab}Q_{ac}(\partial_b v^I,u^J) + T_J^{Ia}\, Q(\partial_a v^I,u^J)$$
$$+ C^{Iab}(u,u')\, \partial_{ab}^2 v^I + f^I$$

where $v^I = \Gamma^\alpha u^I$ and $f^I = f_1^I + f_2^I + \ldots + f_6^I$,

$$f_1^I = [\Gamma^\alpha, \quad]u^I$$

$$f_2^I = T_J^{Iabc}([\Gamma^\alpha,Q_{ab}](\partial_c u^I,u^J) + Q_{ab}(\partial_c u^I,\Gamma^s u^J)$$
$$+ Q_{ab}([\Gamma^\alpha,\partial_c]u^I,u^J))$$

$$f_3^I = T_J^{Ia}([\Gamma^\alpha,Q](\partial_c u^I,u^J) + Q(\partial_c u^I,\Gamma^s u^J)$$

$$+ Q([\Gamma^{\alpha},\partial_a]u^I,u^J))$$

$$f_4^I = T_{JK}^{Iab}\ \Gamma^{\alpha}Q_{ab}(u^J,u^K) + T_{JK}^{I}\ \Gamma^{\alpha}Q(u^J,u^K)$$

$$f_5^I = [\Gamma^{\alpha},C^{Iab}]\partial_{ab}^2 u^I + C^{Iab}[\Gamma^{\alpha},\partial_{ab}^2]u$$

$$f_6^I = \Gamma^{\alpha}\ D^I(u,u')$$

We can also rewrite (4.3) in the form

$$(4.3') \qquad - A^{Iab}\ \partial_{ab}^2 v^I = f^I$$

with

$$A^{Iab} = \eta^{ab} + B^{I[ab]} + C^{Iab}\ , \quad B^{I[ab]} = \frac{1}{2}(B^{Iab} + B^{Iba})$$

and

$$B^{Iab} = (T_J^{I(ac)b}u_c^J + T_J^{Ia}\eta^{bc}u_c^J), \quad T_J^{I(ac)} = (T_J^{Iacb} - T_J^{Icab})$$

According to the definition of [[ ]] in (3.10), we have

$$[[C(t)]]_2 \leq c[u(t)]_2\ ,$$

provided that $[u(t)]_2 < 1$. Therefore according to Lemma 3.2 applied to each equation (4.3′) we derive the following

$$(4.4) \quad \frac{d}{dt}\|v(t)\|_E$$

$$\leq \left(\int (1 + M(t,|x|))^2\ |f(t,x)|^2 dx\right)^{1/2} + c\ \delta\ \frac{1}{t+1}\|v(t)\|_1$$

provided that

$$(4.5) \qquad [u(t)]_2 \leq \delta$$

Here, $\|v(t)\|_E^2 = \sum_{J=1}^{p} \|v^J(t)\|_{E,A_J} \geq m_0\|v(t)\|_{\Gamma,1}^2$ and $\leq (m_0)^{-1}\cdot\|v(t)\|_{\Gamma,1}^2$ for some small, positive m_0. Our main task now is to estimate the integral

$$(4.6) \qquad \left(\int (1 + M(t,|x|))^2\ |f(t,x)|^2\ dx\right)^{1/2}$$

Before doing this we record here two elementary and crude lemmas (see [15] for the sharp statements). For simplicity we will denote, in the remainder of this section, by $\|u(t)\|_k$, $|u(t)|_k$, the previous defined quantities $\|u(t)\|_{\Gamma,k}$, $|u(t)|_{\Gamma,k}$ (see Notation 2.1).

Lemma 4.1. Let u(t,x), v(t,x) be two C^∞ functions, bounded with respect to the expressions appearing below. Then, for any α, $|\alpha| \leq k$

$$\|\Gamma^\alpha(u\,v)(t)\| \leq c(|u(t)|_{k/2}\cdot\|v(t)\|_k + |v(t)|_{k/2}\,\|u(t)\|_k)$$

with k/2 representing the largest integer $\geq$ k/2.

Lemma 4.2. Let $v = (v_1,\ldots,v_p)$, C^∞ functions, bounded with regard to $\|\ \|_k$, and assume that f(v) is uniformly bounded with respect to its derivatives up to order k on $|v| = |v_1| + \ldots + |v_p| \leq 1$. Then, given any 11-index α, $|\alpha| = k$, we have

$$\|\Gamma^\alpha f(v(t))\| \leq c\,\|v(t)\|_k$$

whenever $|v(t)|_{k/2} \leq 1$.

To estimate the integral (4.6) we make the important remark that f_2^I, f_3^I and f_4^I, appearing in the definition of f, can be written as a sum of terms of the type,

$$Q_{ab}(\Gamma^\beta u^J,\Gamma^\gamma u^K),\ Q(\Gamma^\beta u^J,\Gamma^\gamma u^K)$$

for $|\beta| + |\gamma| \leq k$. Indeed this is an immediate consquence of Lemma 1.2. Therefore, by Lemma 1.1

$$\sum_{\ell=2}^{4} |f_\ell^I(t,x)| \leq c\,\frac{1}{1+M(t,|x|)}\sum_{|\beta|+|\gamma|\leq k+1} |\Gamma^\beta u(t,x)|\,|\Gamma^\gamma u(t,x)|$$

and consequently, with the notation of (3.10)

$$\sum_{\ell=2}^{4} \int (1+M(t,|x|))^2\,|f_\ell(t,x)|^2\,dx \leq c\,\frac{1}{t+1}\,[u(t)]_{1+k/2}\cdot\|u(t)\|_{k+1}^2$$

Similarly using Lemmas 4.1 and 4.2,

$$\sum_{\ell=5,6} \int (1+M(t,|x|))^2 |f_\ell(t,x)|^2 \, dx$$

$$\leq c \frac{1}{t+1} [u(t)]^2_{1+k/2} \cdot \|u(t)\|^2_{k+1}$$

provided that $|u(t)|_{(k+1)/2} \leq 1$.

Finally we estimate f_1 by using the commutation properties of the Γ's with and the estimates above. We therefore derive,

$$(4.7) \qquad \int (1+M(t,|x|))^2 |f(t,x)| \leq c \frac{1}{t+1} \delta \cdot \|u(t)\|^2_{k+1}$$

provided, that,

$$(4.8) \qquad [u(t)]_{1+k/2} < \delta < 1$$

Using this in (4.4) we find

$$(4.9) \qquad \frac{d}{dt} \|v(t)\|_E \leq c \ \delta \frac{1}{t+1} (\|u(t)\|_{k+1} + \|v(t)\|_E)$$

provided that (4.8) holds. Since (4.9) holds for every $v = \Gamma^\alpha u$ with $|\alpha| \leq k$ we derive, under the assumption (4.8),

$$(4.10) \qquad \frac{d}{dt} \|u(t)\|_{E,k} \leq c \ \delta \frac{1}{t+1} \|u(t)\|_{k+1} + \|u(t)\|_{E,K}$$

where

$$\|u(t)\|_{E,k} = \sum_{|\alpha| \leq k} \|\Gamma^\alpha u(t)\|_E$$

Differentiating the exprssion $(t+1)^{-\lambda} \|u(t)\|_{E,k}$ and using, $\|u(t)\|^2_{E,k} \geq m_0 \|u(t)\|^2_{k+1}$, for some positive m_0 , we find

$$(4.11) \qquad \frac{d}{dt} ((t+1)^{-\lambda} \|u(t)\|_{E,k}) \leq 0$$

provided that $c\delta < \lambda$, where c is a constant depending only on k and the coefficients of the equation (4.1).

We have therefore proved the following

Proposition 4.1. Let $u = u(t,x)$ be a smooth solution of (4.1) in the time interval $[0,T]$, verifying the assumption

$$(4.12) \qquad [u(t)]_{1+k/2} < \delta < 1$$

for all $t \in [0,T]$, and some integer $k > 0$. Let λ be a positive constant less or equal to 1/2. Then, if δ is sufficiently small,

$$(4.13) \qquad \|u(t)\|_{k+1} \leq c(1+t)^{\lambda} \|u(0)\|_{k+1}$$

for all $t \in [0,T]$.

To estimate $[u(t)]_{(k+1)/2}$ we will use Proposition 2.1 ii. Thus, we have,

$$(4.14) \qquad [u(t)]_{1+k/2} \leq |||u'(t)|||_{5+k/2}$$

where $u' = Du$ and $||| \ |||_k$ is the norm introduced by Notation 2.1.

It only remains to estimate $|||u'(t)|||_{5+k/2}$. To do this we appeal to the corollary to Proposition 2.2. Therefore, in the interval $[0,T]$, the solution u of (4.1), subject to the initial value problem

$$\text{I.V.P.} \qquad u = \varepsilon f(x) , \quad u_t = \varepsilon g(x) \quad \text{at} \quad t = 0 ,$$

satisfies the following,

$$(4.15) \quad |||u'(t)|||_{5+k/2}$$

$$\leq c\,\varepsilon + c \sum_{|\alpha|\leq 6+k/2} \int_0^t ds \int \frac{1}{|x|} \left(1+ \log \frac{t-s+|x|}{|t-s-|x||}\right) |\Gamma^{\alpha}F(s,x)|\, dx$$

To estimate the integrals on the right hand side of (4.14) we note, using Lemma 1.2, that the quadratic terms appearing in the expression of $\Gamma^{\alpha}F$ are of the form

$$Q_{ab}(\Gamma^{\beta}u^{J},\Gamma^{\gamma}u^{K}),\quad Q(\Gamma^{\beta}u^{J},\Gamma^{\gamma}u^{K})$$

with $|\beta| + |\gamma| \leq k/2 + 6$. Therefore using Lemma 1.1 we can estimate them by

$$\sum_{|\beta|+|\gamma|\leq k+1} \left(\frac{1}{1+M(s,|x|)} |\Gamma^{\beta}u| \cdot |\Gamma^{\gamma}u|\right)$$

provided that

$$k/2 + 6 \leq k+1 \tag{4.16}$$

Similarly, the nonquadratic terms can be estimated by

$$\sum_{|\beta|+|\gamma|+|\delta|\leq k+1} |\Gamma^{\beta}u|\ |\Gamma^{\gamma}u|\cdot|\Gamma^{\delta}u|$$

provided that $|u(t)|_{(k+1)/2} \leq 1$ and (4.16) holds. Therefore we find, using also the Cauchy-Schwartz inequality,

$$\int \frac{1}{|x|}\left(1+\log\frac{t-s+|x|}{|t-s-|x||}\right)|\Gamma^{\alpha}F(s,x)|\,dx \tag{4.17}$$

$$\leq c(1+s)^{-1}I(t,s)\left([u(s)]_{(k+1)/2}\right)\cdot[u(s)]_{(k+1)/2}\cdot\|u(s)\|_{k+1}$$

with,

$$I(t,s) = \left(\int \frac{1}{|x|^{2}}\frac{1}{(1+M(s,|x|))^{2}}\left(1+\log\frac{t-s+|x|}{|t-s-|x||}\right)dx\right)^{1/2} \tag{4.18}$$

$$\leq c(1+s)^{-1/3}$$

Consequently, for $k \geq 11$, and $[u(t)]_{k+1/2} \leq 1$, as $t \in [0,T]$,

$$|||u'(t)|||_{5+k/2} \leq c\,\varepsilon + c\int_0^t (1+s)^{-1-1/3}\,\|u(s)\|_{k+1}\,ds \tag{4.19}$$

or, by (4.12), with $0 < \lambda < 1/3$

$$|||u'(t)|||_{5+k/2} \leq c\ \varepsilon$$

for all $t \in [0,T]$. Finally, from (4.13),

(4.20) $$[u(t)]_{1+k/2} \leq c\ \varepsilon$$

for all $t \in [0,T]$, with c a constant depending only on the coefficients of (4.1) and the fixed initial data f,g. We have thus proved the following

Proposition 4.2. Let $u = u(t,x)$ be a smooth solution of (4.1) in the time interval $[0,T]$, verifying the initial conditions

I.V.P. $\qquad u = \varepsilon\ f(x)\ ,\quad u_t = \varepsilon\ g(x)\quad \text{at}\quad t = 0$

with $f,g \in C_0^\infty(\underline{R}^3)$ and a small parameter $\varepsilon > 0$. Then, if ε is sufficiently small we have,

$$[u(t)]_6 < c\ \varepsilon < 1$$

and

$$\|u(t)\|_{12} \leq c\ \varepsilon(1 + t)^{1/3}$$

for all $t \in [0,T]$, with c a positive constant depending on almost 12 derivatives of f,g and the coefficients of the equation (4.1).

The main theorem stated in the introduction is now an immediate consequence of Proposition 4.2 and the classical local existence theorem.

Bibliography

[1] F. John, "Slow-up for quasilinear wave equations in three space dimensions," Comm. Pure Appl. Math. 34, 29-51 (1981).

[2] F. John, "Blow-up of radial solutions of $\Box u = \frac{\partial F(u_t)}{\partial t}$," preprint.

[3] F. John, S. Klainerman, "Almost global existence to nonlinear wave equations in three space dimensions," Comm. Pure Appl. Math., Vol. XXXVII, 443-455 (1984).

[4] S. Klainerman, "Uniform Decay Estimates and the Lorentz Invariance of the Classical Wave Equation," to appear in Comm. Pure Appl. Math., 1985.

[5] S. Klainerman, "Long time behavior of solutions to Nonlinear Evolution Equations," Arch. Rat. Mech. and Anal. 78 (1982) pp. 73-98.

[6] J. Shatah, "Global Existence of Small Solutions to Nonlinear Evolution Equations," J. Diff. Eqts. 46, Dec. 1982.

[7] S. Klainerman, G. Ponce, "Global Small Amplitude Solutions to Nonlinear Evolution Equations," Comm. Pure Appl. Math. 36 (1983) pp. 133-141.

[8] S. Klainerman, "Long time behavior of solutions to nonlinear wave equations," Proc. of the Int. Congress for Mathematicians, Warsaw, 1982.

[9] D. Christodoulou, "Solutions globales des equations de Champ de Yang-Mills," C.R. Acad. Sci. Paris, 293, Series A, pp. 39-42 (1981).

[10] Y.C. Bruhat and D. Christodoulou, "Existence of global solutions of the Yang-Mills Higgs and Spinor Field equations in 3+1 dimensions," Annales de l´ecole normale superieur, 4th series, 14, pp. 481-506 (1981).

[11] D. Christodoulou,"Global solutions of nonlinear hyperbolic equations for small data," preprint.

[12] S. Klainerman, "L^∞-L^1 estimates for solutions to the classical wave equations in 3-dimensions," Comm. Pure Appl. Math. 37, 1984, pp. 269-288.

[13] L. Tartar, Lecture Notes at the University of Wisconsin, Madison, 1976, Math. Res. Ctr. Report No. 1584.

[14] C. Morawetz, "Notes on the decay and scattering for some

hyperbolic problems," Soc. Ind. Appl. Math., Philadelphia.

[15] L. Hörmander, "On Sobolev spaces associated with Some Lie Algebras," preprint.

Department of Mathematics
Courant Institute of Mathematical Sciences
New York University
New York, New York 10012

Lectures in Applied Mathematics
Volume 23, 1986

WAVE PROPAGATION IN A BUBBLY LIQUID

Russel E. Caflisch
Michael J. Miksis
George C. Papanicolaou
Lu Ting[1]

ABSTRACT: The macroscopic behavior of a liquid containing many small gas bubbles can be analyzed under various special assumptions. The system has a sound speed which depends singularly on the volume fraction, and it is dispersive. For disturbance frequencies near the resonant frequency, a set of nonlinear macroscopic equations is derived using Foldy's method. Global existence is proved for the initial value problem for these equations.

1. BUBBLY LIQUIDS - MACROSCOPIC VS. MICROSCOPIC

The subject of this discussion is wave propagation in a liquid containing many small gas bubbles. In particular the focus will be on the macroscopic description of waves, with wavelength much larger than the microscopic size of inter-bubble distance, and on the derivation and properties of this macroscopic theory.

There are at least four levels of description of a bubbly liquid. First at the highest level, a bubbly liquid is fluid flowing through a system of pipes with complicated boundary and

1980 Mathematics Subject Classification. 76T05, 35L75

[1]Research partially supported by Office of Naval Research under Contract No. N00014-81-K-0002.

initial conditions and forces. The second description comes from simpifying this situation and considering a bubbly liquid as an infinite homogeneous continuum propagating some kind of disturbance in pressure and velocity. This description will be referred to as macroscopic. At a finer microscopic level the bubbly liquid consists of gas bubbles surrounded by liquid, undergoing complicated motion on the scale of the bubble size. At an even finer fourth level the two fluids may be considered to be composed of molecules interacting through electromagnetic forces.

The goal here is to derive a macroscopic description from the microscopic description and to analyze the macroscopic theory. This derivation requires an understanding of the microscopic state: The distribution and geometry of the two phases (e.g. uniformly distributed spherical bubbles) and some properties of the local fluid variables (u,p,T) must be known. Furthermore a mathematical analysis is possible only if there is some special correspondence between the microscopic and macroscopic scales:

First there must be clear separation of these two scales; then macroscopic variables are defined by averaging over the microscopic scale. Such averaging inevitably leads to closure problems requiring new constitutive relations. These relations can be found only if the interaction between the two scales is relatively simple. Examples of the macroscopic properties which are sought in such a derivation are sound speed for acoustic waves, description of shock waves, rates of mass and heat transfer and modes of instability. In this paper a system of macroscopic equations for mildly nonlinear acoustic waves is derived and discussed.

The full details of this derivation and analysis are given in [2,3,4]. Additional properties of bubbly liquids and further references may be found in [1,10,15,16]. The authors have benefitted from a number of discussions with Andrea Prosperetti and Don Drew.

2. MACROSCOPIC PHENOMENA

Experimental observations of acoustic waves in bubbly liquids show a number of interesting phenomena [13]. The most important parameters in these experiments are the frequency ω and wavenumber k of the acoustic disturbance and the volume fraction $\beta = n(\frac{4}{3}\pi R^3)$, in which n is the number density of bubbles and R is the bubble radius. We also denote by ρ_ℓ and ρ_g (or c_ℓ and c_g) the densities (or soundspeeds) of liquid and gas, and by $\langle \cdot \rangle$ we mean the volume average.

First of all the effective sound speed $c^* = \omega/k$ (for k not too large) is found to behave singularly near $\beta = 0$. In fact while the sound speed of pure water ($\beta = 0$) is approximately 1500 m/sec and that of pure air is approximately 350 m/sec, a bubbly liquid with $\beta = .2$ has a sound speed c^* of approximately 30 m/sec. Even for $\beta = .005$, $c^* = 160$ m/sec.

The reason for this singular behavior is that sound speeds do not average, but inertia and compressibility do. The effective inertia ρ^* is $\rho^* = \langle\rho\rangle = (1-\beta)\rho_\ell + \beta\rho_g$ and the effective compressibility $\kappa^* = \langle 1/\rho c^2 \rangle = (1-\beta)/\rho_\ell c_\ell^2 + \beta/\rho_g c_g^2$. The effective sound speed c^*, whose square is inversely proportional to the product of compressibility and inertia, is given by

$$1/c^{*2} = \langle 1/\rho c^2 \rangle \langle \rho \rangle . \tag{2.1}$$

Since the density ratio $\rho_\ell/\rho_g = 1000$ is very large, a small volume fraction of gas has a large effect on c^*. In other words water has a large sound speed because it has very little compressibility, air has a large sound speed because it has very little inertia; the mixture has both compressibility and inertia and thus a small sound speed.

This singular behavior in c^* means that there is a large decrease in the sound speed in going from a clear liquid to a bubbly liquid. Such an impedance mismatch leads to very effective reflection which may be used to trap a disturbance and thus has many engineering applications. For example a layer of bubbles is used to shield a dam from the pressure wave generated by an underwater explosion [6]. Bubble layers are also used to mask ship propellor noise.

A further macroscopic phenomena of acoustic waves in a bubbly liquid is the dispersive character $(\omega = \omega(k))$ of the propagation. In fact there is a resonant frequency ω_0 at which waves do not propagate and the bubbles oscillate freely. For example at β = .003 and R = 2 mm, ω_0 = 1500/sec.

3. MICROSCOPIC DESCRIPTION

There are four basic length scales involved in wave propagation in a bubbly liquid. From longest to shortest, they are wavelength λ, average inter-bubble distance d, bubble radius R, and amplitude ΔR of radial oscillations of the bubbles (in anticipation of later assumptions the bubbles are described as spherical). The following two scaling assumptions are vital to this theory:

$$\lambda >> d \qquad d >> R . \tag{3.1}$$

The first of these is necessary for the validity of a continuum macroscopic theory; the second implies that direct bubble interactions are weak. In addition the scaling $\Delta R \cong R$ results in significant nonlinear effects.

The most important dimensionless parameters are the relative inter- bubble distance $\varepsilon = d/\lambda$ and the relative bubble radius $\delta = R/\lambda$; the volume fraction is $\beta = \frac{4}{3}\pi(\delta/\varepsilon)^3$. We introduce a new parameter $\chi = \delta/\varepsilon^3$ which is the product of the dimensionless bubble radius δ and number density $n = \varepsilon^{-3}$ and might be called the effective capacity by analogy to electrostatics.

The microscopic fluid state will be described by the inviscid isentropic Euler equations with an equation of state, i.e.

$$(\rho c^2)^{-1}(p_t + u \cdot \nabla p) + \nabla \cdot u = 0 \tag{3.2}$$

$$\rho \quad (u_t + u \cdot \nabla u) + \nabla p = 0 \tag{3.3}$$

$$EOS(p,\rho) = 0 \tag{3.4}$$

are assumed to hold in each phase with the definition of EOS different from gas to liquid. Across the phase boundaries there are the continuity conditions:

$$p \text{ and } u \cdot n \text{ continuous} \tag{3.5}$$

$$u \cdot n = R_t \quad \text{(for spheres).} \tag{3.6}$$

In the gas we shall use the polytropic equation of state

$$p_g = \tilde{\kappa}\rho_g^{\gamma} = \kappa R^{-3\gamma} \tag{3.7}$$

in which $\rho_g = M/(\frac{4}{3}\pi R^3)$, $\kappa = \tilde{\kappa}(3M/4\pi)^{\gamma}$ and M is the mass of a bubble which is assumed to be constant. In the liquid we shall make the acoustic approximation.

Note that surface tension, heat conductivity, viscosity and mass transfer have been neglected. These could be included, at least in certain regimes [8,12], without qualitative change in the results.

4. LINEARIZED THEORY - DERIVATION OF c^*.

The linearization of equations (3.2), (3.3), (3.5) is the system

$$(\rho c^2)^{-1} p_t + \nabla \cdot u = 0 \tag{4.1}$$

$$\rho u_t + \nabla p = 0 \tag{4.2}$$

which hold in each phase, with the continuity condition

$$p,\ u \cdot n \quad \text{continuous} \tag{4.3}$$

across the boundary. Here ρ and c are taken to be prescribed constants in each phase and the boundaries are constant.

Since p and $u \cdot n$ are continuous, equations (4.1), (4.2) are in natural form in the sense that a volume integration of (4.1), (4.2) would pick up no term from the discontinuities at the phase boundaries. To leading order in β, it is found that the averages $\bar{p}$ and $\bar{u}$ satisfy

$$\langle 1/\rho c^2 \rangle \bar{p}_t + \nabla \cdot \bar{u} = 0 \tag{4.4}$$

$$\langle \rho \rangle \bar{u}_t + \nabla \bar{p} = 0 \tag{4.5}$$

Therefore the effective sound speed is indeed given by

$$1/c^{*2} = \langle 1/\rho c^2 \rangle \langle \rho \rangle \tag{4.6}$$

$$= 1/c_\ell^2 + \beta \rho_\ell / \gamma p_g + O(\beta)$$

in which we have identified $\gamma p_g = \rho_g c_g^2$. Since $\rho_\ell >> \rho_g$ the second term will be dominant unless β is extremely small; the $O(\beta)$ term is uniformly small for small β. Full details of this analysis are presented in [3].

Note that this analysis does not evidence any dispersion. In fact the derivation is only valid for $\omega << \omega_0$. For $\omega = O(\omega_0)$ microscopic structure develops which makes the averaging (4.4), (4.5) invalid. This will be discussed in the next section.

5. NONLINEAR OSCILLATIONS

For frequencies near but below resonance, $\omega < \omega_0$, the bubble oscillations are nonlinear. We make the following scaling assumptions in addition to (3.1):

$$\Delta R/R = O(1) \tag{5.1}$$

$$\Delta p/p = O(1) \tag{5.2}$$

$$u/c^* = O(\delta^2) \tag{5.3}$$

$$\chi = \delta/\varepsilon^3 = O(1) \tag{5.4}$$

$$\zeta = (4\pi^2/3\gamma)(\omega_0/\omega)^2 = O(1) \tag{5.5}$$

The first two of these result in significant nonlinear effects in the bubble oscillation; the third implies that convection is negligible except nearby the bubbles where (5.3) becomes non-uniform and $u = O(\delta c^*)$. Assumptions (5.4) and (5.5) are conditions that the bubbles be well separated and that the frequency be near resonance. Note that under these assumptions the volume fraction is quite small, i.e. $\beta = O(\delta/\varepsilon)^3 = O(\delta^2)$.

The analysis proceeds by asymptotic expansion of the solutions of (3.2) - (3.6). Since the details of that expansion are presented in [2], we shall only describe those physical effects which are found to be significant and the three regions in which the expansions are performed.

A. Bubble Interior. Since the gas has small inertia and order one sound speed $\left(\rho_g/\rho_\ell \ll 1 \text{ and } c_g/c_\ell = O(1)\right)$ spatial non-uniformities within a bubble equilibrate almost instantaneously. Therefore we take the pressure within a bubble to be spatially uniform and the bubbles to be spherical and undergoing purely radial oscillations. As a macroscopic wave propagates through the fluid, p_g and R may vary on the macroscopic length scale and are related by $p_g(x,t) = \kappa\, R(x,t)^{-3\gamma}$. Furthermore since macroscopic convection is negligible, the bubble centers may be assumed motionless.

B. Bubble Neighborhood. On the small length scale R near a

bubble, the liquid may be assumed incompressible but convective effects must be retained. The resulting equations for $|x| > R$ are

$$\nabla \cdot u = 0 \tag{5.6}$$

$$\rho_\ell (u_t + u \cdot \nabla u) + \nabla p = 0 \tag{5.7}$$

with boundary conditions

$$p = p_g(t) \qquad u \cdot n = R_t(t), \text{ on } |x| = R \tag{5.8}$$

$$p = p_\infty(t) \qquad \text{at } |x| = \infty \,. \tag{5.9}$$

This problem, solved first by Rayleigh, has an irrotational solution with $u = \nabla\phi$, $\nabla^2\phi = 0$ in $|x| > R$ solved by

$$\phi = -\frac{R_t R^2}{|x|} + \bar{\phi} \,. \tag{5.10}$$

The pressure is then found by Bernoulli's law, which at $|x| = R$ yields the relation

$$\rho_\ell (RR_{tt} + \tfrac{3}{2}R_t^2) = p_g - p_\infty \tag{5.11}$$
$$= \kappa R^{-3\gamma} - p_\infty$$

known as the Rayleigh equation. In (5.11) p_∞ should be thought of as the pressure in the ambient fluid, and so the Rayleigh equation gives the complete relation between the macroscopic fluid state (described by $p_\ell = p_\infty$) and the microscopic fluid state (described by R).

C. <u>Ambient Fluid</u>. Because of the wide separation of the bubbles ($\delta = O(\varepsilon^3)$, local pressure variations around different bubbles do not interact directly but only through their effect on the ambient fluid. This is the condition for the validity of

Foldy's method [5] which, in the nonlinear generalization used here, yields the following macroscopic equations in dimensionless form:

$$\zeta C^{-2}\bar{p}_t + \nabla \cdot \bar{u} = \chi\theta(\tfrac{4}{3}\pi R^3)_t \tag{5.12}$$

$$\bar{u}_t + \zeta\nabla\bar{p} = 0 \tag{5.13}$$

$$RR_{tt} + \tfrac{3}{2}R_t^2 = \zeta(\kappa R^{-3\gamma} - \bar{p}) . \tag{5.14}$$

In (5.12) - (5.14), $C = c_\ell/c^*$ and $\theta = \theta(x)$ is a dimensionless function measuring the spatial variations in the number of bubbles per unit volume. These equations were first proposed in slightly different form by Van Wijngaarden [14,15] in 1968. The main purpose of our work is to give a formal derivation of the equations, a discussion of their range of validity and an analysis of their properties.

6. PROPERTIES OF THE MACROSCOPIC EQUATIONS

The macroscopic equations can be rewritten, after dropping some constants, as

$$p_t + \nabla \cdot u = (\tfrac{4}{3}\pi R^3)_t \tag{6.1}$$

$$u_t + \nabla p = 0 \tag{6.2}$$

$$RR_{tt} + \tfrac{3}{2}R_t^2 = R^{-3\gamma} - p \tag{6.3}$$

This system has an energy density (recall $\gamma > 1$)

$$E = \tfrac{1}{2}(p^2 + u^2) + 2\pi R^3 R_t^2 + \tfrac{4}{3}\pi(\gamma - 1)^{-1}R^{-3(\gamma-1)} \tag{6.4}$$

satisfying

$$E_t + \nabla \cdot pu = 0 \tag{6.5}$$

The first term in E represents the acoustic energy of the liquid, the second is the kinetic energy of the liquid motion near the bubble neighborhood, the third is the energy in the bubble.

The dispersion relation for linearized solutions of (6.1)-(6.3) with p, u and R near p_0, $u_0 = 0$ and $R_0 = p_0^{1/3\gamma}$ is

$$(c^*)^{-2} = (k/\omega)^2 = 1 + 4\pi R_0/(\omega_0^2 - \omega^2) \tag{6.6}$$

with the resonant frequency ω_0 given by

$$\omega_0^2 = 3\gamma p_0/R_0^2 \,. \tag{6.7}$$

A rigorous nonlinear analysis of (6.1)-(6.3) has been performed in [4] for the initial value and initial boundary value problems in one space dimension. The main result is the following global existence theorem:

Global Existence Theorem (1-D) Suppose p, u, R, R^{-1}, R_t are smooth and bounded at $t=0$ for all x. Then the equations (6.1)-(6.3) for $-\infty < x < \infty$ have a unique solution for all time, with p, u, R, R^{-1}, R_t smooth and bounded at each t. The same is true on $0 < x < \infty$ if $p(x=0,t)$ is specified and is smooth and bounded.

This theorem shows that (6.1)-(6.3) do not describe bubble collapse or blow-up. Resonance does not enter into this theorem since the theorem does not discuss the stationary boundary value problem.

The proof of this theorem uses the following reformulation of(6.1)-(6.3) as a semi-linear system: Denote $f = p+u$, $g = p-u$, $D = RR_t$;then (6.1)-(6.3) is equivalent to

$$f_t + f_x = 4\pi RD \tag{6.8}$$

$$g_t - g_x = 4\pi RD \tag{6.9}$$

$$R_t = R^{-1}D \tag{6.10}$$

$$D_t = -\frac{1}{2}R^{-2}D^2 + R^{-3\gamma} - \frac{1}{2}(f+g) . \tag{6.11}$$

A priori bounds for (6.1)-(6.3) come from the energy density E in (6.4) and the following Gronwall type inequality for the Rayleigh equation (6.3):

Lemma. Suppose that R satisfies (6.3) with $\gamma > 1$, $R(t=0) = R_0 > 0$ *and* $R_t(t=0) = R_1$. *Denote*

$$N(t) = \int_0^t |p(s)|ds. \tag{6.12}$$

Then

$$\begin{aligned} R &\leq c\{1 + t^{2/5} + t^{1/2}N^{1/2}\} \\ R^{-1} &\leq c\{1 + N + t^{1/4}N^{5/4}\}^{2/3(\gamma-1)} \\ |R^{3/2}R_t| &\leq c\{1 + N + t^{1/4}N^{5/4}\} \end{aligned} \tag{6.13}$$

in which c is a constant depending only on R_0, R_1 *and* γ.

7. TRAVELING WAVES AND LONG WAVE APPROXIMATIONS

Let us next construct travelling wave solutions for the system (6.1)-(6.3). We look for one-dimensional solutions that travel with speed c and have the form

$$u = u(y),\ R = R(y),\ p = p(y),\ y = x-ct. \tag{7.1}$$

The wave profiles u, R and p satisfy the ordinary differential equations

$$-cp' + u' = -c(\frac{4}{3}\pi R^3)' \qquad (7.2)$$

$$-cu' + p' = 0 \qquad (7.3)$$

$$c^2(RR'' + \frac{3}{2}R'^2) = R^{-3\gamma} - p . \qquad (7.4)$$

Equations (7.2) and (7.3) are combined and integrated to give

$$p = p(R,c,b) = -\alpha(c)R^3 + b \qquad (7.5)$$

in which b is an arbitrary constant compatible with the requirement that p and R be positive and

$$\alpha(c) = \frac{4}{3}\pi\ (c^{-2} - 1)^{-1}. \qquad (7.6)$$

Equation (7.4) becomes now

$$RR'' = -\frac{3}{2}R'^2 + c^{-2}[R^{-3\gamma} + \alpha(c)R^3 - b] \qquad (7.7)$$

and this can be integrated once to give

$$R^3R'^2 + 2c^{-2}[\frac{1}{3(\gamma-1)}R^{-3\gamma+3} - \frac{\alpha}{6}R^6 + \frac{b}{3}R^3] = \text{constant}. (7.8)$$

We analyze (7.7) around the stationary points $R = R_0$, $R' = 0$ where

$$p_0 = p(R_0,c,b) = R_0^{-3\gamma} . \qquad (7.9)$$

Linearization about R_0 with $R = R_0 + r$ gives

$$r'' = \frac{1}{c^2R_0}(-3\gamma p_0/R_0 + 3\alpha(c)R_0^2)r . \qquad (7.10)$$

The coefficient of r on the right side of (7.10) is the negative of the square of the wavelength k. Recalling the definition (6.7) of the bubble resonant frequency ω_0, k can be expressed as

$$k^2 = c^{-2} \left(\omega_0^2 - 3\alpha(c)R_0 \right) . \tag{7.11}$$

The frequency ω is $\omega^2 = c^2k^2$. Using (7.6) we recognize (7.11) as the dispersion relation (6.6). The constant b is related to R_0 and ω by (7.5) which gives

$$b = (\gamma+1)p_0 - \frac{1}{3} R_0^2\omega^2 . \tag{7.12}$$

The stationary point $(R,R') = (R_0,0)$ is a center if $k^2 > 0$ and a saddle if $k^2 < 0$. There are no spiral points since the system is conservative. There are two branches of such solutions as follows.

(i) If $|c| > 1$ (i.e. $\alpha(c) < 0$) then for every b equation (7.7) has a single stationary point $(R_0,0)$ which is a center. It corresponds to the high speed optical branch of the dispersion relation (6.6) and has $\omega > \sqrt{\omega_0^2 + 4\pi R_0}$ at the center. The invariant quantity (7.8) has a maximum at $(R_0,0)$ and hence all solutions of (7.7) with $R > 0$ are periodic.

(ii) If $c < 1$ (i.e. $\alpha > 0$) and if $b > (\gamma+1)(\alpha/\gamma)^{\frac{\gamma}{\gamma+1}}$ equation (7.7) has two stationary points $(R_1,0)$ and $(R_2,0)$ with $0 < R_1 < R_2$. The point $(R_1,0)$ is a center and has $\omega < \omega_0$ corresponding to the low speed or acoustic branch of the dispersion relation (6.6). The second point $(R_2,0)$ is a saddle point. In this case the function (7.8) goes to $+\infty$ as R goes to zero and to $-\infty$ as R goes to $+\infty$ along the axis $R' = 0$. It has a local minimum at R_1 and a local maximum at R_2. It follows that there are periodic solutions centered around $(R_1,0)$. These solutions are encircled by a solitary wave (homoclinic solution) which has the value $R = R_2$ as y goes to $\pm\infty$. Note that the solitary wave is a wave of bubble compression with $R(y) < R_2 = R(\pm\infty)$, for finite y.

For all other values of c and b there are no stationary points and no periodic or solitary wave solutions.

In the remainder of this section we shall carry out an asymptotic analysis of (6.1)-(6.3), appropriate for weakly nonlinear, slowly varying solution. Our objective is the Boussinesq and Korteweg de Vries approximations for (6.1)-(6.3).

Let ε be a small dimensionless scale factor (not related to the ε of the other sections) and introduce scaled variables as follows

$$x = x'/\sqrt{\varepsilon}\ , \quad t = t'/\sqrt{\varepsilon} \tag{7.13}$$

$$P = p_0 + \varepsilon p'\ , \quad R = R_0 + \varepsilon R'\ , \quad u_1 = \varepsilon u'\ .$$

Here p_0 and R_0 are constants satisfying $p_0 = R_0^{-3\gamma}$. We substitute these into (4.1)-(4.4) and expand in powers of ε keeping only the first two orders to obtain

$$p'_{t'} + \nabla' \cdot u' = 4\pi R_0^2\ (R'_{t'} + 2\varepsilon\ R_0^{-1}\ R'R'_{t'}) \tag{7.14}$$

$$u'_{t'} + \nabla' p' = 0 \tag{7.15}$$

$$\varepsilon\ R_0 R'_{t't'} = p_0(-3\gamma\ R_0^{-1}\ R' + \varepsilon\ \frac{3}{2}\gamma(3\gamma+1)R_0^{-2}R'^2) - p'\ . \tag{7.16}$$

We can now eliminate p' to obtain, with primes omitted, the (modified) Boussinesq system

$$\frac{R_0\omega_0^2}{c^2} R_t - \nabla \cdot u = \varepsilon((3\gamma+1)\ \omega_0^2 - 8\pi R_0)RR_t - \varepsilon\ R_0 R_{ttt} \tag{7.17}$$

$$u_t - \omega_0^2 R_0 \nabla R = -\varepsilon(3\gamma+1)\omega_0^2 R\nabla R + \varepsilon\ R_0 \nabla R_{tt} \tag{7.18}$$

in which c (the low frequency effective sound speed) is defined by $c^{-2} = 1 + 4\pi R_0/\omega_0^2$ and ω_0 by (6.7).

System (7.17), (7.18) is a "good" Boussinesq system in that it is linearly stable. It is interesting to note that if we had eliminated R rather than p in (7.14)-(7.16), the resulting Boussinesq-type system would be unstable. That procedure would involve dropping some terms of order ε^2 in the elimination.

To obtain a KdV equation for R we look for a plane wave solution propagating in a single direction, say in the positive x direction, with speed c. Following the method of Whitham [17], we look for solutions of (7.17), (7.18) of the form

$$R(x,t) = R_1(x-ct) + \varepsilon\, R_2(x,t) \tag{7.19}$$

$$u(x,t) = \frac{\omega_0^2 R_0}{c}\big(R(x,t) + \varepsilon u_2(x-ct)\big)\ . \tag{7.20}$$

Inserting these expressions in (7.17), (7.18) we find that

$$u_2 = -\frac{1}{4R_0}\left[3(\gamma+1)c^2 + 3\gamma - 1\right] R^2 + \frac{c^2}{2\omega_0^2}\left(c^2+1\right) R_{xxx} + O(\varepsilon) \tag{7.21}$$

and the resulting KdV equation for R is

$$R_t + cR_x = \varepsilon\, \frac{3c}{2R_0}(\gamma+1)\left(1-c^2\right) RR_x + \varepsilon\, \frac{c^3}{2\omega_0^2}\left(c^2-1\right) R_{xxx}\ . \tag{7.22}$$

This equation for R was obtained previously by Van Wijngaarden [14,15] and has been used by many authors, e.g. Kuznetsov et. al. [7], Nakoryakov [9], Pokusaev [11].

8. BIBLIOGRAPHY

1. Batchelor, G.K., Compression waves in a suspension of gas bubbles in liquid, Fluid Dynamics Transactions 4 (1969) 425-445.
2. Caflisch, R.E., Miksis, M.J., Papanicolaou, G.C. and Ting, L., Effective equations for wave propagation in bubbly liquids, JFM submitted, 1984.
3. Caflisch, R.E., Miksis, M.J., Papanicolaou, G.C. and Ting L.,

Wave Propagation in Bubbly Liquids at Finite Volume Fraction, JFM, submitted 1984.

4. Caflisch, R.E., Global existence for a nonlinear theory of bubbly liquids, Comm. Pure Appl. Math., to appear 1985.

5. Carstensen, E.L. and Foldy L.L. Propagation of sound through a liquid containing bubbles. JASA 19 (1947) 481-501.

6. Domenico, S.N., Acoustic wave propagation in air-bubble curtains in water, Geophysics 47 (1982) 345-353.

7. Kuznetsov, V.V., Nakoryakov, V.E., Pokusaev, B.G. and Shreiber, I.R. Propagation of perturbations in a gas-liquid mixture. JFM 85 (1978) 85-96.

8. Miksis, M.J. and Ting, L. Nonlinear radial oscillations of a gas bubble including thermal effects, JASA 1984, to appear.

9. Nakoryakov, V.E., Shryber, I.R., and Gasenko, V.G. Moderate-strength waves in the liquids containing gas bubbles, Fluid Mechanics (Sov.) 10 (1981) 51-66.

10. Plesset, M.S. and Prosperetti, A. Bubble dynamics and cavitation. Ann. Rev. Fl. Mech. 9 (1977) 145-185.

11. Pokusaev, B.G., Konabel'Nikov, A.V. and Pribaturin, N.A.. Pressure waves in a liquid containing gas bubbles, Fluid Mechanics, Soviet Research, Vol. 10, No. 2, (1981) 67-93.

12. Prosperetti, A. Bubble phenomena in sound fields, Ultrasonics, 22 (1984) 69-77 and 115-124.

13. Silberman, E. Sound velocity and attenuation in bubbly mixtures measured in standing wave tubes, JASA 29 (1957) 925-933.

14. Van Wijngaarden, L. On equation of motion for mixtures of liquid and gas bubbles, JFM 33 (1968) 465-474.

15. Van Wijngaarden, L. One-dimensional flow of liquids containing small gas bubbles, Ann. Rev. Fl. Mech. (1972) 369-394.

16. Wallis, G.B. One dimensional two-phase flow, McGraw-Hill 1969.

17. Whitham, G.B. Linear and nonlinear waves, Wiley 1974.

Courant Institute of Mathematical Sciences
New York University
New York, NY 10012

Present address of Miksis:
Department of Mathematics
Duke University
Durham, NC 27706

Lectures in Applied Mathematics
Volume 23, 1986

STABILITY IN SYSTEMS OF CONSERVATION LAWS WITH DISSIPATION

Robert L. Pego[1]

Consider a hyperbolic system of conservation laws in one space dimension,

$$u_t + f(u)_x = 0, \ u \in \mathbb{R}^m \tag{1}$$

The 'viscosity method' for existence and uniqueness questions associated with solutions (usually discontinuous) of the Cauchy problem for this system is to consider the limit $\nu \to 0$ in a related parabolic system, which we write

$$u_t + f(u)_x = \nu(B(u)u_x)_x \tag{2}$$

Thus, for some fixed initial data $u_0(x)$, let $u^\nu(x,t)$ be the solution of (2) with

$$u^\nu(x,0) = u_0(x) \tag{3}$$

One would like to show that $u^\nu \to u$ as $\nu \to 0$ where u solves (1) and $u(x,0) = u_0(x)$.

Especially relevant examples of the "viscosity matrix" B are $B = I$ in general, and, say for the compressible Navier-Stokes equations of gas flow, $B(u)$ as determined by physical viscosity and heat conduction. (In that case, B is singular.)

1980 Mathematics Subject Classifications: 35K55
35L65

[1]This material was based upon work supported by the National Science Foundation under Grant No. DMS 84-01614.

How is the limit affected by the choice of B ? In particular: Can one identify those B for which one can expect the limit to exist? In that case, can one isolate a simple "entropy condition" which identifies those weak solutions of (1) which are viscous limits?

In [3] Majda and Pego identified and studied the class of constant matrices B yielding convergence of the viscosity method in L^2 for the constant coefficient equation linearized about $u = \bar{u}$,

$$u_t + Au_x = \nu Bu_{xx} \tag{4}$$

where A is the Jacobian matrix $f_u(\bar{u})$. When (1) is strictly hyperbolic, so that A has distinct real eigenvalues, one finds that generically such B are characterized, among other things, by the following condition:

(5)
A constant $c > 0$ exists so that
any eigenvalue κ of the symbol
$$P(\xi) = -\nu\xi^2 B - i\xi A$$
of equation (3) must satisfy
$\operatorname{Re} \kappa \leq -c|\xi|^2$ for all real ξ .

(Such a condition was also obtained by Smoller and Taylor [8].) Condition (5) is a linearized stability condition for (2) at $\bar{u}$ and when it is satisfied, the viscosity matrix B is called strictly stable (at $\bar{u}$). Majda and Pego also performed an analysis of the shock structure problem for (1) for small amplitude shocks. (When are simple shock wave solutions of (1) the viscous limit of smooth traveling wave solutions of (2)?) They showed that for all strictly stable viscosity matrices B , a single entropy condition due to T.-P. Liu selects those simple shocks which are viscous limits.

Thus the strictly stable viscosity matrices emerge as natural objects for study. In the remainder of this paper we will discuss a nonlinear problem naturally associated with

Condition (5), namely: Fix $\nu > 0$, for instance $\nu = 1$. Suppose B is strictly stable at $\bar{u}$, so the constant state is linearly stable by Condition (5). Is $\bar{u}$ nonlinearly stable to perturbations in equation (2)? That is, if $u(x,t)$ is a solution of (2) with $u(x,0) = \bar{u} + v_0(x)$ with v_0 small, does $u - \bar{u} \to 0$ as $t \to \infty$? (Without loss of generality, we may take $\bar{u} = 0$.)

For the compressible Navier Stokes equations in three space dimensions, this question was answered affirmitavely (for small perturbations in $H^4(\mathbb{R}^3)$) by Matsumura and Nishida in [4] and [5]. Related problems have been treated by Klainerman and Ponce [2] and Shatah [7]. Curiously, the present problem is more delicate for its being posed in one space dimension. (The decay rate is slower.) Nevertheless, one may answer in the affirmative and obtain a rate of decay for small, smooth solutions. We assume that $A = f_u(0)$ has in distinct real eigenvalues, and that $B(u)$ is smooth and strictly stable at $\bar{u} = 0$.

THEOREM 1. Fix $\nu = 1$, and let $u_0 \in L_1 \cap H^6$. There exist positive constants ε and C such that if

$$E_0 = \|u_0\|_{L^1 \cap H^6} < \varepsilon$$

then a solution $u(x,t)$ to the system (2) exists in $C([0,\infty), H^6)$ and satisfies, for some small $\sigma > 0$

i) $\|u(t)\|_{L^2} \leq CE_0\,(1+t)^{-1/4}$

ii) $\|u_x(t)\|_{L^2} \leq CE_0(1+t)^{-3/4 + 2\sigma}$

iii) $\|u(t)\|_{L^\infty} \leq CE_0(1+t)^{-1/2 + \sigma}$

REMARKS

1. Six derivatives of u_0 must be small in the proof I will outline, valid for smoothly varying $B(u)$. By invoking local regularity results for the parabolic system (2), fewer derivatives may be required.

2. The L^∞ decay rate iii) is a consequence of i) and ii), using the inequality

$$\|u\|_{L^\infty} \le C \|u\|_{L^2}^{1/2} \|u_x\|_{L^2}^{1/2}$$

which is a special case of the Gagliardo-Nirenberg inequality (see [1]).

We will give a fairly complete sketch of the proof, which is closely related to the techniques of Matsumura and Nishida [4]. The basic idea is to obtain a <u>time-uniform</u> bound in H^6 on existing solutions in terms of the size E_0 of the initial data, then to apply a local existence theorem and a simple induction argument to deduce global existence of the solution. The decay rates i) and ii) are 'built in' to the time-uniform bound and follow as a consequece of global existence.

<u>PROPOSITION 1</u> (Time-uniform bound)

There exist positive ε_1 and C_1 independent of T such that: If a (smooth) solution $u(x,t)$ of (2) exists on $\mathbb{R} \times [0,T]$ satisfying

$$\|u(t)\|_{H^6} < \varepsilon_1 \quad \text{for} \quad 0 \le t \le T \ ,$$

then in fact

$$\|u(t)\|_{H^6} \le C_1 \|u_0\|_{L^1 \cap H^6} \quad \text{for} \quad 0 \le t \le T \ .$$

The ingredients of the proof of the time-uniform bound are:

1) Standard energy estimates for $u_x(t)$ in H^5. Here a bound for $\|u_x(t)\|^2_{H^5}$ is obtained in terms of only $\|u_x(0)\|^2_{H^5}$ and $\int_0^t \|u_x(s)\|^2_{L^2}\,ds$.

2) Decay estimates in L^2 for $u(t)$, $u_x(t)$. These estimates bound the time integral above and complete the time-uniform bound on $\|u(t)\|_{H^6}$. They are based on decay rates for the linearized equation (4) and the variation of parameters expression for the solution. Because of the slow rate of decay in one dimension, it is necessary to estimate decay rates for u_{xx}, u_{xxx} and u_{xxxx}. The estimates involve nonlinear terms with two more derivatives than that, which accounts for the use of six derivatives. Below, we use the notation $D^j u$ for $\partial^j u/\partial x^j$.

PROPOSITION 2 (Energy estimates)

There exist positive ε_2 and C_2 such that a (smooth) solution $u(x,t)$ of (2) on $\mathbb{R}\times[0,T]$ satisfying

$$\|u(t)\|_{H^6} < \varepsilon_2$$

must satisfy, for $0 \le t \le T$,

$$(E_j)\qquad \|D^j u(t)\|^2 + \int_0^t \|D^{j+1}u(s)\|^2\,ds \le$$

$$\le C_2\left[\|D^j u(0)\|^2 + \int_0^t \|D^j u(s)\|^2\,ds\right]$$

For the proof, we first make a linear change of coordinates so that $(u,B(c)u) \ge c|u|^2$ for some $c > 0$ and all vectors $u \in \mathbb{R}^m$. This can be done since (2) is parabolic. Below, integrals are taken with respect to x unless otherwise indicated. Integrating by parts we necessary, we calculate

$$\frac{d}{dt}\int \frac{1}{2}\, D^j u(t)\ ^2 = \int (D^j u, D^j (B(u)u_x - f(u))_x)$$
$$= - \int (D^{j+1}u, B(o)D^{j+1}u)$$
$$+ \int (D^{j+1}u, D^j((B(o)-B(u))u_x + f(u)))$$
$$- c\int |D^{j+1}u|^2 + \frac{c}{2}\int |D^{j+1}u|^2$$
$$+ \frac{1}{4c}\int |D^j((B(o)-B(u))u_x)|^2$$
$$+ \int |D^j f(u)|^2$$

Using Moser's calculus inequalities (see (A 1-2) of the Appendix), and the presumed bound on $\|u\|_{L^\infty}$ implied by ε_2, this is

$$- \frac{c}{2}\int |D^{j+1}u|^2 + C\left[\|u\|^2_{L^\infty}\|D^{j+1}u\|^2 + \|u_x\|^2_{L^\infty}\|D^j u\|^2\right]$$
$$+ C\ \ D^j u\ ^2$$

So if $\|u(t)\|_{L^\infty}$ is sufficiently small and $\|u_x\|_{L^\infty}$ is bounded uniformly on $[0,T]$, we may integrate from 0 to t to obtain (E_j).

In order to obtain the time-uniform bound of Proposition 1, we will bound $\int_0^t \|D^j u(s)\|^2\, ds$ by initial data for appropriate j, namely $j = 1$, because from linear analysis (below) we expect $\|u_x(t)\| \leq CE_0(1+t)^{-3/4}$, which is square-integrable in time. Combining $(E_1) + C_2(E_2) + \ldots + C_2^5(E_6)$ we may then write

$$\|u_x(t)\|^2_{H^5} \leq C(\|u_x(0)\|^2_{H^5} + \int_0^t \|u_x(s)\|^2\, ds \tag{6}$$

To obtain decay estimates, we introduce depending on a solution $u(x,t)$ to (2) on $\mathbb{R}\times[0,T]$, a quantity

$$M(T) = \max_{0\leq j\leq 4}\ \sup_{0\leq t\leq T}\ (\|D^j u(t)\|\,(1+t)^{\alpha_j}) \tag{7}$$

where as it turns out, $\alpha_j = \frac{1}{4}, \frac{3}{4} - 2\sigma, 1 - 2\sigma, \frac{1}{2} - \sigma, \frac{1}{2} - \sigma$ for $j = 0,1,2,3,4$ respectively, for some fixed $\sigma > 0$ sufficiently small. For convenience, we also define

$$E(T) = \sup_{0 \le t \le T} \|u(t)\|_{H^6}$$

PROPOSITION 3 (Decay estimates)

There exist positive ε_3 and C_3 so that if a (smooth) solution $u(x,t)$ to (2) exists on $\mathbb{R} \times [0,T]$ satisfying

$$E(T) < \varepsilon_3$$

then

$$M(T) \le C_3(E_0 + M^2 + ME)$$

Hence if ε_3 is sufficiently small,

$$M(T) \le 2C_3 E_0 .$$

Starting with a variation of parameters formula for the solution, the parts of the proof are:

a) Establishing L^1 to L^2 decay rates for solutions of the linearized equation (4) .

b) Estimating nonlinear terms via Moser inequalities, then in terms of M^2 and ME .

c) Establishing that the constants arising in these estimates, which are integrals over time, are independent of T .

To begin, we write equation (2) in the form

$$u_t = Lu + N(u)_x \tag{8}$$

where

$$Lu = B(0)u_{xx} = Au_x$$

$$N(u) = (B(u)-B(0))u_x - (f(u) - Au)$$

(We may assume $f(0) = 0$.) Denote the solution of the initial value problem

$$(9)\qquad \begin{aligned} v_t &= Lv \\ v(x,0) &= v_0(x) \end{aligned}$$

at time t by $v(t) = S(t)v_0$. Then the solution to equation (2) or (8) with initial value $u_0(x)$ may be expressed by the variation of parameters formula

$$(10)\qquad u(t) = S(t)u_0 + \int_0^t S(t-s)N(u(s))_x\, ds$$

Lemma 1 (Smooth) solutions of the linear equation (9) satisfy the decay estimates

$$(11)\qquad \| D^j v(t) \|_{L^2} \le C(1+t)^{-1/4 - j/2}(\| v_0 \|_{L^1} + \| D^j v_0 \|_{L^2})$$

A sketch of the proof appears in the Appendix. Applying these decay estimates to the variation of parameters formula, we find

$$(12)\qquad \| D^j u(t) \|_{L^2} \le C(1+t)^{-1/4 - j/2}(\| u_0 \|_{L^1} + \| D^j u_0 \|_{L^2}) + C \int_0^t (1+t-s)^{-1/4 - j/2}(\| N(u(s))_x \|_{L^1} + \| D^{j+1} N(u(s)) \|_{L^2})\, ds$$

(This estimate will not be used for $j = 0$. See below.)

Lemma 2 The nonlinear term $N(u)$ satisfies the estimates (assuming $\| u \|_{L^\infty} \le C$)

$$\| N(u)_x \|_{L^1} \le C \| u \|_{H^1} \| Du \|_{H^1}$$

$$\| D^j N(u) \|_{L^2} \le C \| u \|_{1,\infty} \| D^j u \|_{H^1}$$

$$\| N(u) \|_{L^1} \le C \| u \|_{L^2} \| u \|_{H^1}$$

Here $\|u\|_{1,\infty} = \|u\|_{L^\infty} + \|Du\|_{L^\infty}$. The proof of this lemma is based on Moser's inequalities (see Appendix). Thus

$$\|D^j u(t)\|_{L^2} \le C(1+t)^{-1/4 - j/2}(\|u_0\|_{L^1} + \|D^j u_0\|_{L^2})$$
$$+ C\int_0^t (1+t-s)^{-1/4 - j/2}(\|u\|_{H^1}\|Du\|_{H^1}$$
$$+ \|u\|_{1,\infty}\|D^{j+1}u\|_{H^1})(s)\, ds .$$

We will estimate these last terms in terms of M^2 for $j = 1$ and 2 and in terms of $M^2 + ME$ for $j = 3$ and 4. From the inequalities

$$\|D^j u(s)\| \le C(1+s)^{-\alpha_j} M(T) \quad \text{for} \quad s \le T$$
$$\|u\|_{L^\infty} \le C(\|u\|_{L^2}\|Du\|_{L^2})^{1/2}$$

we obtain

$$\|u\|_{1,\infty} \le C(1+w)^{-1/2 + \sigma}M(T) \text{ for } s \le T .$$

For $j = 1$ and 2, then, we obtain, for $t \le T$,

$$(1+t)^{\alpha_j}\|D^j u(t)\| \le CE_0 +$$
$$+ CM^2(T)\int_0^t (1+t)^{\alpha_j}(1+t-s)^{-1/4 - j/2}((1+s)^{-1/4 - 3/4 + 2\sigma}$$
$$+ (1+s)^{(1/2 - \sigma)2})\, ds .$$

For $j = 3$ and 4,

$$(1+t)^{1/2 - \sigma}\|D^j u(t)\| \le CE_0 +$$
$$+ C\int_0^t (1+t)^{1/2 - \sigma}(1+t-s)^{-1/4 - j/2}[(1+s)^{-1}M^2(T) +$$
$$+ (1+s)^{-1/2 + \sigma} M(T)E(T)]\, ds$$

In terms of the quantity

$$(13) \qquad I(\alpha,\beta,\gamma)(t) = (1+t)^{\alpha} \int_0^t (1+t-s)^{-\beta}(1+s)^{-\gamma}\, ds$$

these bounds read

$$(1+t)^{\alpha_j} \| D^j u(t) \| \le CE_0 + CI(\alpha_j, \tfrac{1}{4}, \tfrac{j}{2}, 1-2\sigma)(t) M^2(T) \qquad j = 1,2$$

$$(14) \quad (1+t)^{1/2-\sigma} \| D^j u(t) \|_{L^2} \le CE_0 + CI(\tfrac{1}{2}-\sigma, \tfrac{1}{4}+\tfrac{j}{2}, 1-2\sigma)(t) M^2(T)$$

$$+ CI(\tfrac{1}{2}-\sigma, \tfrac{1}{4}+\tfrac{j}{2}, \tfrac{1}{2}-\sigma)(t) M(T) E(T)$$

$$j = 3,4$$

Lemma 3 Let α, β, γ be positive. In order that $I(\alpha,\beta,\gamma)(t) \le C$ independent of $t > 0$, it is necessary that $\alpha \le \min\{\beta,\gamma,\beta+\gamma-1\}$ and necessary and sufficient that, in addition, if $\beta = 1$ or $\gamma = 1$, then $\alpha < \beta+\gamma-1$.

This lemma is proved using the simple estimates $(1+t)/2 < 1+s < 1+t$ for $s \in (t/2,t)$, and $(1+t)/2 < 1+t-s < 1+t$ for $s \in (0,t/2)$. Applying Lemma 3 to (14) with values of α_j as specified in (7), it follows that

$$(15) \qquad \sup_{0 \le t \le T} (1+t)^{\alpha_j} \| D^j u(t) \| \le CE_0 + C(M^2 + ME)(T)$$

for $j = 1,2,3,4$. The estimate for $j = 0$ utilizes the special fact that $S(t-s)D\,N(u(s) = DS(t-s)N(u(s))$. Thus we directly estimate (10) by

$$\| u(t) \|_{L^2} \le C(1+t)^{-1/4} \| u_0 \|_{L^1 \cap L^2} +$$

$$+ C \int_0^t (1+t-s)^{-3/4} \| N(u(s)) \|_{L^1 \cap L^2}\, ds$$

Applying Lemmas 2 and 3 we obtain

$$(16) \qquad \begin{aligned} (1+t)^{1/4} \| u(t) \| &\le CE_0 + CI(\tfrac{1}{4}, \tfrac{3}{4}, \tfrac{1}{2}) M^2(T) \\ &\le CE_0 + CM^2(T) \quad \text{for } t \le T . \end{aligned}$$

Combining (15) and (16), Proposition 3 is proved. The time uniform bound of Proposition 1 then follows from (6). The proof of Theorem 1 now rests on a local existence theorem and an induction argument.

Theorem 2 (Local existence)

Let $u_0 \in H^m$ where $m \geq 6$. Given $M > 0$, there exist positive T_0 and C_m so that if $\|u_0\|_{H^6} < M$, then a solution $u(x,t)$ to the system (2) exists in $C([0,T_0],H^m)$ satisfying

$$\|u(t)\|_{H^m} \leq C_m \|u_0\|_{H^m} \quad \text{for} \quad 0 \leq t \leq T_0$$

Furthermore, the solution $u(t)$ depends continuously on u_0 in H^m .

We will not give a proof of this theorem here. By continuous dependence, the time-uniform bound of Proposition 1 holds for all solutions in $C([0,T],H^6)$.

We conclude by sketching the induction argument for Theorem 1. Take ε_1, C_1 from Proposition 1 and C_6 from Theorem 2. Let $\varepsilon = \varepsilon_1/C_1C_6$. Assume $\|u_0\|_{L^1 \cap H^6} < \varepsilon$ and that the solution $u(t)$ of (2) exists on $[0,T]$ satisfying $\|u(t)\| \leq C_1\varepsilon$. (E.g., this is true with $T = 0$.) Then by local existence the solution may be continued to the interval $[0,T+T_0]$ satisfying $\|u(t)\|_{H^6} \leq C_6C_1\varepsilon < \varepsilon_1$. But the time independent bound guarantees $\|u(t)\|_{H^6} \leq C_1\varepsilon$ on $[0,T+T_0]$. Global existence with $\|u(t)\|_{H^6} \leq C_1\varepsilon$ follows.

Appendix

Lemma 2 follows from Moser's inequalities [6]

$$\text{(A.1)} \qquad \|D^j(fg)\|_{L^r} \leq C(\|f\|_{L^p} \|D^j g\|_{L^q} + \|D^j f\|_{L^q} \|g\|_{L^p})$$

where $\frac{1}{r} = \frac{1}{p} + \frac{1}{q}$, and

(A.2) $$\| D^j H(u) \|_{L^2} \le CC_0 \| D^j u \|_{L^2}$$

so long as $(1 + \| u \|_{L^\infty})^{j-1} \sum_{k=1}^{j} \sup_u |D_u^k H| < C_0$.

Write the nonlinear term in the form $N(u) = H_1(u)u_x + H_2(u)u$ where $H_i(u)$ is a smooth matrix with $H_i(0) = 0$, and apply Moser's inequalities.

Lemma 1 may be proved using the Fourier transform: Let $P(\xi) = -\xi^2 B - i\xi A$ be the symbol of equation (4) or (9) . The solution of (9) satisfies $D^j v(\xi,t) = (i\xi)^j e^{tP(\xi)} \hat{v}_0(\xi)$. Because B is strictly stable, it may be shown that

$$|e^{tP(\xi)}| \le Ce^{-c\xi^2 t} \quad \text{for all real } \xi \ .$$

This follows easily from the analysis in the proof of Theorem 2.1 of [3] , which characterizes strictly stable viscosity matrices. Then

$$\| D^j v(t) \|_{L^2}^2 \le C \int \xi^{2j} e^{-2c\xi^2 t} d\xi \, \| \hat{v}_0 \|_{L^\infty}^2$$
$$\le Ct^{-1/2-j} \| v_0 \|_{L^1}^2 \ .$$

Use this inequality for $t \ge 1$ and the inequality $\| D^j v(t) \|_{L^2} \le C \| D^j v_0 \|_{L^2}$ for $t \le 1$ to establish Lemma 1 .

Acknowledgement. The author thanks T. Beale and T. Nishida for many discussions.

BIBLIOGRAPHY

1. A. Friedman, Partial Differential Equations, Holt, Rinehart and Winston, New York, 1969.

2. S. Klainerman and G. Ponce, "Global small amplitude solutions to nonlinear evolution equations", Comm. Pure Appl. Math. 36 (1983), 133-141.

3. A. Majda and R. Pego, "Stable viscosity matrices for systems of conservation laws", to appear in J. Diff. Eqns.

4. A. Matsumura and T. Nishida, "The initial value problem for the equations of motion of compressible viscous and heat-conductive fluids", Proc. Jap. Acad. 55A (1979), 337-342.

5. A. Matsumura and T. Nishida, "The initial value problem for the equations of motion of viscous and heat conductive gases", J. Math. Kyoto Univ. 20 (1980), 67-104.

6. J. Moser, "A rapidly convergent iteration method and nonlinear partial differential equations I", Ann. Scuola Norm. Sup. Pisa 20 (1966), 265-315.

7. J. Shatah, "Global existence of small solutions to nonlinear evolution equations", J. Diff. Eqns. 46 (1982), 409-425.

8. J. Smoller and M. Taylor, "Wave front sets and the viscosity method", Bull. Amer. Math. Soc. 79 (1973), 431-436.

DEPARTMENT OF MATHEMATICS
UNIVERSITY OF MICHIGAN
ANN ARBOR, MICHIGAN 48109

Lectures in Applied Mathematics
Volume 23, 1986

ON THE CONVERGENCE OF THE VISCOSITY METHOD FOR THE SYSTEM OF NON-LINEAR (1-D) ELASTICITY

Michel Rascle

ABSTRACT. We consider the particular hyperbolic system of non-linear (1D) elasticity (resp. visco-elasticity)

$$(S)\quad \begin{cases} u_t-(\sigma(v))_x=o, \sigma'(v) > o \\ v_t-u_x=o \end{cases} \qquad (\text{resp.} (S_\varepsilon) \begin{cases} u^\varepsilon_t-(\sigma(v^\varepsilon))_x=\varepsilon u^\varepsilon_{xx}, \varepsilon \to o_+ \\ v^\varepsilon_t-u^\varepsilon_x=o \end{cases}$$

Let $w^\varepsilon=w(u^\varepsilon,v^\varepsilon), z^\varepsilon=z(u^\varepsilon,v^\varepsilon)$ be the Riemann Invariants. For the system (S), we have the following conjecture :

$$(C)\qquad w^\varepsilon \rightharpoonup w^*, \; z^\varepsilon \rightharpoonup z^*, \text{ and } w^\varepsilon z^\varepsilon \rightharpoonup w^* z^* (\text{weakly in some } L^p \text{ space})$$

which is a (difficult) conjecture of "compensated compactness with variable coefficients". We prove that, if this conjecture is true, then the viscosity method is convergent when $\varepsilon \to o_+$. The proof simply consists in exhibiting a sequence of negative functions (due to the Cauchy - Schwarz inequality) whose weak limit is positive : therefore the limit is zero and the convergence is strong. Moreover, the linearly degenerate case is simply the case of equality in the Cauchy - Schwarz inequality.

1980 Mathematics Subject classification 35L65.

1 - INTRODUCTION

We are concerned with the general problem of the convergence of approximate solutions to 2x2 quasi-linear hyperbolic systems of conservation laws. We consider here the viscosity method, which approximates the Cauchy Problem :

$$(\mathcal{P}) \quad \begin{cases} u_t+(f(u))_x = o & (1.1) \\ u(x,o)=u_o(x) \quad , \; u:Q = \mathbb{R} \times \mathbb{R}_+ \to \mathbb{R}^N & (1.2) \end{cases}$$

by

$$(\mathcal{P}_\varepsilon) \quad \begin{cases} u^\varepsilon_t+(f(u^\varepsilon))_x=\varepsilon Du^\varepsilon_{xx} & (1.1_\varepsilon) \\ u^\varepsilon(x,o)=u_o(x) & (1.2) \end{cases}$$

where $\varepsilon \to o_+$, D is a positive diffusion matrix, and $f:\mathbb{R}^N \to \mathbb{R}^N$ is a smooth function.

Let us recall that the system (1.1) is said strictly hyperbolic if all the eigenvalues of the Jacobian matric $A(u)=f'(u)$ are real and distinct for all u, and that an eigenvalue $\lambda_k(u)$ of $f'(u)$ associated to a right eigenvector $r^k(u)$ is said linearly degenerate iff $\nabla\lambda_k(u).r^k(u) \equiv o$, and genuinely non linear iff, on the contrary, $\forall u, \nabla\lambda_k(u).r^k(u)\neq o$.

Here and in the sequel, $\nabla = (\frac{\partial}{\partial u_1}, --, \frac{\partial}{\partial u_N})$ is the gradient vector in the phase space, and the dot denotes the scalar product in $\mathbb{R}^N$.

Let us recall also that a scalar function $w_k(u)$ is called a Riemann Invariant of system (1.1), in the sense of Lax, associated to λ_k, iff

$$\forall u, \; \nabla w_k(u).r^k(u)=o \qquad (1.3)$$

For 2x2 system, this implies that $\nabla w_k(u)$ is a left eigenvector of $f'(u)$, associated to $\lambda_j(u)$, $j=3-k$. Multiplying both members of (1.1) by $\nabla w_k(u)$, we then obtain :

$$\frac{\partial}{\partial t} w_k(u)+\lambda_j(u) \frac{\partial}{\partial x} w_j(u)=o, \ j=3-k \tag{1.4}$$

for smooth solutions of (1.1). Finally, a couple of scalar functions $(\eta(u,q(u))$ is a pair <u>(entropy, flux)</u> iff

$$\nabla\eta(u)f'(u) \equiv \nabla q(u) \tag{1.5}$$

which implies

$$\frac{\partial}{\partial t} \eta(u) + \frac{\partial}{\partial x} q(u) = o \tag{1.6}$$

for smooth solutions of (1.1). It is well known that the classical <u>Entropy Condition</u> for Lax [13] requires that an admissible weak solution of (1.1) satisfies

$$\frac{\partial}{\partial t} \eta(u) + \frac{\partial}{\partial x} q(u) \leqslant o \tag{1.6'}$$

for all convex entropy (if there exists) associated to the flux q. (This condition is formally justified by the viscosity method).

The crucial point to prove the convergence of a sub-sequence (u^ε) to an admissible solution of $(\mathscr{P})$ is classically to take the weak-limit of non-linear tems : assuming for simplicity that the family $\{u^\varepsilon\}$satisfies a uniform (w.r.t.ε)L^∞ estimate (1), we deduce that, at least for a sub-sequence,

(1) Actually, the only case where one can prove such an estimate is the very particular case where (1.1) admits convex <u>bounded</u> invariant regions defined by the Riemann invariants of (1.1) see chueh, Conley and Smoller [ ].

$$u^{\varepsilon} \rightharpoonup u \ , \ f(u^{\varepsilon}) \rightharpoonup f^{*} \text{ in } (L^{\infty}(Q))^{N} \text{ weak-star} \tag{1.7}$$

and the problem is obviously to prove that

$$f^{*} = f(u) \tag{1.8}$$

The classical way to prove (1.8) is the compactness method, see e.g. Lions [14]. For hyperbolic systems, unfortunately, it is and open question to obtain a BV-estimate on the family $\{u^{\varepsilon}\}$, and therefore the compactness method fails for this problem.

Thus, during several years, there was no progress on the viscosity method, and the only approximate method known to be convergent was the Glimm scheme (with small initial data), precisely because, in his celebrated Theorem, J. Glim [8] was able to prove a BV estimate for the family of approximate solutions he constructed.

However, in the late seventies, an alternative approach, based on the compensated compactness, was proposed by L. Tartar [20] : in some cases, on can prove that any pair (entropy, flux) of system (1.1) satisfies :

$$\frac{\partial}{\partial t}\eta(u^{\varepsilon}) + \frac{\partial}{\partial x} q(u^{\varepsilon}) \in \text{compact sef of } H^{-1}(Q), \tag{1.9}$$

(where $H^{-1}(Q)$ is the dual space of the Sobolev space $H^{1}_{0}(Q)$ and is equipped with its strong topology). Then if (η', q') is another pair, the compensated compactness implies :

$$\eta(u^{\varepsilon})q'(u^{\varepsilon}) - \eta'(u^{\varepsilon})q(u^{\varepsilon}) \rightharpoonup \eta^{*}q'^{*} - \eta'^{*}q^{*} \tag{1.10}$$

in the distribution sense, where η^{*}, q'^{*} etc... are obviously the weak limits of subsequences $\eta(u^{\varepsilon})$, $q'(u^{\varepsilon})$ etc...

Let us express (1.10) in terms of Young measures, see Young [21], Tartar [20]. If, for any $\varepsilon > 0$, u^ε takes its values in a compact set K of $\mathbb{R}^N$, then there exists a sub-sequence (u^ε) and a family of probability measures $\nu_{x,t}$ on $\mathbb{R}^N$, with supports in the closure $\overline{K}$, such that

$$\forall g \in \mathcal{C}(\mathbb{R}^N; \mathbb{R}),\ g(u^\varepsilon) \longrightarrow g^* \text{ in } L^\infty(Q) \text{ weak star}$$

with

$$g^*(x,t) = \langle \nu_{x,t}, g(.) \rangle \text{ a.e. in } Q \tag{1.11}$$

With these notations, we can rewrite (1.10) under the form

$$\langle \nu_{x,t}, \eta q' - \eta' q \rangle = \langle \nu_{x,t}, \eta \rangle . \langle \nu_{x,t}, q' \rangle - \langle \nu_{x,t}, \eta' \rangle . \langle \nu_{x,t}, q \rangle \tag{1.12}$$

Now, a 2x2 strictly hyperbolic system of conservation laws admits a very large family of pairs (η,q). Therefore, this family of equations (1.12) provides much information about the measure $\nu_{x,t}$, and one can hope that this information is sufficient to identify $\nu_{x,t}$ and (at least in the genuinely non linear case) to prove that actually

$$\nu_{x,t} = \delta_{u(x,t)} \text{ (a point-mass)} \tag{1.13}$$

which is equivalent to the strong convergence (since $\forall i=1,\ldots,N$, $u_i^\varepsilon \rightarrow u_i$ and $(u_i^\varepsilon)^2 \rightarrow (u_i)^2$).

This program was proposed by L. Tartar [20], and performed by R.J. Di Perna for a general 2x2 strictly hyperbolic and genuinely non linear system [6], and for the (non strictly hyperbolic) isentropic gas dynamics system [7].

It is quite natural that the genuine non linearity of the system plays an important part, since it is quite easy, in the linearly degenerate case, to construct periodic oscillating solutions of (1.1), which converge weakly, but not strongly[(1)].

Here, our approach is different : we consider the particular system of non-linear 1D-elasticity :

$$(S)\qquad \begin{cases} u_t - (\sigma(v))_x = o \quad , \quad \sigma'(v) \geqslant c > o \text{ (strict hyperbolicity)} \\ \\ v_t - u_x = o \end{cases}$$

that we approximate by visco-elasticity

$$(S\ )\qquad \begin{cases} u_t^\varepsilon - (\sigma(v^\varepsilon))_x = \varepsilon\, u_{xx}^\varepsilon \ , \quad \varepsilon \to o_+ \\ \\ v_t^\varepsilon - u_x^\varepsilon = o \end{cases}$$

For this system, we have a conjecture

(C) The Riemann Invariants $w^\varepsilon = w(u^\varepsilon, v^\varepsilon)$ and $z^\varepsilon = z(u^\varepsilon, v^\varepsilon)$ of system (S) satisfy
$w^\varepsilon \rightharpoonup w^*$, $z^\varepsilon \rightharpoonup z^*$ in a weak sens <u>and</u>
$w^\varepsilon . z^\varepsilon \rightharpoonup w^* . z^*$

This is a conjecture of "<u>compensated compactness with variable directions</u>", since w^ε and z^ε satisfy

$$\begin{cases} \dfrac{\partial w^\varepsilon}{\partial t} + (\sigma'(v^\varepsilon))^{1/2} \dfrac{\partial w^\varepsilon}{\partial x} = \varepsilon\, u_{xx}^\varepsilon \in \text{compact sef of } H^{-1}(Q) \\ \\ \dfrac{\partial z^\varepsilon}{\partial t} - (\sigma'(v^\varepsilon))^{1/2} \dfrac{\partial z^\varepsilon}{\partial x} = \varepsilon\, u_{xx}^\varepsilon \in \text{compact sef of } H^{-1}(Q) \end{cases} \qquad (1.14)$$

(1) Actually, these solutions are note solutions of the Cauchy Problem : the initial data preclude such periodic oscillations.

We shall prove that, if this conjecture is true, then a subsequence $(u^\varepsilon, v^\varepsilon)$ converges weakly to a weak solution (u,v) of (S) for which, if there is no non-trivial interval on which σ is linear, the total energy :

$$E(t)=\int\left(\frac{(u^\varepsilon)^2}{2}+\Sigma(v^\varepsilon)\right)(x,t)dx,\ \Sigma(v)=\int_0^v\sigma(s)ds \qquad (1.15)$$

is a decreasing function of the time.

The scheme of the paper is the following : after this Introduction, and some reminders about system (S) in Section 2, we recall the (classical) a-priori estimates and the conjecture (C) in Section 3, finally we give in Section 4 the results of convergence implied by (C).

We don't give here all the details, for which we refer to [15] [16]. For a discussion of the conjecture (C) and a larger conjecture for a general 2x2 hyperbolic system of conservation laws, we refer to Serre [19], Rascle [16], Rascle and Serre [17].

2 - THE SYSTEM (S)

This well-known system describes the longitudinal vibrations in a rigid bar : let $y(x,t)$ be the position at time t of material point whose initial position was x, let $u=y_t$ be the velocity and $v=y_x-1$ the relative elongation. Then if the tension $\mathcal{T}$ is given by

$$\mathcal{T} = \sigma(v) \quad , \quad \sigma'(v) > o \qquad (2.1)$$

(non-linear elasticity) or by

$$\mathcal{T} = \sigma(v) + \varepsilon v_t \quad , \quad \varepsilon > o \qquad (2.2)$$

(visco-elasticity), and if, for simplicity, we neglect the external forces and assume that the initial density is $\rho_o \equiv 1$, the fundamental law of mechanics can be written

$$u_t - \mathcal{T}_x = o$$

which, with (2.1) (resp.(2.2)) provides system (S) (resp.S_ε).

For simplicity, we consider the case where both ends of the bar are fixed : then the initial boundary value problem is

$$(\mathcal{P})\left\{\begin{array}{lll} (S)\left\{\begin{array}{l} u_t-(\sigma(v))_x=o \\ v_t-u_x=o \end{array}\right. & \text{in } Q=]0,L[\times]0,T[=\Omega\times]0,T[& \\ u|_\Sigma = o & \text{on } \Sigma=\Gamma\times]0,T[\,;\ \Gamma=\partial\Omega & (2.3) \\ (u(x,o),v(x,o))=(u_o(x),v_o(x))\ (u_o(x),o) \text{ in } \Omega & & (2.4) \end{array}\right.$$

which we approximate by

$$(\mathscr{P}_\varepsilon)\begin{cases} (S_\varepsilon)\begin{cases} u^\varepsilon_t - (\sigma(v^\varepsilon))_x = \varepsilon u^\varepsilon_{xx} \\ v^\varepsilon_t - u^\varepsilon_x = o \end{cases} & \text{in } Q \\ u^\varepsilon = o & \text{on } \Sigma \quad (2.3) \\ (u^\varepsilon(x,o),v^\varepsilon(x,o))=(u_o(x),v_o(x)) \ (u_o(x),o) & \text{in } \Omega \quad (2.4) \end{cases}$$

We assume that $\sigma'(v) \geqslant c > o$ ("uniform" hyperbolicity) and that $\sigma(o)=o$. Then the eigenvalues satisfy

$$\lambda_1 \equiv \lambda_1(v) = -(\sigma'(v))^{1/2} \leqslant -c < o < +c \leqslant \lambda_2(v) = +(\sigma'(v))^{1/2} \qquad (2.5)$$

the Riemann invariants are

$$w_1 \equiv w = u - g(v), w_2 \equiv z = u + g(v), g(v) = \int_o^v (\sigma'(s))^{1/2} ds \qquad (2.6)$$

λ_1 and λ_2 are linearly degenerate iff $\sigma''(v)$ o and genuinely non linear iff, on the contrary $\forall v, \sigma''(v) \neq o$.

Finally, the total energy

$$\eta = \frac{u^2}{2} + \Sigma\,(v),\ \Sigma\,(v) = \int_o^v \sigma(s)ds \geqslant c\,\frac{v^2}{2} \qquad (2.7)$$

is a particular entropy of system (S), associated to the flux $q = -u\sigma(v)$.

Remark :

1) For the solution of Problem $(\mathscr{P}_\varepsilon)$, see e.g. Greenberg, Mac Camy and Mizel [9], Andrews [1], Dafermos [5], Kim [11].

2) The same problem is also the (totally unrealistic!) simplified version of the dynamics of cables, see e.g. Cristescu [4], Keyfitz-Kranzer [10]... or the p-system of isentropic gas dynamics in Lagrangian-mass coordinates, see Courant-Friedrichs [3]...

For sake of concreteness, we shall assume that
(H) $\sigma'(v)=|v|^p+\sigma_1'(v)$, $\sigma_1'(v)=o(|v|^p)(v \to \infty), p > o$. We shall use the following Cauchy-Schwarz inequality

$$(g(a)-g(b))^2 \leqslant (a-b)(\sigma(a)-\sigma(b)) \tag{2.8}$$

with equality iff σ is linear between a and b, i.e. iff system (S) is <u>linearly degenerate</u> for v between a and b.

It is easy to prove the following Lemma [15].

LEMMA 1 :

<u>The assymption (H) implies</u>

i) <u>$\forall\alpha > o, \exists c < 1, \exists M / \forall b$ such that</u> $|b| \geqslant M, \forall a / |\frac{a}{b} - 1| \geqslant \alpha$,

$$\underline{(g(a)-g(b))^2 \leqslant c(\alpha(a)-\alpha(b))(a-b)} \tag{2.9}$$

ii) <u>$\forall\beta > o, \exists\alpha > o, \exists M / \forall (a,b)$ such that</u> $|b| \geqslant M$ and $|\frac{a}{b} - 1| \leqslant \alpha$,

$$\underline{(g(a)-g(b))^2 \leqslant \beta(\Sigma(a)+\Sigma(b))} \tag{2.10}$$

(The property (i) means that σ is non-linear at infinity, since if e.g. σ was linear out of a compact set, one should take c=1).

This lemma will be useful to prove the strong convergence of a sub-sequence $(u^\varepsilon, v^\varepsilon)$ to an "admissible" weak solution (u,v) of Problem ( ) (in the genuinely-non linear case), even without L^∞ estimate.

3 - THE A-PRIORI ESTIMATES - THE CONJECTURE

We shall use the following classical spaces $L^q(L^p)=L^q(0,T;L^p(\Omega)), 1 \leqslant p,q \leqslant +\infty$; $W^{1,p}(\Omega)=\{u \in L^p(\Omega)$ /

$\frac{\partial u}{\partial x} \in L^p(\Omega)\}$; $W_o^{1,p}(\Omega)=\{u \in W^{1,p}(\Omega)/u|_\Gamma=o\}$; $W^{-1,p}(\Omega)$ its dual space... all equipped with their usual norms ; $\mathcal{D}(\Omega)$ is the space of test functions, $\mathcal{D}'(\Omega)$ the space of distributions and $\mathfrak{M}(\Omega)$ the space of Radon measures on Ω etc ...

The following results are classical :

PROPOSITION 1 :

Suppose that u_o is smoth and satisfies the compatibility conditions (see e.g. [12]) at a suitable order, for the Problem $(\mathscr{P}_c)$. Then, for any fixed $c > o$,

1) The Problem $(\mathscr{P}_\varepsilon)$ admits a unique smooth solution $(u^\varepsilon, v^\varepsilon)$, globally defined for all time, which satisfies, under the assymption (H) :

$$\|u^\varepsilon\|_{L^\infty(L^2)}+\|\Sigma(v^\varepsilon)\|_{L^\infty(L^1)}+\|g(v^\varepsilon)\|_{L^\infty(L^2)}+$$
$$+\|w^\varepsilon\|_{L^\infty(L^2)}+\|z^\varepsilon\|_{L^\infty(L^2)}+\|\sigma(v^\varepsilon)\|_{L^\infty(L^{(p+2)/(p+1)})} \leqslant c \tag{3.1}$$

(c denotes several constants, independent of ε)

2) Moreover, we have

$$\frac{\partial}{\partial t}\left(\frac{(u^\varepsilon)^2}{2}+\Sigma(v^\varepsilon)\right)+\frac{\partial}{\partial x}(-u^\varepsilon\sigma(v^\varepsilon))=\varepsilon\left(\frac{(u^\varepsilon)^2}{2}\right)_{xx}-\varepsilon(u_x^\varepsilon)^2=$$
$$=\alpha^\varepsilon+\beta^\varepsilon \tag{3.4}$$

which implies

$$\|\sqrt{\varepsilon}\, u_x^\varepsilon\|_{L^2(L^2)} \leqslant c \tag{3.5}$$

$$\alpha^\varepsilon = \left(\varepsilon \frac{(u^\varepsilon)^2}{2}\right)_{xx} \to o \text{ in } \mathcal{D}'(Q), \quad \beta^\varepsilon \leqslant o, \|\beta^\varepsilon\|_{L^\infty(L^1)} \leqslant c \tag{3.6}$$

$$\frac{\partial w^\varepsilon}{\partial t} + (\sigma'(v^\varepsilon))^{1/2} \frac{\partial w^\varepsilon}{\partial x} = \frac{\partial z^\varepsilon}{\partial t} - (\sigma'(v^\varepsilon))^{1/2} \frac{\partial z^\varepsilon}{\partial x} = \varepsilon\, u_{xx}^\varepsilon \to o \tag{3.7}$$

<u>in $H^{-1}(\Omega) = W^{-1,2}(\Omega)$ strong.</u>

The proof is easy, see e.g. the previously mentioned references. The last estimates are based on (3.4), which simply expresses that $\eta = \frac{u^2}{2} + \Sigma(v)$ is convex entropy of (S), associated to the flux $q = -u\,\sigma(v)$. It would be also easy to prove

$$\|\sqrt{\varepsilon}\, v_x^\varepsilon\|_{L^2(L^2)} \leqslant c$$

so that the right-hand side of (3.7) satisfies in fact
$\varepsilon\, v_{xt}^\varepsilon = \varepsilon\, u_{xx}^\varepsilon \to o$ in $L^2(H^{-1}) \cap H^{-1}(L^2)$ strong (see [15]).

<u>Remark</u> :

The well-known result of Chueh, Conley, Smoller [2] does not apply to this problem, therefore one cannot prove a uniform (w.r.t.ε) L^∞ estimate for the solution of $(\mathcal{P}_\varepsilon)$. The only case where one knowns such an estimate is the case where $v.\sigma''(v) \geqslant o$, and where the second equation of (S_ε) is replaced by

$$v_t^\varepsilon - u_x^\varepsilon = \varepsilon\, u_{xx}^\varepsilon \text{ (with the same } \varepsilon \text{ in both equations).}$$

Now, to be concrete, we consider the conjecture

$$(C) \quad \begin{cases} w^\varepsilon \rightharpoonup w^*,\ z^\varepsilon \rightharpoonup z^* \text{ in } L^\infty(L^2) \text{ weak-star, } \underline{\text{and}} \\ \iint_Q w^\varepsilon z^\varepsilon dxdt \to \iint_Q w^* z^* \text{ when } \varepsilon \to o_+ \end{cases}$$

Remark :

1) We could have considered the similar conjecture :

$$w^{\varepsilon}z^{\varepsilon} \to w^{*}z^{*} \text{ in } \mathcal{D}'(Q) \tag{3.8}$$

but (C) is rather weaker, and will provide strong convergences in some L^q spaces, instead of L^q_{loc} spaces if we supposed (3.8).

2) As explained in the Introduction, (C) (or (3.8)) is a conjecture of compensated compactness with variable coefficients, since (3.7) provides informations on the derivatives of w^{ε} and z^{ε} respectively in the directions $(+\sigma'(v^{\varepsilon}(x,t)),1)$ and $(-\sigma'(v^{\varepsilon}(x,t)),1)$ which are transverse since $\sigma'(v) \geqslant c > o$. However, for some 2x2 hyperbolic systems, linearly degenerate for both their eigenvalues, D. Serre and L. Tartar have constructed a counter-example in which $w^{\varepsilon}z^{\varepsilon} \not\rightharpoonup w^{*}z^{*}$.

3) So this conjecture, which was already mentioned in [6], [20] is not true for any 2x2 hyperbolic system (at least if one does not use the informations on the initial data, that compensated compactness can't take into account). For a more precise discussion of a more general conjecture and its consequences, see [19], [17], [16].

Now, we define a weak solution (u,v) of Problem $\mathcal{P}$) as a couple (u,v) such that

$$\iint_Q (u\phi_t - \sigma(v)\phi_x)\, dxdt + \int_\Omega u_o(x)\phi(x,o)dx = o \tag{3.9}$$

and

$$\iint_\Omega (v\psi_t - u\psi_x)\, dxdt + \int_\Omega v_o(x)\psi(x,o)dx = o \tag{3.10}$$

for any couple smooth functions (ϕ,ψ) with compact supports in $\Omega x[0,T[$ and $\overline{\Omega}\, x[0,T[$ respectively.

4 - THE CONVERGENCE WHEN $\varepsilon \to o_+$

We simply summarize the results. For the complete proof, see [15].

THEOREM 1 :

With the previous, we suppose that conjecture (C) is true. Then, when $\varepsilon \to o_+$, one can extract sub-sequences (u^ε) and (v^ε) such that, in $L^\infty(L^q)$ spaces, with the suitable exponents q, we have :

1)
$$\begin{cases} v^\varepsilon \rightharpoonup v,\ g(v^\varepsilon) \rightharpoonup g^* = g(v) & \text{weak-star} \\ w^\varepsilon = w(u^\varepsilon, v^\varepsilon) \rightharpoonup w^* = u - g^* = w(u,v) & \text{"} \\ z^\varepsilon = -z(u^\varepsilon, v^\varepsilon) \rightharpoonup z^* = u + g^* = z(u,v) & \text{"} \end{cases} \tag{4.1}$$

$$\begin{cases} \sigma(v^\varepsilon) \rightharpoonup \sigma^* = \sigma(v) & \text{"} \\ \sigma'(v^\varepsilon) \to \sigma'^* = \sigma'(v) \text{ a.e in } Q \text{ and in } L^\infty(L^{(p+2)/p}) & \text{weak-star} \end{cases} \tag{4.2}$$

2) Therefore (u,v) is a weak solution of Problem ($\mathcal{P}$). Moreover, $\forall\lambda$, let $I(\lambda)$ be the largest interval, may be reduced to $\{\lambda\}$, which contains λ and on which σ is linear. Then the distance

$$d(v^\varepsilon(x,t),\ I(v(x,t))) \to o \quad \text{a.e. in } Q \tag{4.3}$$

In order words, the supports of the Young measures $\nu_{x,t}$ associated to (v^ε) satisfy a.e. in Q :

$$\text{supp } \nu_{x,t} \subset I(v(x,t)) \tag{4.4}$$

3) Suppose now that thene is no nontrivial interval on which σ is linear. Then all the convergences (4.1), (4.2) are strong in the corresponding $L^r(L^q)$ spaces, $\forall r < +\infty$, (with the same exponents q). Moreover

$$\eta^\varepsilon = \frac{(u^\varepsilon)^2}{2} + \Sigma(v^\varepsilon) \to \eta = \frac{u^2}{2} + \Sigma(v) \text{ in } L^r(L^1) \text{ strong, } \forall r < +\infty \quad (4.5)$$

$$q^\varepsilon = -u^\varepsilon \sigma(v^\varepsilon) \to q = -u\,\sigma(v) \text{ a.e. in } Q \quad (4.6)$$

and the total energy

$$E(t) = \int_\Omega \left(\frac{u^2}{2} + \Sigma(v)\right)(x,t)dx$$

is a decreasing function of time.

Sketch of the proof :

1) The crucial point is to prove (4.3), (4.4), which imply

$$g^* = g(v),\ \sigma^* = \sigma(v),\ (\sigma')^* = \sigma'(v) \text{ a.e. in } Q \quad (4.7)$$

and then (4.1), (4.2). For this purpose, observe first that (a slight modification of) the compensated compactness gives

$$\iint_Q (u^\varepsilon)^2 - v^\varepsilon \sigma(v^\varepsilon) dxdt \underset{(\varepsilon \to 0)}{\to} \iint_Q (u^2 - v\sigma^*) dxdt \quad (4.8)$$

Now, the conjecture (C) can be written

$$\iint_Q ((u^\varepsilon)^2 - (g(v^\varepsilon))^2) dxdt = \iint_Q w^\varepsilon z^\varepsilon dxdt \to \iint_Q w^* z^* dxdt =$$

$$= \iint_Q u^2 - (g^*)^2 dxdt \quad (4.9)$$

Substrating (4.9) from (4.8), we obtain

$$\iint_Q ((g(v^\varepsilon))^2 - v^\varepsilon \sigma(v^\varepsilon)) dxdt \to \iint_Q ((g^*)^2 - v\sigma^*) dxdt \quad (4.10)$$

Let us set

$$\chi^\varepsilon = (g(v^\varepsilon) - g(v))^2 - (v^\varepsilon - v)(\sigma(v^\varepsilon) - \sigma(v)) \quad (4.11)$$

We develop χ^ε , integrate over Q, and pass to the limit

$$\iint_Q \chi^\varepsilon dxdt \to \iint_Q [(g^*)^2-2g^*g(v)+(g(v))^2-v\sigma^*+v\sigma^*-v\sigma(v)+v\sigma(v)]\,dxdt=$$

$$=\iint_Q (g^*-g(v))^2 dxdt \geqslant o \qquad (4.12)$$

On the other hand, χ^ε is negative for any ε, thanks to the Cauchy-Schwarz inequality (2.8), and $\chi^\varepsilon(x,t) = o$ iff $v^\varepsilon(x,t) \in I(v(x,t))$. Therefore

$$\chi^\varepsilon \to o \quad \text{in } L^1(L^1) \text{ strong, and } g^*=g(v) \text{ a.e.} \qquad (4.12)$$

from which, it is easy to deduce (4.3). In terms of Young measures we can write (4.12) under the form :

$$< \nu_{x,t}, h(.,v(x,t)) > \underset{df}{=} < \nu_{x,t}, (g(.)-g(v(x,t)))^2 -$$

$$-(.-v(x,t))(\sigma(.)-\sigma(v(x,t)))> = o \qquad (4.13)$$

Therefore, the support of $\nu_{x,t}$ is located in the set

$$I(v(x,t)) = \{\lambda/h(\lambda,v(x,t))=o\}$$

since $\nu_{x,t}$ and $-h$ are positive. This proves (4.4), and (4.7) and (4.2) follow easily, since, for example

$$\sigma^*(x,t)=<\nu_{x,t},\sigma(\lambda)>=<\nu_{x,t},a(x,t)\lambda+b(x,t)> =$$

$$=a(x,t)<\nu_{x,t},\lambda>+b(x,t)=\sigma(v(x,t) \text{ a.e. in } Q$$

Thus, in any case, $\sigma^*=\sigma(v)$, which implies that (u,v) is a weak solution of $\mathscr{P}$).

2) The proof of the (3) would be obvious if we had L^∞ estimates on the solution, since for $I(\lambda)=\{\lambda\}$, (4.3), (4.4) become

$$v^\varepsilon \to v \text{ a.e. in } Q, \text{ i.e. } \nu_{x,t}=\delta_{u(x,t)} \text{ (a point mass) a.e. in } Q$$

whence

$$v^\varepsilon \rightharpoonup v=\langle \nu_{x,t},\lambda \rangle \text{ and } (v^\varepsilon)^2 \rightharpoonup v^2=(\langle \nu_{x,t},\lambda \rangle)^2$$

which implies the strong convergence (observe that the same argument implies, in all cases that, in the (1) :

$$\sigma'(v^\varepsilon) \to \sigma(v) \text{ in } L^r(L^q) \text{ strong } \forall r < +\infty, \forall q < (p+2)/p)$$

Moreover, with these assumptions, the weak solution (u,v) would obviously satisfy the Las Entropy Condition for the uniformly convex entropy η (1).

Here, without L^∞ estimates, it is more complicated : observe that e.g.

$$\eta^\varepsilon= \frac{(u^\varepsilon)^2}{2} +\Sigma(v^\varepsilon) \to \frac{u^2}{2} +\Sigma(v)=\eta, q^\varepsilon=-u^\varepsilon\sigma(v^\varepsilon) \to q=-u\sigma(v) \text{ a.e. in } Q$$

while we only know : $\|\eta^\varepsilon\|_{L^\infty(L^1)} \leqslant c$, and we don't even know if $\|q^\varepsilon\|_{L^\infty(L^1)} \leqslant c$.

So we need some (tedious) technical arguments (see [15]), based on Lemma 1, to prove the (3). Unfortunately the estimates of proposition 1 are too weak to prove that (u,v) satisfies the Lax Entropy Condition, due to the lack of estimates on q^ε (2).

---

(1) In the genuinely nonlinear case, it is well known that the Lax Entropy Condition is satisfied for <u>all</u> convex entropy iff it is satisfied for <u>one</u> strictly convex entropy η.

(2) Actually, one also needs to be careful to apply the theory of Young measures to sequences of functions which are only bounded in L^p, $p < +\infty$, se Schonbek [18]. To apply $\nu_{x,t}$ to the function $h(\lambda,v(x,t))$, one needs to use (4.12), see [15].

In conclusion, we have thus proved that the conjecture (C) implies the convergence of the viscosity method for the nonlinear elasticity system (S), in a very simple way, which is a pleasant generalization of the work of Tartar [20] for scalar equations, and which gives also a very simple interpretation of the linearly degenerate case as the case of equality in the Cauchy - Schwarz inequality. A larger conjecture for general 2x2 hyperbolic systems of conservation laws, given by D. Serre [19] has already permitted to extend the present method, based on the construction of a sequence of negative functions which converges weakly to a positive limit , to the system of isentropic gas dynamics [17].

Therefore, the proof of this (difficult!) conjecture seems to be a crucial point to understand better the convergence of approximate solutions to some 2x2 quasi-linear hyperbolic systems of conservation laws.

BIBLIOGRAPHY

1. G. Andrews .- On the existence of solutions to the Equation $u_{tt}=u_{xxt}+\sigma(u_x)_x$. J. Of Diff. Eq., 35, 2 (1980), 200-231.

2. K.N. Chueh, C.C. Conley and J.A. Smoller .- Positively Invariant Regions for Systems of Nonlinear Equations. Indiana Univ. Math. J., 26, 2 (1977), 373-392.

3. R. Courant, K.O. Friedrichs .- Supersonic Flow and Shock Waves. Interscience, New-York (1948).

4 N. Cristescu .- Spatial motion of elastic plastic strings. J. Mech. Phys. Solids, vol.9 (1961), 165-178.

5 C.M. Dafermos .- The mixed initial boundary value problem for the equations of nonlinear one dimensional visco-elasticity. J. of Diff. Eq.,6 (1969), 71-86.

6. R.J. Di Perna .- Convergence of Approximate Solutions to a class of non-linear hyperbolic systems of equations. Arch. for Rat. Mech. and Anal.

7. R.J. Di Perna .- Convergence of the Viscosity Method for Isentropic Gas Dynamics, Arch. for Rat. Mech. and Anal.

8. J.M. Greenberg, R.C. Mac Camy and V.G. Mizel .- On the existence, uniqueness and stability of solutions to the equation $\sigma'(u_x)u_{xx}+\lambda u_{xxt}=\rho_o u_{tt}$. J. of. Math. Mech., 17 (1968), 707-728.

9. J. Glimm .- Solutions in the large for Non Linear Hyperbolic systems of Equations. Comm. in Pure Appl. Math., 15 (1965), 697-715.

10. B.L. Keyfitz and H.C. Kranzer .- A system of non-strictly hyperbolic conservation laws arising in elasticity theory. Arch. for Rat. Mech. and Anal., 72 (1980), 219-241.

11. J.U. Kim .- Solutions to the Equations of one dimensional visco-elasticity in BV. SIAM J. on Math. Anal. (1983).

12. O.A. Ladyzenskaja, V.A. Solonnikov and N.N. Ural'ceva .- Linear and Quasi-Linear Equations of Parabolic Type. Amer. Math. Soc. Translations, vol.23, A.M.S., Providence, R.I., (1968).

13. P.D. Lax .- Shock Waves and Entropy, in Contributions to Non-Linear Analysis. Zarantonello editor, Academic Press, Yew-York (1971).

14. J.L. Lions .- Quelques méthodes de résolution des problèmes aux limites non linéaires. Dunod-Gauthier-Villars, Paris (1969).

15. M. Rascle .- Perturbations par viscosité de certains systèmes hyperboliques non linéaires, Thèse, Université Lyon 1 (1983).

16. M. Rascle .- To appear

17. M. Rascle and D. Serre .- To appear

18. M.E. Schonbek .- Convergence of Solutions to Non-Linear Dispersive Equations. Comm. in Partial Diff. Eq., 7, 8, (1982), 959-1000.

19. D. Serre .- To appear

20. L. Tartar .- Compensated Compactness and Applications to Partial Differential Equations, in Non Linear Analysis. Heriot-Watt Symposium, volIV, R.J. Knops editor, Pitman (1979), 136-212.

21. L.C. Young .- Lectures on the Calculus of Variations and Optimal Control Theory. W.B. Saunders, Philadelphia, Pa (1969).

ANALYSE NUMERIQUE, UNIVERSITE DE ST-ETIENNE

42023, ST-ETIENNE CEDEX (FRANCE)

Lectures in Applied Mathematics
Volume 23, 1986

THE RIEMANN PROBLEM FOR NONMONOTONE STRESS-STRAIN FUNCTIONS: A "HYSTERESIS" APPROACH

Barbara Lee Keyfitz

ACKNOWLEDGEMENTS. The idea for this paper was formed three years ago when I spent a semester at Duke University and talked to Michael Shearer, who had just completed an analysis of the Riemann Problem for a model of an elastic rod which allowed coexistence of two phases. His solution of an initial-value problem for a problem with an elliptic region seemed remarkable. To shed some light on the well-posedness of the problem, I sketched out the treatment contained in the present talk, which replaced the elliptic region by a degenerate but still hyperbolic one. The difficulty was, and still is, that I could not justify this approach by any appeal to continuum mechanics. But it seems to offer a useful simplification of Shearer's solution, as well as a way of understanding some points in it that had seemed unsatisfactory. There is doubt concerning the correct admissibility conditions for shocks in elasticity and van der Waals fluid equations which model phase transitions, and after some very helpful discussions with Rick James and John Ball I was encouraged to write down some details of my approach. Although this paper may add yet another model to the confusion, I hope it may have some part in resolving it. In any case, I want to thank the people mentioned above for their suggestions. I also want to thank my

---

1980 Mathematics Subject Classification 35L65.
*Research partially supported by NSF grants MCS-80-02751 and MCS-81-03441.

frequent collaborator, Herb Kranzer. The results here are largely based on our joint work.

1. Background. There has been interest recently in models for physical phenomena which involve a system of conservation laws which is not everywhere hyperbolic but is, formally, elliptic in an open region of phase space. That this happens in steady fluid flow in the transonic regime is well known [12] and appropriate boundary conditions for well-posedness have been discussed [12, 13]. The models we are concerned with here, however, involve time-dependent problems, and classical theory predicts that the Cauchy problem may be illposed. This need not be the case, however. In [14], Shearer showed that for a particular class of mixed-type conservation laws which model phase changes in fluids or in elastic bars, the Riemann Problem could be uniquely solved if one made a particular set of assumptions on what constituted admissible shocks. Shearer's approach was closely tied to some work of Keyfitz and Kranzer [7] which solved the Riemann Problem for a model equation which was hyperbolic except along a curve in phase space. Our solution displayed some features that were new in this context: shock curves in phase space that were detached from the local shock curves near the originating point, and also shocks which satisfied the classical Lax entropy condition [9] only with a weak inequality. Such shocks are known in the Russian continuum mechanics literature as "Jouguet shocks" (see [8]), from the analogy with shocks in combustion models [2]. In any case, the waves in [7] satisfy standard admissibility criteria, being the limits of viscosity profiles, and also Liu's entropy condition for systems [10].

Shearer [14] showed that one could solve the Riemann problem for systems in which the degenerate curve widened to an open region of elliptic points by allowing the shocks introduced in [7] and adding shocks of one other type: stationary shocks which did not satisfy the classical Lax entropy condition or Liu's condition but could be justified on other mathematical grounds. Shearer's solution was also even more complex to describe than ours, as it included 17 regions in which the solution of the Riemann problem was defined differently [14, 15].

The purpose of this paper is to show that one can solve the Riemann Problem using the construction of [7] for a system of conservation laws obtained from the mixed-type system by altering the flux function in the elliptic region so that the system is no longer elliptic there but degenerate hyperbolic. This generalizes the solution found in [7]. Our solution is qualitatively similar to Shearer's, but has the advantages of being simpler to describe (we are back to the 12 regions of [7]) and also of stipulating an admissibility criterion for shocks that is equivalent to all the others (i.e., the Lax entropy condition, Liu's entropy condition and the existence of artifical viscosity profiles). Our solution also involves stationary discontinuities, but very few, and they appear as contact discontinuties rather than shocks. Thus we feel that this approach lends support to the validity of Shearer's solution.

The main difficulty with the model studied here is that we cannot claim for it a rational derivation from principles of continuum mechanics. I can offer it only on the mathematical grounds that it indicates a class of conservation laws for which the Riemann problem has a unique solution which can be described in straightforward terms, and that it suggests why an equation of mixed type may lead to a well-posed problem. It is clearly related to the models studied by Shearer, Slemrod and James [4, 5, 14, 17]. I am quite optimistic that a rational derivation can be given. I also hope that this model may help resolve the question of what constitutes an admissibility criterion for shocks.

The basic equation that leads to a system of mixed type of the form we are considering here is

$$w_{tt} = \sigma(w_x)_x \tag{1.1}$$

where $w(x,t)$ is the position coordinate at time t of a point on a rod that had position x in a prestressed reference configuration. The reference configuration is supposed to have unifrom density, cross-section and stress. We will work with the coordinates

$$\begin{aligned} u &= w_x \\ v &= w_t \quad . \end{aligned} \tag{1.2}$$

Here u is the local stretch $+1$ and v the local velocity. The function $\sigma(u)$ is the constitutive or stress-strain relation. The materials which admit phase changes of the type we are considering here are characterized by a nonmonotone function σ. We will make the standard assumptions that σ is twice continuously differentiable and that there exist values a and b with $a < b$ such that $\sigma' > 0$ for $u < a$ and $u > b$, and $\sigma' < 0$ in (a,b). We will also assume that $\sigma'' < 0$ for $u < a$ and $\sigma'' > 0$ for $u > b$, although this is just for convenience, and could be modified as discussed in section 5. To get global solutions (also for convenience) we will also assume that $|\sigma|$ grows faster than linearly as $|u| \to \infty$. These are basically Shearer's assumptions (i) - (iii) in [14], although we do not need as much control on u in (a,b). A typical function σ is sketched in Figurc 1.1. The points c, where $\sigma(c) = \sigma(b)$ and d, where

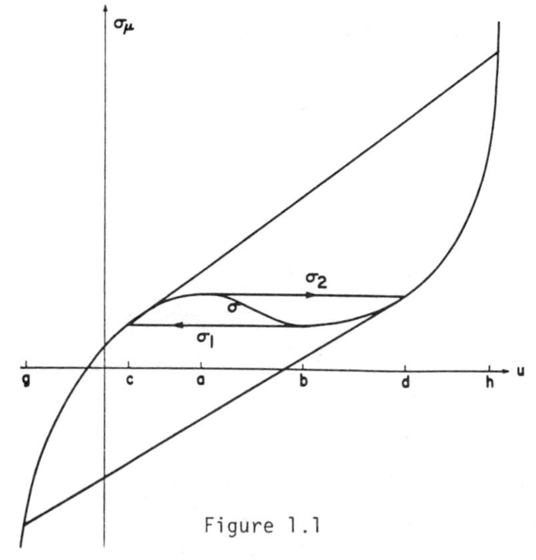

Figure 1.1

$\sigma(a) = \sigma(d)$, will be important in what follows. We also define, for reference, the points g and h where the tangents to σ at d and c respectively meet σ again.

Writing (1.1) as a system in the standard way,

$$\begin{aligned} u_t &= v_x \\ v_t &= \sigma(u)_x, \end{aligned} \tag{1.3}$$

we see that (1.3) is hyperbolic with characteristic speeds $\pm\sqrt{\sigma'(u)}$ for $u < a$ and $u > b$, but is of elliptic type in the strip (in uv phase space) $a < u < b$.

In this paper, we modify the stress-strain relation σ in the interval $c < u < d$ as follows.

Define

$$\sigma_1(u) = \begin{cases} \sigma(u) & \text{if } u \le c \text{ or } u \ge b \\ \sigma(c) & \text{if } c < u < b \end{cases} \tag{1.4}$$

and

$$\sigma_2(u) = \begin{cases} \sigma(u) & \text{if } u \le a \text{ or } u \ge d \\ \sigma(a) & \text{if } a < u < d \end{cases}, \tag{1.5}$$

and replace σ in (1.3) by

$$\sigma_\mu = \sigma_\mu(u, \operatorname{sgn} u_x) = \begin{cases} \sigma_2(u) & \text{if } \operatorname{sgn} u_x > 0 \\ \sigma_1(u) & \text{if } \operatorname{sgn} u_x < 0 . \end{cases} \tag{1.6}$$

Since $u_x = w_{xx}$, one could think of (1.6) as incorporating some higher-order effects into the constitutive relation, but the physical nature of these effects is unclear. Classical visco-elasticity models, as in Dafermos's paper [3], use a stress function $\sigma(w_x, w_{xt})$ which is smoothly differentiable and increasing in the second argument. Andrews and Ball [1] assume σ is linear in w_{xt}, as does James [5] in studying small viscosity limits. Slemrod [17], who considers (1.1) as a model for a van der Waals fluid, looks also at a viscosity-capillarity model proposed by Korteweg which replaces the stress function by

$$\sigma(u) + D(u)u_x^2 - C(u)u_{xx} + \mu(u)u_t$$

which, although very general, still does not include the specific sort of dependence we find useful. More recently, James [6] has suggested including temperature effects in the model, which replaces (1.1) by a larger system of equations; again it is not clear whether our model might approximate such a system.

Superficially, our function σ_μ, as indicated schematically in Figure 1.1, bears some resemblence to the multi-valued functions sometimes used to describe hysteresis phenomena in elasticity, but it should be emphasized that the phenomenon modelled here is not hysteresis and that the use of the word in the title of this paper is meant to describe the picture of σ_μ, not the constitutive assumptions.

2. Solution of the Riemann Problem.

We consider

$$u_t = v_x \tag{2.1}$$

$$v_t = (\sigma_\mu)_x \tag{2.2}$$

$$(u(x,0),v(x,0)) = \begin{cases} U_\ell & x \leq 0 \\ U_r & x > 0 \end{cases}, \tag{2.3}$$

where σ_μ is given by equation (1.6) for σ as in Figure 1.1.

To solve the Riemann problem we construct, in the u-v plane, curves separating the open regions to which a point U_0 can be joined by a succession of shocks, contact discontinuities and rarefaction waves, with U_0 on the left. These regions fill the plane uniquely. The construction is similar to [7]: it is simpler in some respects because (2.1) is linear; however, the separating curves are not as smooth. The basic types of waves can be seen by considering a couple of cases: if $u_0 > d$ (note that we shall use the convention $U_j = (u_j, v_j)$ throughout), then $\sigma_\mu \equiv \sigma_1$ when we are considering points U to which U_0 may be joined by a single wave. Thus (2.1) is a non-strictly hyperbolic system with a parabolic degeneracy in the strip $c \le u \le b$, and a discontinuity in the characteristic speeds

$$\lambda_{1,2} = \mp \sqrt{\sigma_\mu'(u)} \tag{2.4}$$

at $u = c$. As in [7] we find that S_2 and R_1^+ are finite branches, while there is a detached branch S_1^*. Figure 2.1

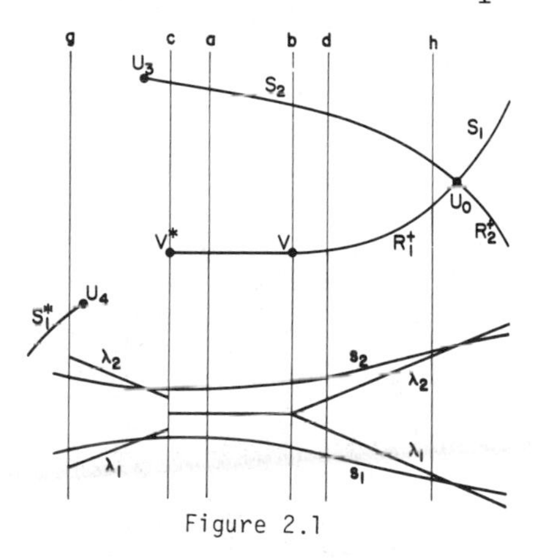

Figure 2.1

illustrates these curves and the associated shock speeds and characteristic values. We use the notation

$$s_{1,2}(u,u_0) = \mp \sqrt{(\sigma_\mu(u) - \sigma(u_0))/(u-u_0)}.$$

The curve denoted R_1^+ is a composite curve: the points on the horizontal part are joined to V by contact discontinuities with zero speed. Also the curve is only C^1 at V. Other significant points are u_3, where

$$s^2(u_0,u_3) \equiv \frac{\sigma_\mu(u_0)-\sigma_\mu(u_3)}{u_0-u_3} = \sigma_\mu'(u_3) \tag{2.5}$$

and u_4 where

$$s^2(u_0,u_4) \equiv \frac{\sigma_\mu(u_0)-\sigma_\mu(u_4)}{u_0-u_4} = \sigma'_\mu(u_0) \quad . \tag{2.6}$$

If $u_0 < h$ we define u_3 to be c; u_4 is defined for all $u_0 > b$, and $u_4 \to c$ as $u_0 \to b$ since σ'_1 is continuous at b.

Note that the shocks admitted here satisfy both the weakened Lax entropy condition [7] and the Liu entropy condition [10].

When $b < u_0 < d$, the left side of the picture is the same, but now $\sigma_\mu = \sigma_2$ for $u > u_0$, and we have the situation shown in Figure 2.2. The equation for U_4 is found from the Rankine-Hugoniot condition, $s[u] = -[v]$; $s[v] = -[\sigma_\mu]$, so, for s_1,

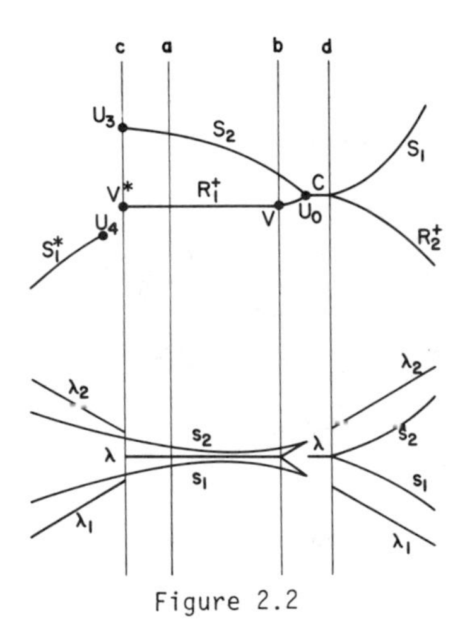

Figure 2.2

where we take the negative square root,

$$[v] = v - v_0 = -s_1[u] = (u-u_0)\sqrt{(\sigma(u)-\sigma(u_0))/(u-u_0)} \tag{2.7}$$

Now as $u_0 \to b$, $u_4 \to c$ and $v_4 \to v_0$ since $\sigma(b) = \sigma(c)$, so $U_4 \to V^*$.

The third different case is $a \le u_0 \le b$. The picture is particularly simple in this case: there are no detached branches and there is a sort of symmetry. This is indicated in Figure 2.3.

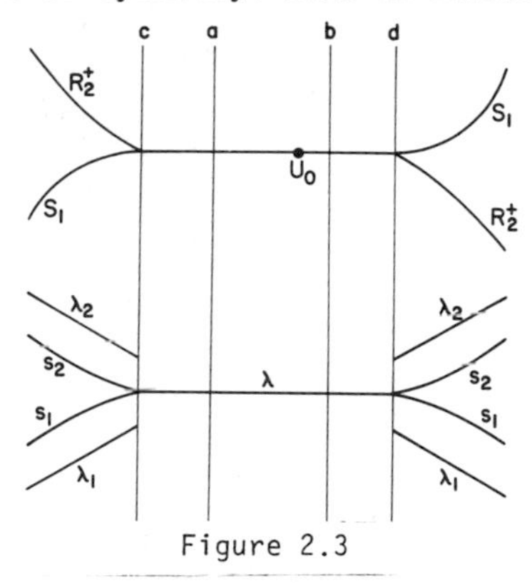

Figure 2.3

The salient point for solving the Riemann problem is that, for $u_0 > b$, if $U_1 \in R_1^+(U_0)$ with $u_1 > b$, then the speed of the 1-shock joining U_1 to $U_4(U_1)$ is greater than or equal to the head speed of the rarefaction joining U_0 to U_1. Hence the system can support two waves of negative speed. Similarly, any point on the curve $R_2^+(U_3)$ can be joined to U_3 by a rarefaction with tail-speed greater than or equal to the speed of the $U_0 - U_3$ shock. As in [7, 14], we let $E(U_0)$ denote the curve $U_4(U_1)$ as U_1 ranges through $R_1^+(U_0)$, and let $J(U_0)$ denote the curve $U_3(U_1)$ as U_1 ranges through $R_1^+(U_0) \cup S_1(U_0)$, and let $J_1(U_0)$ be the curve $U_3(U_1)$ as U_1 ranges through $E(U_0) \cup S_L^*(U_0)$. For the three different ranges of U_0 we divide the plane into 12 regions as indicated in Figures 2.4, 2.5 and 2.6. Within each

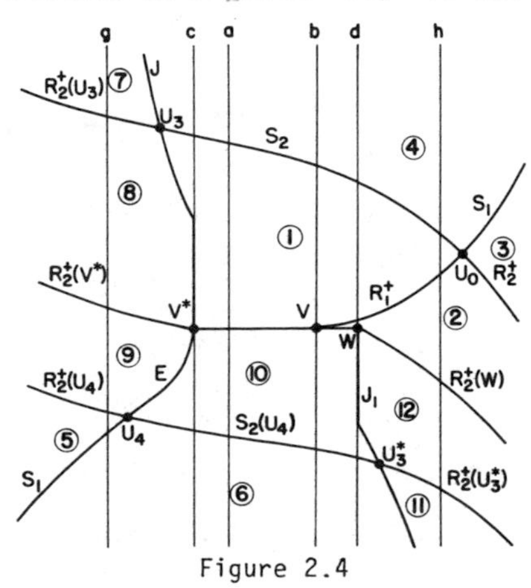

Figure 2.4

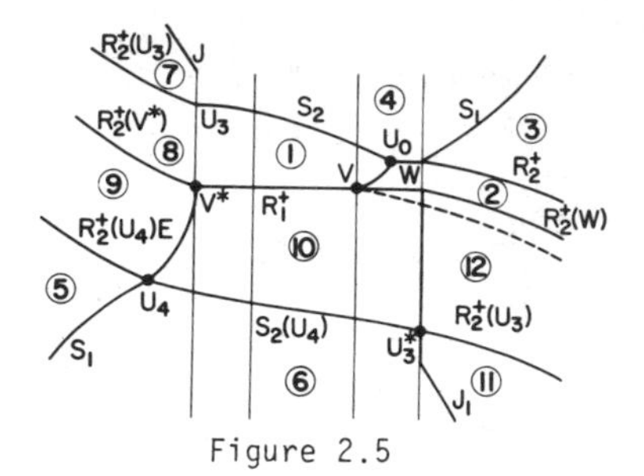

Figure 2.5

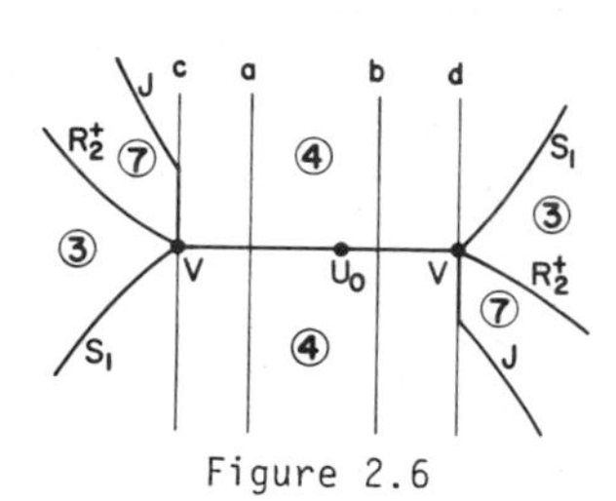

Figure 2.6

region, the structure of connecting waves is as follows:

Region

① $:U_1 \in R_1^+,\quad U_2 \in S_2(U_1)$

② $:U_1 \in R_1^+,\quad U_2 \in R_2^+(U_1)$

③ $:U_1 \in S_1,\quad U_2 \in R_2^+(U_1)$

④ $:U_1 \in S_1,\quad U_2 \in S_2(U_1)$

⑤ $:U_1 \in S_1^*,\quad U_2 \in R_2^+(U_1)$

⑥ $:U_1 \in S_1^*,\quad U_2 \in S_2(U_1)$

⑦ $:U_1 \in S_1(U_0),\ U_2 \in S_2(U_1):\ U_2 = U_3(U_1);\ U_r \in R_2^+(U_2)$

⑧ $:U_1 \in R_1^+(U_0),\ U_2 \in S_2(U_1):\ U_2 = U_3(U_1);\ U_r \in R_2^+(U_2)$

⑨ $:U_1 \in R_1^+(U_0),\ U_2 \in S_1^*(U_1):\ U_2 = U_4(U_1) \in E(U_0);\ U_r \in R_2^+(U_2)$

⑩ $:U_1 \in R_1^+(U_0),\ U_2 \in S_1^*(U_1):\ U_2 = U_4(U_1) \in E(U_0);\ U_r \in S_2(U_2)$

⑪ $:U_1 \in S_1^*(U_0),\ U_2 \in S_2(U_1):\ U_2 = U_3(U_1) \in J_1;\ U_r \in R_2^+(U_2)$

⑫ $:U_1 \in R_1^+(U_0),\ U_2 \in S_1^*(U_1):\ U_2 = U_4(U_1) \in E(U_0);$

$U_3 \in S_2(U_2):\ U_3 = U_3(U_2) \in J_1;\ U_r \in R_2^+(U_3)$.

To establish the validity of this solution, we need certain properties of the curves E, J and J_1. These properties, analogous to Propositions 3.1-3.3 of [7], can be proved the same way, except for the observation that the curves J and J_1 fail to be smooth at one point. The fact that E is vertical at V^* is Proposition 3.1 of [14]. Note that E is identical to Shearer's function.

3. Viscosity Profiles for Shocks. One advantage of this approach to solving the Riemann problem is that all admissible shocks have (artificial) viscosity profiles: that is to say, when (1.3) is replaced by a perturbed system

$$U_t + F(U)_x = U_{xx} \tag{3.1}$$

and $U = U(\xi)$, where $\xi = (x-st)/\varepsilon$, there is a trajectory joining any two points U_0, U_1 that may be connected by an admissible shock of speed s. We will refer to this as the Conley-Smoller condition [18]. The construction here is very similar to [7]. This contrasts with the more complicated situation in Shearer's model: Shearer imposes a "chord condition" which is the same as Liu's entropy condition, and shows, in [15], that it is different from the Conley-Smoller condition. There is a second complication in Shearer's solutions, that the stationary shocks do not satisfy the entropy condition or have profiles. In our approach, the stationary shocks are treated as contact discontinuities. They will be discussed in Section 4. Here we prove

Theorem 3.1. Let U_0 be any point in R^2 and let $U_1 \in S_2(U_0) \cup S_1(U_0) \cup S_1^*(U_0)$. Then there is a unique solution $U(\xi)$ of (3.1) such that $U(-\infty) = U_0$, $U(+\infty) = U_1$.

Proof. For definiteness, assume $u_0 > a$. We shall see that the shock trajectory is monotone in u and hence we may fix $\sigma_\mu = \sigma_1$ or σ_2 for the construction. The cases that do not involve a phase transition are standard; we shall assume, then, $u_0 \geq b$,

$u_1 < b$, so $\sigma = \sigma_1$ for $u \le u_0$. The usual substitutions in (3.1), and an integration, yield

$$\begin{aligned} u' &= -v - su + A \equiv f_1(u,v) \\ v' &= -\sigma_1(u) - sv + B \equiv f_2(u,v) \end{aligned} \tag{3.2}$$

where $A = v_0 + su_0 = v_1 + su_1$ and $B = \sigma_1(u_0) + sv_0 = \sigma_1(u_1) + sv_1$. Now a quick calculation shows that the only zeros of (f_1, f_2) for $u \le u_0$ are at (u_0,v_0) and (u_1,v_1). Furthermore we get a negatively invariant region precisely when $s = s_1 < 0$ and $U_1 \in S_1^*(U_0)$ and a positively invariant one when $s = s_2 > 0$ and $U_1 \in S_2(U_0)$ (including the end points in both cases). The first case is shown in Figure 3.1, the second in 3.2.

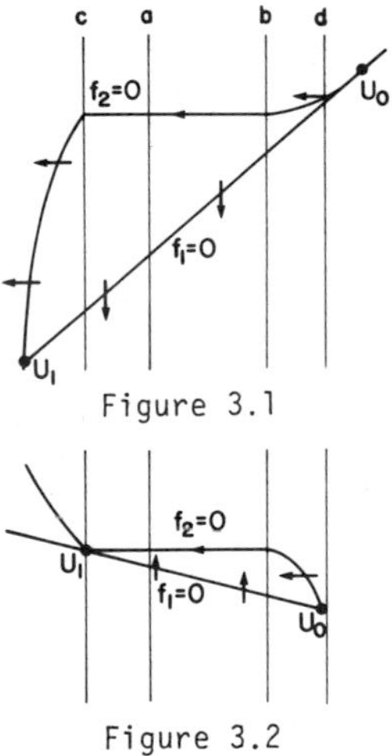

Figure 3.1

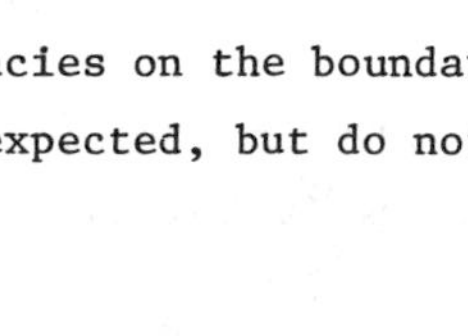

Figure 3.2

Note the degeneracies on the boundary of the invariant region, which might be expected, but do not complicate the proof. We omit the details.

4. Stability of Stationary Waves. Our construction involves two kinds of stationary waves. One type, which does not involve a phase change, appears to be an artifact of the choice of σ_μ made in (2.3). For example, when $b \le u_0 \le d$, every point in region

(2) and some points in region (4) contain stationary discontinuties in the solutions to the Riemann Problem. This is to be expected since σ is affine in an open set in this case, and there stationary solutions have the same status as jumps in a linear equation. The modification of the solution within the same hyperbolic component (same phase) is somewhat unsatisfactory; however, it is interesting that if the definition of σ_μ is modified so that the local solution to the Riemann Problem (for pairs such that the entire solution has a single phase) is unchanged, then there will be more than one solution for some Riemann problems. Specifically, $R_2^+(V)$ would become the dotted line in Figure 2.5 and points between $R_2^+(V)$ and $R_2^+(W)$ could be joined to U_0 in two ways, one involving a phase transition and one not. The definition of σ in this case might be taken as

$$\sigma_\nu = \sigma_\nu(u,u_0) = \begin{cases} \sigma_1(\mu), & u_0 \geq b \\ \sigma_2(u), & u_0 \leq a \\ \sigma_\mu(u, \operatorname{sgn}(u-u_0)), & a \leq u_0 \leq b. \end{cases}$$

The phenomenon of nonuniqueness in the original model to which this is closely related is currently being studied by Shearer and is described in [16].

Stationary jumps that involve a phase transition (i.e. one of u_0 and u above b, the other below a) do not occur in any open region of our construction. If $u_0 > d$, as in Figure 2.5, then all points U in $R_2^+(V^*)$, including V^*, will join U_0 by a sequence involving a stationary shock from V to V^*. For points close to $R_2^+(V^*)$ in region (8), this shock will be replaced by a shock of small positive speed, and for points in region (9) by a shock of small negative speed joining points close to V and V^*. These moving shocks have viscosity profiles as described in section 3, but there is no viscosity profile for the stationary shock because U_0 is a degenerate critical point of the vector field.

Finally, we observe that when u_0 is between a and b, the solution still contains stationary shocks only for U on the segment V^*V, and for U in the complement of this line segment we again have only moving shocks. As before, the stationary shocks have no viscosity profiles, but are the limits of shocks with viscosity profiles.

5. Comparison with other solutions. As emphasized throughout, one merit of the approach taken here is that the solution of the Riemann problem is easier to construct than Shearer's, yet shows the same qualitative features. In Figure 5.1 we reproduce Figure

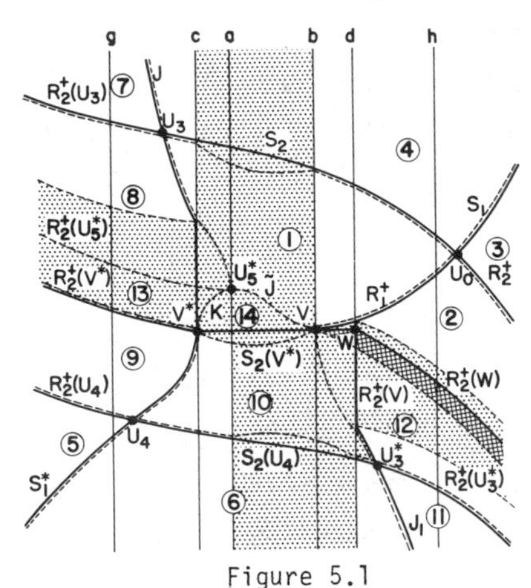

Figure 5.1

2.4 with Shearer's division of the plane superimposed in dotted lines. The shaded area indicates the region where the solution constructed here differs from Shearer's. In most places the difference is qualitative. For example, in the part of our region (8) corresponding to Shearer's (13) the difference is that one of our intermediate states is on J, with $u = c$, while in Shearer's constuction the corresponding state is on K, with $c < u < a$ and there is a stationary shock. We do not have a stationary shock in this case, but a shock with slow, positive speed. In the cross-hatched part of region (12) however, we predict a phase transition and Shearer predicts none. An explanation for this was given in section 4.

It should also be mentioned that the treatment of phase boundaries discussed in this paper, which is consistent with Shearer's (in the sense just described) is, like Shearer's, inconsistent with James's definition of admissible phase boundaries. To be precise, James, in [4] and [5] allows only stationary and moving phase boundaries in which the Lax entropy condition is violated: for a given $u_0 > b$, for example, he would require $u_3 > u > u_4$ for the state U. In fact, this is equivalent to his definition of a phase boundary as a discontinuity that moves more slowly than the sound speed both ahead of and behind the discontinuity. It is standard in shock-wave theory to rule out such discontinuities by arguments concerning well-posedness [18], but James states that one should perhaps not expect a unique solution in the class of functions admitting phase boundaries. It would be impossible to decide on the relative validity of Shearer's and James's solutions to a given Riemann problem without appeal to many other considerations. Suffice it to say, in this paper, that when U_0 and U are two states in opposite phases, say $u_0 > b$ and $u < a$, then our solution, at least for U in regions (7), (8) or (9), appeals to intermediate states on J or E. Such phase boundaries, which satisfy our admissibility criteria in a weak sense (these are the Jouguet points), also satisfy James's criteria in a weak sense. (It is not clear that James would admit these solutions.) The same holds in region (12). However, the solution of our problem (and Shearer's solution) in regions (5), (6) and (11) involves points on $S_1^*(U_0)$. This entire branch is apparently rejected by James.

Another treatment of a related problem involving phase changes in a van der Waals fluid is proposed by Slemrod. Typically, the convexity of the function σ in his problem is different from Figure 1. Like James, Slemrod admits certain shocks that violate the standard entropy condition, and therefore his solutions are quite different from those found here and in Shearer's study [14]. It should be emphasized that the changes in convexity of σ within the hyperbolic region could be handled by Shearer's method or the one proposed here: as is known from the work of Liu on strictly hyperbolic systems without convexity conditions, [11], it is sufficient to replace the shock and rarefaction curves by composite wave curves. The construction, although complicated in detail, is straightforward in principle.

6. Conclusions. We have offered here an alternative method of constructing a unique centered solution to the Riemann Problem for a pair of conservation laws modelling phase changes in elastic bars or fluids. Our construction is based upon Shearer's approach, of admitting in the solution only shocks that satisfy classical entropy conditions, but it incorporates into the model a change that much simplifies the construction and, in addition, makes equivalent the standard tests of admissibility (which give inequivalent results for Shearer's solution). Under our hypotheses, we construct a solution to the Riemann problem which agrees with Shearer's qualitatively everywhere and in detail over much of the plane. Further work must be done to reconcile this solution with solutions proposed by James and by Slemrod, and also to give a rational derivation of our model.

7. References

1. G. Andrews and J. M. Ball, "Asymptotic behavior and changes of phase in one-dimensional nonlinear viscoelasticity", Jour. Diff. Eqns., 44 (1982) 306-341.

2. R. Courant and K. O. Friedrichs, Supersonic Flow and Shock Waves, Wiley-Interscience, New York, 1948.

3. C. M. Dafermos, "The mixed initial-boundary value problem for the equations of nonlinear one-dimensional viscoelasticity", Jour. Diff. Eqns., 6 (1969), 71-86.

4. R. D. James, "Co-existent phases in the one-dimensional static theory of elastic bars", Arch. Rat. Mech. Anal., 72 (1979), 99-140.

5. ___________, "The propagation of phase boundaries in elastic bars", Arch. Rat. Mech. Anal. 73 (1980), 125-158.

6. ___________, "A relation between the jump in temperature across a propagating phase boundary and the stability of solid phases", Jour. Elast. 13 (1983), 357-378.

7. B. L. Keyfitz and H. C. Kranzer, "The Riemann problem for a class of hyperbolic conservation laws exhibiting a parabolic degeneracy", Jour. Diff. Eqns. 47 (1983), 35-65.

8. A. G. Kulikovskiy and E. I. Sweshnikova, Nonlinear deformation waves in previously stressed elastic media, in Nonlinear Deformation Waves, ed. U. Nigul and J. Engelbrecht, Springer-Verlag, Berlin/Heidelberg/ New York, 1983, 293-298.

9. P. D. Lax, "Hyperbolic Systems of Conservation Laws and the Mathematical Theory of Shock Waves", Conf. Board Math. Sci., 11, SIAM, 1973.

10. T. P. Liu, "The Riemann problem for general 2×2 conservation laws", Trans. Amer. Math. Soc. 199 (1974), 89-112.

11. ____________, "Admissible solutions of hyperbolic conservation laws", Amer. Math. Soc. Memoirs, #240, Providence, 1981.

12. C. S. Morawetz, Nonlinear Waves and Shocks, Tata Institute Notes, Springer, Berlin/Heidelberg/New York, 1981.

13. S. J. Osher, "Boundary value problems for equations of mixed type I, The Lavrent'ev-Bitzadze model", Comm. Partial Diff. Eqns. 2 (1977), 499-547.

14. M. Shearer, "The Riemann problem for a class of conservation laws of mixed type", Jour. Diff. Eqns. 46 (1982), 426-443.

15. __________, "Admissibility criteria for shock wave solutions of system of conservation laws of mixed type", Proc. Royal Soc. Edin. A. 93 (1983), 233-244.

16. __________, "Nonuniqueness of admissible solutions of Riemann initial value problems for a system of conservation laws of mixed type", preprint, 1984.

17. M. Slemrod, "Admissibility criteria for propagating phase boundaries in a van der Waals fluid", Arch. Rat. Mech. Anal. 81 (1983), 301-315.

18. J. A. Smoller, Shock Waves and Reaction-Diffusion Equations, Springer, New York, 1983.

DEPARTMENT OF MATHEMATICS
UNIVERSITY OF HOUSTON - UNIVERSITY PARK
HOUSTON, TEXAS 77004

Lectures in Applied Mathematics
Volume 23, 1986

THE RIEMANN PROBLEM FOR A VAN DER WAALS FLUID

Harumi Hattori[1]

1. INTRODUCTION

The purpose of this paper is to discuss the applicability of the entropy rate admissibility criterion proposed by Dafermos [1], [2] to the Riemann problem for a van der Waals fluid. The system is given by

$$\begin{aligned} v_t - u_x &= 0 \\ u_t + p(v)_x &= 0 \end{aligned} \qquad (1.1)$$

where u, v, and p are the velocity, the specific volume, and the pressure of the fluid, respectively. This system expresses the one dimensional flow of a compressible fluid in Lagrangian coordinates. For the isothermal flow of an ideal gas, $p(v) = \frac{R\theta}{v}$, where R is a universal constant, and θ is the constant absolute temperature. In this case, the system (1.1) is hyperbolic. On the other hand, the state equation of a van der Waals fluid is given by

[1]Supported by EPSCOR and a Senate Grant from West Virginia University.

$$p(v) = \frac{R\theta}{v-b} - \frac{a}{v^2} \quad , \tag{1.2}$$

where a and b are the characteristic constants of the fluid. Thus, when the temperature is sufficiantly low, $p'(v)$ is positive, and thus the system (1.1) is elliptic on a certain inteval (α,β), while $p'(v)$ is negative on (b,α)(liquid phase) as well as on (β,∞) (vapor phase) on which (1.1) is hyperbolic.

The entropy rate admissibility criterion was originally proposed for hyperbolic systems of conservation laws of the form

$$w_t + f(w)_x = 0 \ . \tag{1.3}$$

It is well known that in general the weak solutions (bounded measurable functions which satisfy (1.3) in the sense of distributions) are not unique. In order to select a physically relevant solution, various admissibility criteria have been proposed. The entropy rate admissibility criterion, one of these criteria, roughly says that for the admissible solution, the entropy decreases with the highest rate.

Nonuniqueness of weak solutions to (1.1) also arises in the above nonhyperbolic case, so admissibility criteria have to be postulated for that case. So far a physically motivated criterion is the viscosity-capillarity criterion proposed by Slemrod [3],[4]. His argument is based on the capillarity effect of the fluid, which is an extension of the result of Serrin [5]. We shall show that the entropy rate admissibility criterion is, also, applicable to the Riemann problem of the above nonhyperbolic system. Specifically, we apply this criterion to a small class of solutions consisting of constant states separated by a backward wave, a forward wave, and a phase boundary. We should notice that this criterion is applied in a limited sense.

This paper consists of 6 sections. In Sections 2, 3, and 4, we discuss the preliminaries necessary to the subsequent sections, namely, the van der Waals fluid, the Riemann problem, and the entropy rate admissibility criterion. We treat, in Section 5, the possible solutions which join two constant states lying in different phases by backward waves, phase boundaries, and forward waves. In Section 6, we discuss a special Riemann problem and show that the well-known Maxwell construction is admissible according to the entropy rate criterion.

2. VAN DER WAALS FLUID

Consider a van der Waals fluid in which the state equation is given by (1.2). In Fig. 1 we sketch a few isotherms of (1.2) for

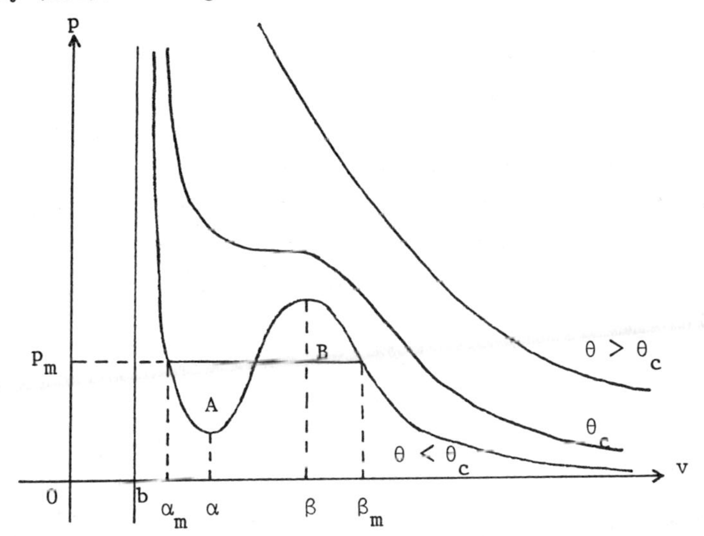

Figure 1.

different values of θ. If the temperature is below the critical temperature $\theta_c = \frac{8a}{27bR}$, the isotherm is not monotone so that the system (1.1) is nonhyperbolic. Since we do not need to restrict attention to the specific equation (1.2), we will be assuming in what follows that $p(v)$ is any function satisfying the following conditions:

(C1) $p_v(v) < 0$ on $(b,\alpha) \cup (\beta,\infty)$,

(C2) $p'(\alpha) = p'(\beta) = 0$,

(C3) $p'(v) > 0$ on (α,β) ,

(C4) $p(v) > 0$,

(C5) $p''(v) > 0$ on $(b,v_*) \cup (v^*, \infty)$,

$$(\alpha < v_* < \beta, \ \beta < v^* < \infty) ,$$

(C6) $p''(v) < 0$ on (v_*, v^*) .

The intervals (b,α) and (β,∞) will be called the α-phase (liquid phase) and the β-phase (vapor phase), respectively. The horizontal line segment for which areas A and B are equal is called the Maxwell line. We denote the pressure at the Maxwell line by p_m. The values of v in the α-phase and the β-phase at which the pressure is equal to p_m are denoted by α_m amd β_m, respectively. The fluid may behave in a totally different way at pressures just below and just above p_m. Various properties change discontinuously in transition. This observation motivates the following definitions; the homogeneous solution $v(x)=$ constant is

(S1) stable if $b < v < \alpha_m$ or $\beta_m < v$.

(S2) neutrally stable if $v=\alpha_m$ or $v=\beta_m$.

(S3) metastable if $\alpha_m < v \leq \alpha$ or $\beta \leq v < \beta_m$.

(S4) unstable if $\alpha < v < \beta$.

The state (S3) may be observed, yet a small perturbation will change the phase drastically. The above argument is based on the Gibbs relation; see Fermi [6].

3. RIEMANN PROBLEM AND ELEMENTARY WAVES

The Riemann problem is a special initial value problem in which the initial condition is given by

$$(u,v)(x,0) = \begin{cases} (u_0,v_0) & x < 0 \\ (u_1,v_1) & x > 0 \end{cases} , \tag{3.1}$$

where the right hand side are constants. In this paper we, furthermore, require the following:

(i) v_0 is in the α-phase and v_1 is in the β-phase, and they are near the Maxwell line, (3.2)

(ii) $p_0 = p_1$, $u_0 = u_1$.

Therefore, the phase change takes place at $x=0$ in the initial data. Since both the system (1.1) and the initial data (3.1) are invariant under the transformation $(x,t) \to (\gamma x, \gamma t)$, $\gamma > 0$, the solution of (1.1), (3.1) is a function of $\frac{x}{t}$, i.e., it consists of a fan of waves that emanate from the origin and propagate with individual speeds. Specifically, we assume that the solutions consist of three different types of elementary waves, namely, a rarefaction wave, a shock, and a phase boundary and are given as Fig. 2.

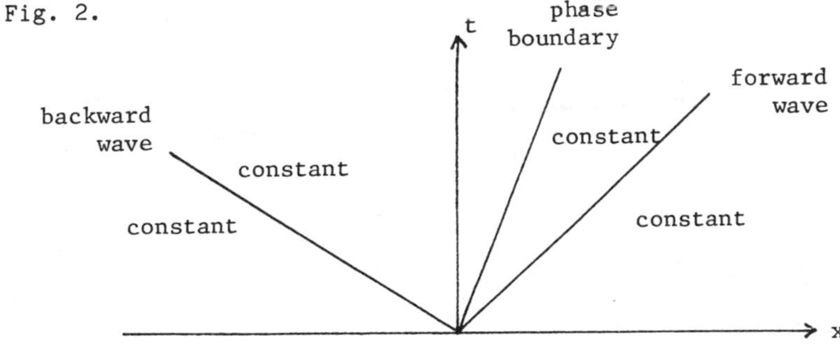

In this figure a wave means either a rarefaction wave or a shock and a forward (backward) wave means a wave with a positive (negative) speed. We note that there might be solutions of other types, for instance having more than one phase boundary (odd numbers) between the forward and backward waves and that if v_0 and v_1 are away from α_m and β_m, the connection may be as in Fig. 3

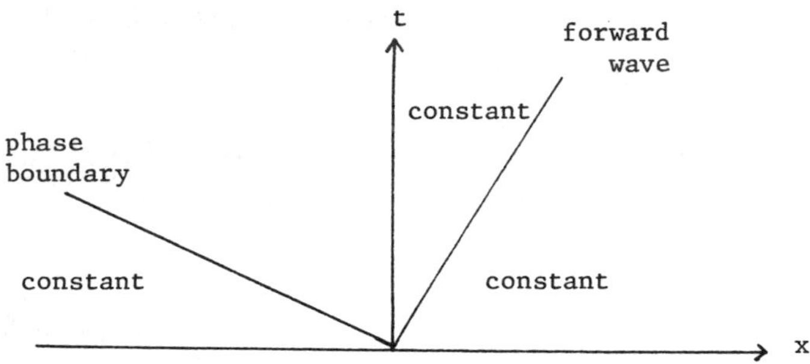

with phase boundary connecting the α-phase and the β-phase as shown in Fig. 4.

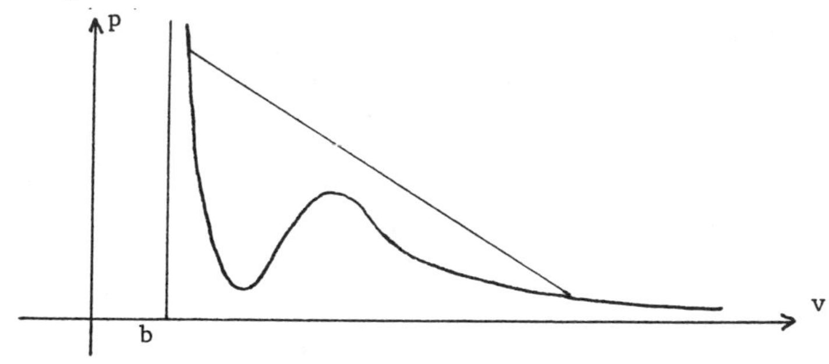

However, we do not treat these cases in this paper.

It is now in order to explain the three types of waves briefly:

(a) Rarefaction wave; this is a wedge in the x-t plane. If two constant states (u_-,v_-) and (u_+,v_+) are adjacent to a forward (backward) rarefaction wave on the left and right, the relation

$$u_+ - u_- = (\mp)\int_{v_-}^{v_+} \sqrt{-p'(w)}\,dw \tag{3.3}$$

is satisfied. As characteristic curves expand in the rarefaction wave region, we have the following relation between v_- and v_+ if $p'' \neq 0$.

$$\begin{aligned} &\text{forward rarefaction wave:} && v_- > v_+ \quad \text{if} \quad p'' > 0, \\ & && v_- < v_+ \quad \text{if} \quad p'' < 0, \\ &\text{backward rarefaction wave:} && v_- < v_+ \quad \text{if} \quad p'' > 0, \\ & && v_- > v_+ \quad \text{if} \quad p'' < 0. \end{aligned} \tag{3.4}$$

(b) Shock; this is a jump discontinuity across which the Rankine-Hugoniot condition is satisfied. This condition is given by

$$\begin{aligned} \sigma[u_+ - u_-] &= [p(v_+) - p(v_-)], \\ \sigma[v_+ - v_-] &= -[u_+ - u_-], \end{aligned} \tag{3.5}$$

where σ is the speed of propagation of the shock, and (u_-,v_-) and (u_+,v_+) are the states on the left and the right of the shock. Solving (3.5) for σ and substituting it in (3.5b), we obtain

$$u_+ - u_- = (\mp)\sqrt{-\frac{p_+ - p_-}{v_+ - v_-}}\,(v_+ - v_-) \tag{3.6}$$

as the relation for the forward (backward) shock. Applying an appropriate admissibility criterion such as the energy (entropy) rate criterion (this will be discussed in the next section), if $p'' \neq 0$, we obtain the relation

$$\begin{array}{lll} \text{forward shock:} & v_- < v_+ \quad \text{if} \quad p'' > 0, \\ & v_- > v_+ \quad \text{if} \quad p'' < 0, \\ \text{backward shock:} & v_- > v_+ \quad \text{if} \quad p'' > 0, \\ & v_- < v_+ \quad \text{if} \quad p'' < 0. \end{array} \tag{3.7}$$

(c) Phase boundary; this is, also, a jump discontinuity across which the Rankine-Hugoniot condition is satisfied. The phase is same across a shock, but on the other hand, the phase change takes place across a phase boundary.

From (3.3) and (3.6), it is clear that if we fix $(u_-,v_-)((u_+,v_+))$, the set of (u_+,v_+) $((u_-,v_-))$ forms one parameter family of states connected to $(u_-,v_-)((u_+,v_+))$, on the right (left), by a rarefaction wave or a jump discontinuity. These sets are called forward (backward) wave curves or phase boundary curves, as appropriate. In this paper the forward wave curve means that set of (u_-,v_-) connected to a given (u_+,v_+) by a forward shock or a forward rarefaction wave (or a combination of them) and the backward wave curve means the set of (u_+,v_+) connected to a given (u_-,v_-) by a backward shock or a backward rarefaction wave.

4. ENTROPY RATE ADMISSIBILITY CRITERION

A convex function $\eta(w)$ is called an entropy for (1.3), with entrophy flux $q(w)$, if

$$\eta(w)_t + q(w)_x = 0$$

holds identically for any smooth vector field $w(x,t)$ which satisfies (1.3), i.e., if

$$\sum_{j=1}^{n} \frac{\partial\eta}{\partial w_j} \cdot \frac{\partial f_j}{\partial w_k} = \frac{\partial q}{\partial w_k}, \quad k=1,\ldots,n.$$

Dafermos [1], [2] has proposed the entropy rate admissibility criterion to choose an admissible weak solution. A solution $w(x,t)$ will be called admissible if there is no solution $w(x,t)$ with the property that for some τ $[0,T]$, $w(x,t)=w(x,t)$ on $(-\infty,\infty)x[0,\tau]$ and $D_+H_w(\tau) < D_+H_w(\tau)$, where

$$H_w(t) = \int_{-\infty}^{\infty} \eta(w(x,t))dx.$$

For system (1.1) we employ the mechanical energy $\frac{1}{2}u^2 + \int^{v}(-p(w))dw$ as the entropy. The corresponding entropy flux is given by $up(v)$. For the above choice, the rate of entropy decay is

$$D_+H_w(\tau) = \sum_{\substack{\text{jump}\\ \text{discontinuities}}} \sigma\left\{ -\frac{1}{2}(p(v_R)+p(v_L))(v_R-v_L) + \int_{v_L}^{v_R} p(w)dw \right\}, \tag{4.1}$$

where v_R and v_L are the values of v on the right and the left of the jump discontinuity. Since the mechanical energy is employed as the entropy, it may be more appropriate here to refer the criterion as the energy rate admissibility criterion.

5. POSSIBLE SOLUTIONS TO THE RIEMANN PROBLEM AND COMPARISON OF THE ENERGY RATE.

In this section, we study possible solutions to the Riemann problem, i.e., possible connections of the states (u_0,v_0) and (u_1,v_1) in (3.1) by means of shock waves, rarefaction waves, and phase boundaries (see Fig. 2). We denote by (u_L,v_L) the intermediate constant state on the left of the phase boundary and by (u_R,v_R) the intermediate constant state on the right of the phase boundary. Then, v_L is in the α-phase and v_R is in the β-phase.

As the backward and forward waves may be either shocks or rarefaction waves, and there are backward and forward phase boundaries, there are 8 possible combinations of connections, namely,

(1)	B.R. - F.P.B. - F.R.,	(5)	B.R. - B.P.B. - F.R.,
(2)	B.R. - F.P.B. - F.S.,	(6)	B.R. - B.P.B. - F.S.,
(3)	B.S. - F.P.B. - F.R.,	(7)	B.S. - B.P.B. - F.R.,
(4)	B.S. - F.P.B. - F.S.,	(8)	B.S. - B.P.B. - F.S..

Here, for example, B.S., F.P.B., and F.R. stand for "backward shock", "forward phase boundary", and "forward rarefaction wave", respectively. Therefore, (1) means that (u_0,v_0) is joined to (u_L,v_L) by a backward rarefaction wave, then (u_L,v_L) is joined to (u_R,v_R) by a forward phase boundary, and (u_R,v_R) is joined to (u_1,v_1) by a forward rarefaction wave. We derive the differential equations which are satisfied by (1) - (8). As p'' changes sign in the β-phase, we may have a combination of shock and rarefaction wave for the forward wave. However, to avoid complication we treat the above cases only.

We start from case (1) (and (5)). In this case u_0 is joined to u_L by a backward rarefaction wave, hence,

$$u_L = u_0 + \int_{v_0}^{v_L} \lambda(w)dw, \tag{5.1}$$

where $\lambda(w)=\sqrt{-p'(w)}$. Then, u_1 is joined to u_R by the forward (backward for (5)) phase boundary, namely,

$$u_R = u_L \ (\mp)\sqrt{\frac{p_R-p_L}{v_R-v_L}}\,(v_R-v_L). \tag{5.2}$$

Also, u_R is joined to u_1 by the forward rarefaction wave, therefore,

$$u_1 = u_R - \int_{v_R}^{v_1} \lambda(w)dw. \tag{5.3}$$

From (5.1) - (5.3), we deduce

$$u_0 + \int_{v_0}^{v_L} \lambda(w)dw \ (\mp)\sigma_P(v_R-v_L) - \int_{v_R}^{v_1} \lambda(w)dw = u_1, \tag{5.4}$$

where $\sigma_P = \sqrt{\frac{p_R-p_L}{v_R-v_L}}$. If we differentiate (5.4) with respect to v_L, regarding v_R as a function of v_L, we obtain the differential equation

$$\frac{dv_R}{dv_L} = \frac{(\lambda_L(\pm)\sigma_P)^2}{(\lambda_R(\mp)\sigma_P)^2} \tag{5.5}$$

where $\lambda_L = \sqrt{-p_L'}$ and $\lambda_R = \sqrt{-p_R'}$. In the same manner, we can derive the following differential equations:

$$\text{case (2)((6)):} \quad \frac{dv_R}{dv_L} = \frac{\sigma_F(\lambda_L(\pm)\sigma_P)^2}{(\sigma_F(\mp)\sigma_P)(\lambda_R^2(\mp)\sigma_F\sigma_P)}, \tag{5.6}$$

$$\text{case (3)((7)):} \quad \frac{dv_R}{dv_L} = \frac{(\sigma_B(\pm)\sigma_P)(\lambda_L^2(\pm)\,\sigma_B\sigma_P)}{\sigma_B(\lambda_R(\mp)\sigma_P)^2}, \tag{5.7}$$

$$\text{case (4)((8)):} \quad \frac{dv_R}{dv_L} = \frac{\sigma_F(\sigma_B(-)\sigma_P)(\lambda_L^2(-)\sigma_B\sigma_P}{\sigma_B(\sigma_F(\mp)\sigma_P)(\lambda_R^2(\mp)\sigma_F\sigma_P)}. \tag{5.8}$$

As σ_B, σ_F, σ_P, λ_L, and λ_R satisfy the relations

$$0 \leq \sigma_P < \sigma_B < \lambda_L, \quad 0 \leq \sigma_P < \sigma_F < \lambda_R,$$

$\frac{dv_R}{dv_L}$ has a positive sign in all cases. In particular, if $\sigma_P=0$, namely $p_L = p_R$, then

$$\frac{dv_R}{dv_L} = \frac{\lambda_L^2}{\lambda_R^2} = \frac{p_L'}{p_R'} \tag{5.9}$$

Next, we consider how the energy rate varies as a function of v_L. The energy rates for the backward shock, for the backward (forward) phase boundary, and for the forward shock are given, respectively, by

$$E_B = \sigma_B \left\{\frac{1}{2}(v_L - v_0)(p_L + p_0) - \int_{v_0}^{v_L} p(w)dw\right\},$$

$$E_p = (\pm)\sigma_P\left\{\frac{1}{2}(v_R - v_L)(p_R + p_L) - \int_{v_L}^{v_R} p(w)dw\right\}, \tag{5.10}$$

$$E_F = \sigma_F\left\{\frac{1}{2}(v_1 - v_R)(p_R + p_1) - \int_{v_R}^{v_1} p(w)dw\right\}.$$

Differentiating (5.10) with respect to v_L, regarding v_R as a function of v_L, we obtain

$$\frac{dE_B}{dv_L} = -\frac{1}{4\sigma_B}(\sigma_B^2 - \lambda_L^2)\left(3p_L - p_0 - \frac{2\int_{v_0}^{v_L} p(w)dw}{v_L - v_0}\right),$$

$$\frac{dE_p}{dv_L} = (\mp)\frac{1}{4\sigma_P}\left\{\frac{dv_R}{dv_L}(\sigma_P^2 - \lambda_R^2)\left(3p_R - p_L - \frac{2\int_{v_L}^{v_R} p(w)dw}{v_R - v_L}\right)\right.$$

$$\left. -(\sigma_P^2 - \lambda_L^2)\left(3p_L - p_R - \frac{2\int_{v_L}^{v_R} p(w)dw}{v_R - v_L}\right)\right\}, \tag{5.11}$$

$$\frac{dE_F}{dv_L} = -\frac{\frac{dv_R}{dv_L}}{4\sigma_F}(\sigma_F{}^2-\lambda_R{}^2)(3p_R-p_1-\frac{2\int_{v_1}^{v_R} p(w)dw}{v_R - v_1}).$$

Set

$$A = 3p_R - p_L - \frac{2\int_{v_L}^{v_R} p(w)dw}{v_R-v_L}, \quad B = 3p_L-p_R - \frac{2\int_{v_L}^{v_R} p(w)dw}{v_R-v_L},$$

$$C = 3p_L - p_0 - \frac{2\int_{v_0}^{v_L} p(w)dw}{v_L-v_0}, \quad D = 3p_R - p_1 - \frac{2\int_{v_1}^{v_R} p(w)dw}{v_R-v_1}.$$

Then, they satisfy the following relations:

$$A > 0, \ B > 0 \quad \text{if} \quad p_R = p_L > p_m,$$

$$A < 0, \ B < 0 \quad \text{if} \quad p_R = p_L < p_m,$$

$$A = B = 0 \quad \text{if} \quad p_R = p_L = p_m, \tag{5.12}$$

$$\frac{dA}{dv_L} = \frac{dB}{dv_L} = -2\lambda_L{}^2 \quad \text{at} \quad p_R = p_L = p_m,$$

$$\lim_{v_L \to v_0} C = 0, \quad \lim_{v_R \to v_1} D = 0.$$

6. SPECIAL RIEMANN PROBLEM

In this section we discuss a rather special Riemann problem. Specifically, we treat (1.1) and (3.1) with (3.2), and shall show the following two theorems.

THEOREM 6.1. Suppose v_R obeys one of the differential equations (5.5), (5.6), (5.7), or (5.8), as appropriate. Then, as v_L

approaches v_o, $\frac{dE}{dv_L}$ approaches a negative number if $p_0 < p_m$, a positive number if $p_0 > p_m$, and zero if $p_0 = p_m$.

PROOF. From (5.11a,c) we easily see that $\frac{dE_B}{dv_L}$ and $\frac{dE_F}{dv_L}$ approach zero as v_L approaches v_0. Therefore, it remains to examine the sign of $\frac{dE_P}{dv_L}$ as v_L approaches v_0. Combining the differential equations (5.5) - (5.8) with (5.11b), and taking the limit of (5.11b) as v_L approaches v_0, we obtain

from (5.5) $$\lim_{v_L \to v_0} \frac{dE_P}{dv_L} = -\frac{\lambda_0}{2\lambda_1}(\lambda_1 A + \lambda_0 B),$$

from (5.6) $$\lim_{v_L \to v_0} \frac{dE_P}{dv_L} = -\frac{\lambda_0}{4\lambda_1}\{\lambda_0(A+B)+2\lambda_1 A\},$$

from (5.7) $$\lim_{v_L \to v_0} \frac{dE_P}{dv_L} = -\frac{\lambda_0}{4\lambda_1}\{\lambda_1(A+B)+2\lambda_0 B\},$$

and from (5.8) $$\lim_{v_L \to v_0} \frac{dE_P}{dv_L} = -\frac{\lambda_0}{4\lambda_1}(\lambda_0+\lambda_1)(A+B).$$

Using the relations in (5.12), we conclude that the limit of $\frac{dE}{dv_L}$ as v_L approaches v_0 is negative if $p_0 < p_m$, positive if $p_0 > p_m$, and zero if $p_0 = p_m$. Q.E.D.

As a consequence of Theorem 6.1 we deduce the following

COROLLARY 6.1. If $u_0=u_1$ and $p_0=p_1 < p_m$ or $p_0=p_1 > p_m$, the connection between (u_0,v_0) and (u_1,v_1) by the stationary phase boundary $((u_0,v_0)=(u_L,v_L),\ (u_1,v_1)=(u_R,v_R))$ is not admissible according to the energy rate admissibility criterion.

The case where $p_0 = p_1 = p_m$ is more delicate, yet we can show the following

THEOREM 6.2. Suppose $u_0=u_1$ and $p_0=p_1=p_m$. Then, the connection between (u_0,v_0) and (u_1,v_1) by the stationary phase boundary is admissible in the sence that it minimizes the energy rate for v_L close to v_0.

PROOF. If we draw the backward wave curve through (u_0,v_0) and the forward wave curve through (u_1,v_1) in the u-v plane, there are three possibilities, depicted in Fig. 5, 6, 7, depending on the value of v^*(the v coordinate of the inflection point in the β-phase). Since the proof is similar in all three cases, we shall only consider the case of Fig. 5 and shall show that $\frac{dE}{dv_L}$ is positive if $v_L > v_0$ and negative if $v_L < v_0$, for v_L close to v_0.

Using that $\frac{dv_R}{dv_L}$ is positive, we find that the connection between (u_0,v_0) and (u_1,v_1) is of type (1) if $v_L > v_0$, and of type (8) if $v_L < v_0$ and v_L is close to v_0. If $v_L \ll v_0$, the forward wave may be a combination of a shock and a rarefaction wave (see Greenberg [7] for the details).

In case (1), $\frac{dv_R}{dv_L}$ is given by (5.5) and $\frac{dE}{dv_L}$ is

$$\frac{dE}{dv_L} = \frac{1}{4\sigma_P}\left\{\frac{dv_R}{dv_L}(\sigma_P{}^2-\lambda_R{}^2)A - (\sigma_P{}^2-\lambda_L{}^2)B\right\} \tag{6.1}$$

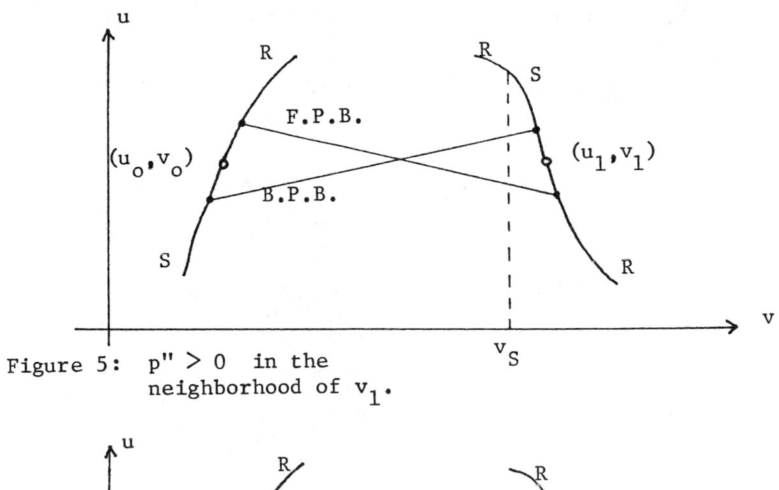

Figure 5: $p'' > 0$ in the neighborhood of v_1.

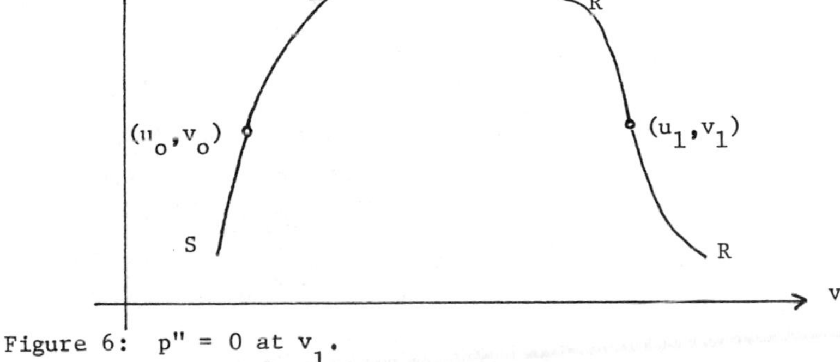

Figure 6: $p'' = 0$ at v_1.

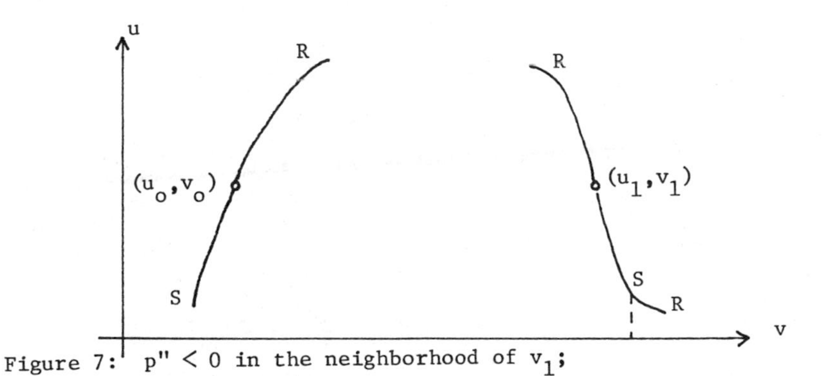

Figure 7: $p'' < 0$ in the neighborhood of v_1;

Combining (5.5) and (6.1), we find

$$\frac{dE}{dv_L} = -\frac{\lambda_L+\sigma_P}{\lambda_R-\sigma_P}\{(\lambda_L+\lambda_R)(A+B)-(v_R-v_L)(\lambda_L\lambda_R+\sigma_P{}^2)\sigma_P\} \qquad (6.2)$$

For $v_L > v_0$, since $A+B < 0$, $\frac{dE}{dv_L}$ in (6.2) is positive.

In case (8), $\frac{dv_R}{dv_L}$ is given by (5.8) and $\frac{dE}{dv_L}$ is

$$\frac{dE}{dv_L} = -\frac{1}{4\sigma_B}(\sigma_B{}^2-\lambda_L{}^2)C - \frac{1}{4\sigma_F}\frac{dv_R}{dv_L}(\sigma_F{}^2-\lambda_R{}^2)D$$

$$+\frac{1}{4\sigma_P}\{\frac{dv_R}{dv_L}(\sigma_P{}^2-\lambda_R{}^2)A+(\sigma_P{}^2-\lambda_L{}^2)B\}.$$

It is no longer easy to find a relation like (6.2). We expand (6.3) in series of (v_L-v_0) and examine the lowest order terms. Then,

$$\frac{dE_S}{dv_L} = \frac{\lambda_0}{4}\{p_0'' \quad +p_1'' \quad (\frac{\lambda_0}{\lambda_1})^5\}(v_L-v_0)^2+o((v_L-v_0)^2), \qquad (6.4)$$

$$\frac{dE_P}{dv_L} = -\lambda_0{}^2(v_R-v_L)\sigma_P+\lambda_0{}^3(1+\frac{\lambda_0}{\lambda_1})(v_L-v_0)+o((v_L-v_0)), \qquad (6.5)$$

where E_S and E_P denote the energy rate of the shocks and the phase boundary, respectively. From (6.4) and (6.5) we find that the orders of the lowest order terms are different, and that the lowest order terms in (6.5) are negative for $v_L < v_0$. This indicates that $\frac{dE}{dv_L}$ is negative if $v_L < v_0$ and close to v_0.

Combining case (1) and (8), we infer that the energy rate attains a relative minimum at $v_L=v_0$. Q.E.D

REMARK 6.1. If we assume that E is locally convex near $v_L=v_0$ on the basis of Theorem 6.2, Theorems 6.1 and 6.2 seem to indicate the following behavior. If $u_0=u_1$ and $p_0=p_1 > p_m$, v_L will be greater than v_0 because $\frac{dE}{dv_L}$ is negative at $v_L=v_0$. Then, as $\frac{dv_R}{dv_L}$ is positive at $v_L = v_0$, from Fig. 5, 6, and 7 we see $u_L > u_R$. Using the Rankine-Hugoniot conditions for the phase boundary, namely,

$$\sigma[u_R-u_L] = [p_R-p_L],$$

$$\sigma[v_R-v_L] = -[u_R-u_L]$$

we conclude that σ is positive. Hence, the phase boundary will move forwards. On the other hand if $u_0=u_1$ and $p_0=p_1 < p_m$, the phase boundary will move backwards by the similar argument.

REMARK 6.2. As a matter of fact, if $u_0=u_1$, $p_0=p_1 > p_m$, $p''(v_1) > 0$ (Fig.7), and v_0 is close to α_m, we can justify Remark 6.1. In this case $\frac{dE}{dv_L} = \frac{dE_P}{dv_L}$ and is given by (6.2). As we increase v_L, (A+B) in (6.2) will change sign from positive to negative. We denote by v_L the first point at which A+B=0, provided a solution exists for $\frac{dv_R}{dv_L}$ on the interval $[v_0,\alpha_m]$. This v_L should be less than α_m, because when $v_L=\alpha_m$, (A+B) is already negative (observe that $p_R < p_L$). Therefore, $\frac{dE}{dv_L}$ changes sign negative to positive before v_L. In this case the state v_L is stable and will take over the vapor state v_R in finite time (the state v_R may be stable or metastable).

ACKNOWLEDGEMENT: I would like to thank Professor Constantine M. Dafermos and Professor Marshall Slemrod for their comments and encouragement.

REFERENCES

[1] Dafermos, C.M., The entropy rate admissibility criterion for solutions of hyperbolic conservation laws, J.Diff.Eqs.14(1973), 202-212.

[2] Dafermos, C.M., The entropy rate admissibility criterion in thermoelasticity, Rediconit della Classe di Science Fisiche, Serie VII, LVII (1974), 113-119.

[3] Slemrod, M. Admissibility criteria for propagating phase boundaries in a van der Waals fluid, Arch. Rat. Mech. Anal. 81(1983), 301-315.

[4] Slemrod, M., Dynamic phase transitions in a van der Waals fluid, to appear in J. Diff. Eqs.

[5] Serrin, J., Phase transitions and interfacial layers for van der Waals fluids, Proc. of SAFA IV Conference, "Recent Methods in Nonlinear Analysis and Appliations", Naples, 1980; A. Cambora, S. Rionero, C. Sbordone, and C. Trombelli, editors.

[6] Fermi. E., Thermodynamics, Dover: New York(1956).

[7] Greenberg, J.M., On the elementary interactions for the quasilinear wave equation, Arch. Rat. Mech. Anal. 43(1971), 325-349.

DEPARTMENT OF MATHEMATICS
WEST VIRGINIA UNIVERSITY
MORGANTOWN, WV 26506

Lectures in Applied Mathematics
Volume 23, 1986

THE GEOMETRY OF CONTINUOUS GLIMM FUNCTIONALS

Michelle Schatzman

ABSTRACT. Let $u_t + f(u)_x$, $x \in \mathbf{R}$, $t \geq 0$, be a strictly hyperbolic system of conservation laws ; the characteristic fields are either genuinely non-linear or linearly degenerate. There exists a functional $\mathcal{F}$ which is equivalent to the total variation on $\mathbf{R}$, and such that, if u is a piecewise C^1 entropic solution of the system, and the initial total variation is small enough, then $t \to \mathcal{F}(u(t))$ decreases. This functional is analogous to the one defined by Glimm [5]. The uniqueness of the solution of Riemann problem is a consequence. These results are generalized to piecewise Lipschitz continuous functions.

1. INTRODUCTION. Given a function f of class C^3 from an open set $\mathcal{U}$ of $\mathbf{R}^N$ to $\mathbf{R}^N$, we say that the system

$$(1) \qquad u_t + f(u)_x = 0, \; x \in \mathbf{R}, \; t \geq 0,$$

is strictly hyperbolic if the derivative of f with respect to its argument, denoted Df, has N distinct real eigenvalues ; these eigenvalues are arranged in increasing order, as follows :

$$(2) \qquad \lambda_1(u) < \lambda_2(u) < \dots < \lambda_N(u).$$

Then, for each u, Df has a basis of eigenvectors

$$(3) \qquad Df(u)\, r_i(u) = \lambda_i(u)\, r_i(u),$$

and a basis of eigenlinear forms

$$(4) \qquad \ell_i(u)\, Df(u) = \ell_i(u)\, \lambda_i(u).$$

1980 Mathematics Subject Classification. 35 L 65.

We assume that r_i and ℓ_i are C^2 in u and satisfy

(5) $\ell_i(u)\ r_j(u) = \delta_{ij}$.

It is well known that solutions of (1) develop discontinuities. Therefore, we must consider weak solutions, i.e. solutions in the sense of distributions. To ensure some hope of uniqueness, a weak solution must satisfy an admissibility criterion, at the singularities.

The problem we consider here is the following : does there exist a functional $\mathcal{F}$, defined on a suitable subset of $BV(\mathbf{R})^N$, the space of functions of bounded variation on $\mathbf{R}$, with values in $\mathbf{R}^N$, such that

(6) $\mathcal{F}$ is equivalent to the total variation

(7) if u is an admissible solution of (1), $\frac{d}{dt}\ \mathcal{F}(u(t)) \leq 0$.

We shall give a positive answer to this question, when u is piecewise continuously differentiable, away from points where singularities meet, appear, or disappear. The admissibility condition is a slightly refined version of Lax entropy condition, which always holds if the shocks are weak enough. The total variation of the initial data must be small enough ; for the result to hold, u must take its values in a small enough ball.

In other words, if u belongs to an adequate functional class, and satisfies an admissibility criterion, L^∞ estimates on u and BV estimates on the initial data imply BV estimates for all timo.

In the construction [5] of a solution of (1), Glimm defines a functional F on piecewise constant functions from $\mathbf{R}$ to $\mathcal{U}$; this functional is equivalent to the total variation ; if $u(n\Delta t)$ is the piecewise constant function obtained after n iterations of Glimm's scheme, with an arbitrary (not necessarily equidistributed) sampling, then $F(u(n\Delta t))$ decreases with respect to n.

Formally, the functional $\mathcal{F}$ constructed here is the result of a passage to the limit on F, as the discretization step tends to zero.

However, it is not possible to obtain the result described here by a passage to the limit on Glimm solutions, because we lack a better convergence than weak BV convergence, and a general uniqueness result. Therefore, we must perform direct computations on the solutions of (1).

This article differs from (1) in the following respects : the detailed computations are not given, a more geometric approach is offered, and under the heading "esprit de l'escalier", a generalization to Lipschitz continuous functions is offered. This generalization came after I sadly described to D. Wagner all the reasons why I was unable to generalize the results ; it relies on a trace theorem for Lipschitz continuous functions which satisfy (1).

2. REVIEW OF THE SOLUTION OF RIEMANN PROBLEM. The so-called Riemann problem is the Cauchy problem for (1), with discontinuous initial data

$$u(x,0) = \begin{cases} u^{\ell} & \text{on } (-\infty,0), \\ u^{r} & \text{on } (0,\infty), \end{cases} \tag{8}$$

where u^{ℓ} and u^{r} are given states, i.e. elements of $\mathcal{U}$.

Riemann problem is solved by taking advantage of the symmetries : the transformation $(x,t) \to (kx,kt)$, for $k > 0$, leaves the problem invariant. Therefore, one seeks solutions of the form

$$u(x,t) = u(x/t) \; ; \tag{9}$$

nevertheless, it is not obvious, a priori, that any solution of (1), (8) is self-similar.

The first self-similar solution of(1), (8) is due to P. Lax, [9], under the assumption that u^{ℓ} and u^{r} are close enough, and that the characteristic fields are genuinely nonlinear, i.e.

$$D\lambda_i(u)\, r_i(u) \neq 0, \; \forall \; u \in \mathcal{U}. \tag{10}$$

The sign of (10) is usually normalized by requiring that

$$D\lambda_i(u)\, r_i(u) > 0. \tag{11}$$

There is another condition under which it is easy to solve Riemann problem : the i-th characteristic field is linearly degenerate if

$$(12) \qquad D\lambda_i(u)\, r_i(u) = 0, \ \forall\, u \in \mathcal{U} .$$

If neither (11) nor (12) is satisfied, see [11,12] for the solution of the Riemann problem ; nevertheless, I will assume that for every i, either (11) or (12) is satisfied, and I shall denote by LD the set of indices of linearly degenerate characteristic fields, and by GNL the complementary set of indices of genuinely non-linear characteristic fields. We shall need now the following notations : let

$$(13) \qquad A(u,v) = \int_0^1 Df(u + tv)\, dt ;$$

if u and v are close enough, or if there is convex entropy functional which acts as a symmetrizer [6], $A(u,v)$ has N real eigenvalues, denoted as

$$(14) \qquad \lambda_1(u,v) \leqq \lambda_2(u,v) \leqq \dots \leqq \lambda_N(u,v),$$

a basis of real eigenvectors,

$$(15) \qquad A(u,v)\, r_i(u,v) = \lambda_i(u,v)\, r_i(u,v),$$

and a basis of real eigenlinear forms

$$(16) \qquad \ell_i(u,v)\, A(u,v) = \lambda_i(u,v)\, \ell_i(u,v).$$

The Rankine-Hugoniot condition is satisfied by weak solutions of (1) across a discontinuity : if $x = s(t)$ parameterizes a discontinuity of u, and if u is of class C^1 on either side of this discontinuity, we denote by u and u^r the respective limits of u from the left and the right ; the transmission condition across the singularity is

$$f(u^r(s,.)) - f(u^\ell(s,.)) = \dot{s}\, (u^r(s,.) - u^\ell(s,.)),$$

or, using the clasical shorthand $[v] = v^r - v^\ell$,

$$[f(u)]\, (s,.) = \dot{s}\, [u]\, (s,.) .$$

This condition can be written alternatively as follows : there exists an index k such that

$$(17) \qquad \ell_i(u^\ell,u^r)\, (u^\ell - u^r) = 0, \ \forall\, i \neq k, \text{ and } \ \dot{s} = \lambda_k(u^\ell,u^r).$$

With these notations, a discontinuity satisfying (17) is termed admissible, if

(18) either $k \in LD$, or $k \in GNL$ and $\lambda_k(u\ ) > \lambda_k(u^\ell,u^r) > \lambda_k(u^r)$.

Let now $\mathcal{U}_1$ be a subset of $\mathcal{U}$ such that, for u and v in $\mathcal{U}_1$, A(u,v) has a basis of real eigenvectors.

A k-wave is a function of bounded variation on **R** with values in $\mathcal{U}_1$, such that, for all $i \neq k$

$$(19)\qquad \lim_{\xi'\uparrow\xi,\xi''\downarrow\xi} \frac{\ell_i(u(\xi'),u(\xi''))\ (u(\xi') - u(\xi''))}{|u(\xi') - u(\xi'')|} = 0\ .$$

In particular, if u is piecewise absolutely continuous, so that

$$\frac{du}{d\xi} = g\ dx + \sum_s \delta(.-\xi_s)\ [u](\xi_s),$$

then u is a k-wave if, for all $i \neq k$,

$$(20)\qquad \left|\begin{array}{l}(\ell_i(u)\ g)(\xi) = 0,\ \forall\, \xi \notin \bigcup_s \{\xi_s\}; \\ \\ \ell_i(u(\xi_s-0),u(\xi_s+0))\ [u](\xi_s) = 0,\ \forall\ s.\end{array}\right.$$

A k-wave will be called admissisble if (18) holds at discontinuity points. If u is a function of time also, it is a k-wave if (19) holds for all time.

One solves Riemann Problem by observing first that a function of the form u(x,t) = v(x/t) solves (1) iff

(21) $f(v)^{\cdot} - \xi\dot{v} = 0$ in the sense of distributions.

If, moreover, v is a k-wave, $\dot{v}$ is proportional to $r_k(v)$ at smooth points, and [v] is proportional to $r_k(v(\xi_s-0),v(\xi_s+0))$ at discontinuous points.

For every u in , and every k in {1,...,N} , P.D. Lax [9] constructs a one parameter family of states u^r such that there exists an admissible k-wave v satisfying (21) and the boundary conditions

(22) $u(-\infty) = u^\ell\ ,\ u(+\infty) = u^r$.

Denote this one-parameter family by

$$u^r = \Phi_k(\varepsilon_k, u^\ell).$$

The parameterization can be chosen so that Φ_k is of class C^2, with bounded third derivatives, and

$$(23) \qquad \Phi_k(0,u^\ell) = u^\ell, \quad \partial\Phi_k/\partial\varepsilon_k\big|_{\varepsilon_k=0} = r_k(u^\ell).$$

Riemann problem is then solved by a sequence of admissible k-waves, $k = 1, 2, \ldots, N$; define a mapping Φ from a neighborhood of $(0,u^\ell)$ in $\mathbf{R}^N \times \mathcal{U}$ to $\mathcal{U}$ by

$$(24) \qquad \Phi(\varepsilon,u^\ell) = \Phi_N(\varepsilon_N,\Phi_{N-1}(\varepsilon_{N-1}, \ldots,\Phi_1(\varepsilon_1,u^\ell))).$$

Then, if u^r is close enough to u^ℓ, the implicit functions theorem shows that there exists a unique ε in a neighborhood of 0 such that

$$(25) \qquad u^r = \Phi(\varepsilon,u^\ell).$$

When (25) holds, we denote

$$(26) \qquad \varepsilon = E(u^\ell,u^r) \; ;$$

ε_i is usually called the strength of the i-wave from u^ℓ to u^r.

Even if the characteristic fields are neither linearly degenerate nor genuinely non-linear, Riemann problem is solved by exhibiting one-parameter families of states u^r which are reached from u^ℓ by a self-similar k wave, and then by using the implicit function theorem, if the data are close enough ; otherwise it is a global problem.

It is important to observe that the self-similar solution of Rieman problem is constant outside of the cone $\lambda_1(u^\ell)t \leqq x \leqq \lambda_N(u^r)t$.

3. REVIEW OF SOME IDEAS IN GLIMM'S CONSTRUCTION. In the random choice method, [5], the first step is to replace the initial condition u_0 by a piecewise constant function on intervals of length Δx. At each point $i\Delta x$, the data are locally the data of a Riemann problem, and if Δt is small enough, the waves which solve two neighboring Riemann problem do not have time to interact. Therefore, one solves (1) exactly on $\mathbf{R} \times [0,\Delta t]$. As the exact solution obtained at

time Δt is not piecewise constant, the next step is to create new piecewise constant data by random sampling [5], with a simplification due to A. Chorin [1], and a deterministic version due to Liu, which is convenient to read in [13].

In order to show the stability of the scheme, J. Glimm introduces two functionals defined on all sequences $u = (u_i)_{i \in \mathbf{Z}}$, where u_i belongs to an open set $\mathcal{U}_2$ such that Riemann problem can be solved for all states u^ℓ and u^r in $\mathcal{U}_2$. The first functional is the total strength of the waves in u

$$L(u) = \sum_{i,k} |E_k(u_i, u_{i+1})|. \tag{27}$$

To define the second functional, let, for ε and η in $\mathbf{R}^N$

$$\Delta(\varepsilon,\eta) = \sum_{k>n} |\varepsilon_k \, \eta_n| + \sum_{k \in GNL} |\varepsilon_k \, \eta_k| - \varepsilon_k^+ \, \eta_k^+ \tag{28}$$

where $r^+ = (r + |r|)/2$. Then the total interaction potential is

$$Q(u) = \sum_{i<j} \Delta(E(u_i,u_{i+1}), E(u_j,u_{j+1})). \tag{29}$$

Clearly, if $\mathcal{U}_2$ is small enough, L is equivalent to the total variation of u, and Q can be estimated by the square of the total variation.

The next important step of Glimm construction is to show that, if the total variation of u, at the initial time is small enough, there exists a constant M such that $n \to L(u(n\Delta t)) + M\, Q(u(n\Delta t))$ decreases with respect to n. This does not depend on the sampling.

The main tool to do this is an interaction estimate : there exists an open set $\mathcal{U}_3$ included in $\mathcal{U}_2$, and a constant K, such that, for all u, v and w in $\mathcal{U}_3$,

$$\|E(u,w) - E(u,v) - E(v,w)\| \leq K \, \Delta(E(u,v),E(v,w)). \tag{30}$$

Here, $\|\varepsilon\| = \sum_i |\varepsilon_i|$.

4. DEFINITION OF THE CONTINUOUS GLIMM FUNCTIONALS. For a function u in BV($\mathbf{R}$), let

(31) $$\frac{du}{d\xi} = \nu + \sum_s \delta(.-\xi_s)\ [u](\xi_s)$$

be the decomposition of the measure $du/d\xi$ into the sum of a diffuse measure ν, and an atomic measure. We can always write

(32) $$\nu = z\,\rho\,,$$

where ρ is a scalar measure, and z is a vector-valued function, belonging to $L^1_{loc}(\mathbf{R},\rho)$; for instance,

$$\rho = \sum_{i=1}^{N} |\rho_i|\,.$$

If u takes its values in $\mathcal{U}_3$, we let

(33) $$M_k(u) = (\ell_k(u)z)\,\rho + \sum_s (.-\xi_s)\ E_k(u(\xi_s-0),u(\xi_s+0)).$$

If $\nu = z'\,\rho'$ is another decomposition, (33) defines the same measure scalar measure $M_k(u)$. We define

(34) $$\mathcal{L}_k(u) = \int |\,M_k(u)\,|\ ,$$

(35) $$\mathcal{L}(u) = \sum_k \ {}_k(u).$$

If we let the interaction potential between i-waves and j-waves be

(36) $$\Delta_{ij}(\varepsilon,\eta) = \begin{vmatrix} |\varepsilon_i\ \eta_j| \text{ if } i > j, \\ |\varepsilon_i\ \eta_i| - \varepsilon_i^+\eta_i^+ \text{ if } i = j \in GNL, \\ 0 \text{ otherwise,} \end{vmatrix}$$

then,

(37) $$\mathcal{Q}_{ij}(u) = \int_{x<y} \Delta_{ij}(M_k(u)(x),\ M_k(u)(y)\ ),$$

(38) $$\mathcal{Q}(u) = \sum_{i,j} \mathcal{Q}_{ij}(u).$$

It is easy to check that, if u is piecewise constant, the respective definitions of and coincide with the definitions of L and Q given in (28) and (29). If u is piecewise continuously differentiable,

(39) $$M_k(u) = \ell_k(u)\ u_x\ dx + \sum_s \delta(.-\xi_s)\ E_k(u(\xi_s-0),\ u(\xi_s+0).$$

5. APPROXIMATE CONSERVATION LAWS FOR WAVE STRENGTH . CASE 1 : NO SINGULARITIES. To estimate the evolution in time of the continuous Glimm functionals, one observes that the strength of waves satisfy an approximate conservation law. We start with the smoth regions. If we assume, for simplicity, that u is of class C^2 in some region of $\mathbf{R} \times \mathbf{R}^+$, and satisfies (1), we can multiply (1) by $\ell_k(u)$ on the left, and by $\operatorname{sgn}(\ell_k(u)u_x)$, and differentiate the expression obtained, with respect to space :

$$(\ell_k(u)\, u_t\,)_x + (\lambda_k(u)\, \ell_k(u)\,)_x = 0.$$

But,

$$(\ell_k\, u_t)_x = (\ell_k u_x)_t + D\ell_k\, (u_x \otimes u_t - u_t \otimes u_x),$$

where we have used the notation

$$D\ell_l(u)\; v \otimes w = \lim_{\tau \to 0} \tau^{-1}\, (\ell_k(u+\tau v) - \ell_k(u))\; w\ .$$

Therefore, as $\ell_k\, u_t$ vanishes whenever $\ell_k u_x$ changes sign, we have

$$|\ell_k\, u_x|_t + (\lambda_k |\ell_k\, u_x|)_x = R_k(u,u_x)\, \operatorname{sgn}(\ell_k\, u_x), \tag{40}$$

with

$$R_k(u,u_x) = D\ell_k\, (u_t \otimes u_x - u_x \otimes u_t).$$

It is convenient to express R_k more explicitly : using the decomposition

$$u_t = -\sum_i \lambda_i\, (\ell_i\, u_x)\, r_i, \quad u_x = \sum_n\, (\ell_n\, u_x)\, r_n\ ,$$

we obtain

$$R_k(u,u_x) = \sum_{n>i} D\ell_k\, (r_i \otimes r_n - r_n \otimes r_i)\ \ell_i\, u_x\ \ \ell_n\, u_x\ . \tag{41}$$

The above calculus is very close to the one used by F. John in [8].

Unless N = 2, and the eigenlinear forms are the gradients of the Riemann invariants, the expression R_k does not vanish identically. Otherwise, it is always possible to find u such that

$$\sum_k \int_{\mathbf{R}} R_k(u,u_x)\, \operatorname{sgn}(\ell_k\, u_x)\, dx > 0.$$

The k-characteristic curve is defined by the equation

$$\frac{\partial X_k}{\partial t}(x,t,s) = \lambda_k(u(X_k(x,t,s),s))$$

$$X_k(x,t,t) = x.$$

Let us estimate the variation of the strength of the k-waves in a characteristic tube :

$$\frac{d}{ds}\int_{X_k(x',t,s)}^{X_k(x'',t,s)} |\ell_k\ u_x|\ dx\Big|_{s=t} = (|\ell_k(u)u_x|\ \lambda_k(u))(x,t)\Big|_{x'}^{x''} +$$

$$+ \int_{x'}^{x''} [R_k\ \mathrm{sgn}(\ell_k u_x) - (\lambda_k(u)\ |\ell_k(u)u_x|)_x]\ dx =$$

$$= \int_{x'}^{x''} R_k\ \mathrm{sgn}(\ell_k(u)\ u_x)\ dx.$$

If we define

$$\Lambda_{ac} = \sum_{i>n} (\lambda_i - \lambda_n)\ |\ell_i u_x|\ |\ell_n u_x|, \tag{42}$$

it can be proved that there exists a constant C_k depending on $\mathcal{U}_3$ such that

$$|R_k| \leqq C_k\ \Lambda_{ac}, \quad \text{for all } u \text{ with values in } \mathcal{U}_3, \tag{43}$$

and therefore,

$$\frac{d}{ds}\int_{X_k(x',t,s)}^{X_k(x'',t,s)} |\ell_k u_x|\ dx\Big|_{s=t} \leqq C_k \int_{x'}^{x''} \Lambda_{ac}\ dx. \tag{44}$$

From the results of Douglis [4] and Hartman and Wintner [7], we know that C^1 solutions can be approximated by C^2 solutions, so that (44) still holds if u is of class C^1, away from the singularities.

6. APPROXIMATE CONSERVATION LAWS FOR WAVE STRENGTH . CASE 2 : ONE SINGULARITY.

What happens to the strength of the waves across a shock ? We consider only the strength of the waves inside a characteristic tube, and the width of a characteristic tube varies discontinuously when the tube encounters a discontuinuity. In particular, if a j-characteristic tube crosses a k-singularity, we can evaluate this discontinuity in the width of the tube as follows : if the k-shock

(45) $$x = x_0 + \sigma(t - t_0), \quad \sigma = \lambda_k(u^\ell, u^r),$$

separates two regions where u is constant and equal to u^ℓ on the left and u^r on the right, the characteristic in the right region have the equation

(46) $$x = x' + \lambda_j^r (t - t_0) .$$

Assume that $j > k$; a j-characteristic starting at (x',t) in the right region is bent when it crosses the shock at time

$$t' = t_0 + (x' - x_0)(\sigma - \lambda_j^r)^{-1} ,$$

and the new equation of the characteristic in the left region is

$$x = x_0 + \sigma(t' - t_0) + \lambda_j^\ell (t - t').$$

Another j-characteristic starting in the right region at (x'',t) would have the equation in the left region

$$x = x_0 + \sigma(t - t_0) + \lambda_j^\ell \quad (t - t''),$$

with

$$t'' = t_0 + (x'' - x_0) (\sigma - \lambda_j^r).$$

Therefore, the width of a characteristic tube is multiplied by $(\sigma - \lambda_j^\ell)(\sigma - \lambda_j^r)^{-1}$, across a singularity, and the relevant quantity to be considered here is the jump $[(\dot{s} - \lambda_j) |\ell_j \; u_x|]$.

Now, if we consider a k-characteristic tube, and a k-singularity, the geometric entropy condition implies that the k-characteristic do not cross the k-singularity ; from the point $(s(t),t)$ on the singularity, there are two backward k-characteristics and no forward k-characteristics. Therefore, the curvilinear sector limited by the two backward k-characteristics is a good generalization of a characteristic tube, in the spirit of [2]. The variation of the total strength of the k-waves in this tube is the sum of the variation of the strength of the singularity, and of a quantity which accounts for the shrinking of the characteristic tube to a point. Thus, the relevant quantity is, according to the same analysis as above,

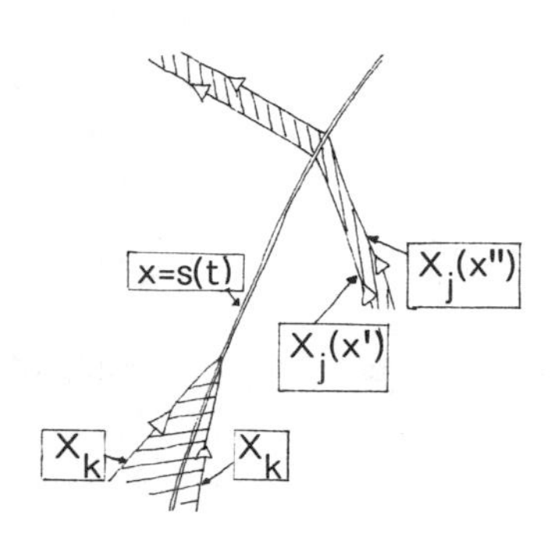

$$\frac{\partial |E_k|}{\partial t} - [(\dot{s} - \lambda_k)\, |\ell_k\, u_x|].$$

This holds if the k-th characteristic field is genuinely nonlinear ; if it is linearly degenerate, the k-characteristics do not cross a k-singularity, and $\dot{s} = \lambda_k^r = \lambda_k^\ell$, so that there is no change in width in the characteristic tube.

To estimate all these quantities, I use a "differential Glimm estimate", which describes the interaction of infinitesimal waves with finite waves.

Let $x = s(t)$ separate two smooth regions, and let

$$u_1 = u^\ell(s(t+h),t+h),$$
$$u_2 = u^\ell(s(t),t),$$
$$u_3 = u^r(s(t),t),$$
$$u_4 = u^r(s(t+h),h) \;;$$

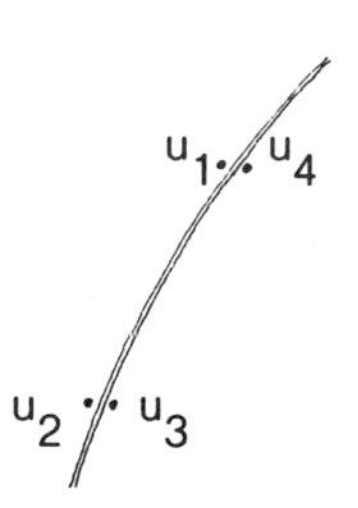

applying twice Glimm's interaction estimate (30), and passing to the limit as h tends to zero, one obtains [14, Corollary 3.4],

$$(47)\qquad \left|\frac{\partial E_k}{\partial t} - [\ell_k u_x(\dot{s} - \lambda_k)]\right| + \sum_{i\neq k} [\ell_i u_x(\dot{s} - \lambda_i)] \leq K\, \Lambda_s |E_k|,$$

where

$$(48)\qquad \Lambda_s = \left|\begin{array}{ll} \sum_{i>k} \{(\lambda_i-\dot{s})\,|\ell_i u_x|\}^\ell + \sum_{i<k} \{(\dot{s}-\lambda_i)\,|\ell_i u_x|\}^r & \text{if } k \in LD \;; \\ \sum_{i\geq k} \{(\lambda_i-\dot{s})\,|\ell_i u_x|\}^\ell + \sum_{i\leq k} \{(\dot{s}-\lambda_i)\,|\ell_i u_x|\}^r & \text{if } k \in GNL. \end{array}\right.$$

If we ask for a reinforced entropy condition

$$(49)\quad \left| \begin{array}{l} * \text{ either } k \in LD \\ * \text{ or } k \in GNL \text{ and} \\ \quad \left| \begin{array}{l} \lambda_k(u^\ell) > \dot{s} > \lambda_k(u^r) ; \\ \max(\lambda_i(u^\ell),\lambda_i(u^r)) < \dot{s}, \ \forall\, i < k ; \\ \min(\lambda_i(u^\ell),\lambda_i(u^r)) > \dot{s}, \ \forall\, i > k ; \end{array}\right. \end{array}\right.$$

then, not only is Λ_s always greater than or equal to zero, but it is possible to estimate what I called previously the "relevant quantities" in terms of Λ_s, with the help of (47). In particular, we obtain

$$|[(\lambda_j - \dot{s})|\ell_j u_x|]| \leq |[(\lambda_j - \dot{s})\ell_j u_x]|$$

$$\left|\frac{\partial |E_k|}{\partial t} + [(\dot{s} - \lambda_k)|\ell_k u_x|]\right| \leq \left|\frac{\partial E_k}{\partial t} + [(\lambda_k - \dot{s})\ell_k u_x]\right| .$$

Let the singularities of a solution u of (1) be described by

$$I_p \to \mathbf{R} \times \mathbf{R}^+, \qquad t \to s_p(t),$$

with s_p of class C^1, and u of class C^1 in a neighborhood, on either side of $\Gamma_p = \{(x,t) \;/\; t \in I_p, \ x = s_p(t)\}$. We define a singular measure

$$<\tilde{\Lambda}_s,\psi> = \sum_p \int_{I_p} \{|E_{k(p)}|\Lambda_{s,p}\}\psi(s_p,.)\, dt,$$

where k(p) is the index of the family to which the singularity s_p belongs, and $\Lambda_{s,p}$ is the expression defined by (48), relative to s_p. With these notations, the folowing holds for a classical characteristic tube :

$$(50)\quad \left| \begin{array}{l} \displaystyle\int_{x'}^{x''} |M_k(u(t))| \leq K_k \Big[\int_{X_k(x',t,s)}^{X_k(x'',t,s)} |M_k(u(s))| + \\ \displaystyle + \int_s^t \int_{X_k(x',t,\tau)}^{X_k(x'',t,\tau)} (\Lambda_{ac} + \tilde{\Lambda}_s)\,(x,\tau) \quad , \quad \forall\, s < t . \end{array}\right.$$

If the sides of the tube are generalized characteristics, i.e. contain segments of shock curves, one must be careful to consider in the above formula only minimal backward characteristics on the left, and maximal backward characteristics on the right ; this is easily deduced by a continuity argument. For (50) to hold, the characteristic tube must not contain a meeting point of singularities. If there is a meeting point of singularities, one more term must be added in (50), under a reasonable assumption on those points. This assumption will be expressed as the existence of a limit, in an adequate topology ; the study of this topology is the object of next paragraph.

7. WHAT IS A GOOD BV TOPOLOGY FOR OUR PURPOSES ?

Consider what the usual BV convergences allow or forbid : let u^n be a sequence of vector valued functions on BV which converge in the norm topology, i.e.

$$\|u - u^n\|_{BV} = \int |d(u^n - u)/dx| + |u^n(-\infty) - u(-\infty)| \to 0 ;$$

if u has discontinuities, then the size of the discontinuities of u^n converges to the size of the discontinuities of u, and moreover, the abscissa of the largest N discontinuities of u^n is identical to the abscissa of the largest N discontinuities, for $n \geqq n_0(N)$. Therefore, the norm convergence is too strong to allow $u(.+\frac{1}{n})$ to converge to u as n tends to infinity, as soon as u has a discontinuity.

On the other hand, if the sequence u^n converges weakly, i.e.

$$<\frac{du^n}{dx} - \frac{du}{dx}, \psi> \to 0, \ \forall \psi \in C_0^0(\mathbf{R}) ; \ u^n(-\infty) - u(-\infty) \to 0,$$

then overshoot, undershoot and splitting are perfectly permissible, and this has unpleasant consequences on nonlinear functionals of u ; for instance, let

$$u^n(x) = -1 \text{ if } x<0, \ 0 \text{ if } 0 < x < n^{-1}, \ \ 1 \text{ if } n^{-1} < 1 ,$$

and define a nonlinear functional on BV by

$$N(u) = \int |\frac{d}{dx} u^2| ;$$

then, it is clear that $N(u^n) = 2$, which does not converge to $N(u^\infty)$, where u^∞ is the weak limit of u^n.

In the present study, we want a topology such that $\mathcal{L}_k$ is continuous with respect to u ; this is achieved if we define a distance on BV as follows : let

(51) $\Psi = \{\psi \in W^{1,1}(\mathbf{R}) / \psi^{-1} \in W^{1,1}(\mathbf{R}), \psi' \geq 0\}$.

Then, for u and v in BV(**R**), let

(52) $d(u,v) = \inf\{\|u - v \circ \psi\|_{BV} + \|\psi - Id\|_{1,1} + \|\psi^{-1} - Id\|_{1,1} / \psi \in \Psi\}$

where

$$\|\psi\|_{1,1} = \int_{\mathbf{R}} (|\psi| + |\psi'|)\, dx.$$

The space BV is complete for this distance, and if u takes its values in $\mathcal{U}_3$, it is easy to check that the mapping $u \to M_k(u)$ is continuous on BV with the distance d. There is an obvious analogous definition of d on BV(a,b), for any real a and b.

We make the following requirement on u at the points (x_q, t_q) where singularities meet, appear or disappear : we scale u at (x_q, t_q) by letting

(53) $u_q^{\varepsilon}(X,T) = u(x_q + \varepsilon X, t_q + \varepsilon T)$.

We shall assume that

(54) There exist $T_- < 0$, and $T_+ > 0$ such that $u^{\varepsilon}(.,T_+)$ & $u^{\varepsilon}(.,T_-)$ are Cauchy sequences for d.

If (53) holds, u_q^{ε} converges to a self similar function u_q^0 ; if u is admissible, so is u_q^0 ; for T>0, u_q^0 is the self-similar solution of the Riemann problem with data $u^{\ell} = u(x_q - 0, t_q)$, $u^r = u(x_q + 0, t_q)$.

8. APPROXIMATE CONSERVATION LAWS FOR WAVE STRENGTH . CASE 3 : SINGULARITIES MEET, APPEAR OR DISAPPEAR.

Assumption (54) allows to define

$$\mathcal{L}_q^{\pm} = \lim \mathcal{L}(u_q^{\varepsilon}(.,T_{\pm})\big|_{[-A,A]}) \ ; \ \mathcal{Q}_q^{\pm} = \lim \mathcal{Q}(u_q^{\varepsilon}(.,T_{\pm}))\big|_{[-A,A]}) \ ;$$

here [-A,A] is any large finite interval. As u^0 solves Riemann problem for T > 0, $\mathcal{Q}_q^+ = 0$; an n-fold Glimm estimate [14, proposition 3.2], and the corresponding passage to the limit as n tends to infinity [14, corollary 4.1] imply that

$$(55)\qquad \left|\begin{array}{l}\lim\limits_{h\downarrow 0}\ [\ \mathcal{L}(u(t_q+0)\big|_{[x_q-h,x_q+h]}) - \mathcal{L}(u(t_q-0)\big|_{[x_q-h,x_q+h]})\ ] \leq \\ \leq\ K\ Q_q^-\ \exp(K\mathcal{L}_q^-).\end{array}\right.$$

If we introduce a measure $\underline{\Lambda}$ on $\mathbf{R} \times \mathbf{R}^+$, which will be naturally called the cancellation measure [11],

$$(56)\qquad \underline{\Lambda} = \Lambda_{ac}dxdt + \sum_p \bar{\Lambda}_{s,p}\delta(. - s_p) + \sum_q \exp(K\mathcal{L}_q)Q_q\ \delta(.-x_q,.-t_q),$$

then the essential result is that, if $x_1 < x_2$, and if the minimal backward k-characteristic through (x_1,t) and the maximal backward k-characteristic through (x_2,t) do not go through a meeting point of singularities, then

$$(57)\qquad \int_{x_1}^{x_2} |M_k(u(t))| \leq \int_{X_k(x_1,t,s)}^{X_k(x_2,t,s)} |M_k(u(s))| + K_k \int_s^t \int_{X_k(x_1,t,s)}^{X_k(x_2,t,\ )} \underline{\Lambda}\ .$$

9. ESTIMATES ON Q ; STATEMENT OF THE RESULT. Estimating the variation of Q with respect to time is essentially a geometric problem. Consider, for instance Q_{jk}, with $j > k$; then, with some abuse of notation,

$$Q_{jk}(u(t)) = \int_{x<y} |M_j(u(x,t))|\ |M_k(u(y,t))|\,dx\ dy.$$

It is natural to compare $M_j(u(x,t))$ and $M_j(u(X_j(x,t,s),s))$, for $t > s$, or rather, the strength of j-waves at the two ends of a j-characteristic tube. Similarly, one compares the strength of the k-waves at the ends of a k-characteristic tube, and the difference can be expressed in terms of the cancellation measure $\underline{\Lambda}$.

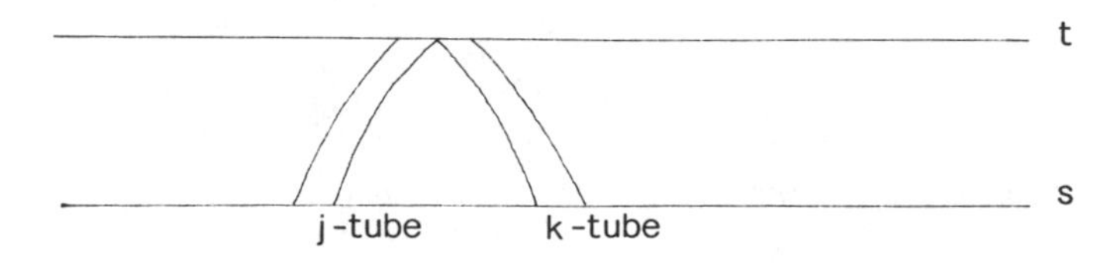

In the course of this process, a piece of Q_{jk} (u(s)) is not accounted for : the piece which is integrated on the set

$$\{(x,y) \;/\; \text{for some } z,\ X_j(z,t,s) < x < y < X_k(z,t,s)\}\ .$$

This piece disappears because close enough j-characteristics and k-characteristics intersect ; see the above figure. In smooth regions, the size of the lost piece of Q_{jk} is

$$\int_{x<y<x+(\lambda_j-\lambda_k)\,\delta t} |\ell_j u_x\|\ell_k u_y|\ dxdy \sim \delta t \int(\lambda_j-\lambda_k)\ |\ell_j u_x\|\ell_k u_x|dx$$

$$\sim\ \delta t \int \Lambda_{ac}\ dx.$$

For the other cases, one can perform analogous computations, and therefore, the variation of Q is estimated by $(K\mathcal{L} - 1)\ \underline{\Lambda}$. We summarize the result and its conditions of validity as follows :

(58) u takes its values in an open set $\mathcal{U}_3$, where Glimm's estimate (30) holds.

(59) There exist curves Γ_p, parameterized by t

$$I_p \to \mathbf{R} \times \mathbf{R}^+,\quad t \to s_p(t),$$

where I_p is open in $\mathbf{R}^+$, $s_p \in C^1(I_p)$, the Γ_p are disjoint, and any compact of $\mathbf{R} \times \mathbf{R}^+$ intersects at most a finite number of the Γ_p's.

We denote

$$S_1 = \bigcup_p \Gamma_p\ ,\ S_2 = \bigcup_p (\overline{\Gamma}_p \setminus \Gamma_p).$$

(60) u is of class C^1 in $\mathbf{R} \times \mathbf{R}^+ \setminus (S_1 \cup S_2)$.

(61) u, u_x and u_t have limits along $\mathbf{R} \times \mathbf{R}^+ \cap \{x \lessgtr s_p(t)\}$ as x' and t' tend to $s_p(t)$, t, for all t in I_p, and all p.

(62) At every point of S_2, condition (54) is satisfied.

Then, we have the following result :

THEOREM 1. Under conditions (58) - (62), if u is a solution of (1), which satisfies the admissibility condition (49), and if $\mathcal{L}$ and $\mathcal{Q}$ are defined respectively by (35) and (38), there exists a positive measure $\underline{\Lambda}$ on $\mathbf{R} \times \mathbf{R}^+$ such that

$$\frac{d}{dt}\mathcal{L}(u(t)) \leq \int \underline{\Lambda}(t)\, dx \tag{63}$$

$$\frac{d}{dt}\mathcal{Q}(u(t)) \leq \int \underline{\Lambda}(t)\, dx \quad (K\mathcal{L} - 1), \tag{64}$$

where K is a constant which depends on $\mathcal{U}_3$ and f.

Arguing as in [5], we have the corollary :

COROLLARY 2. If $\mathcal{L}(u_0)$ is small enough, there exists a constant M such that $(\mathcal{L} + M\,\mathcal{Q})(u(t))$ decreases.

This means, of course that L^∞ estimates on u, and BV estimates on the initial data imply BV estimates on u(t), for all t.

11. UNIQUENESS FOR RIEMANN PROBLEM.

We shall sketch the proof of the following theorem :

THEOREM 3. Assume that u satisfies the assumptions of Theorem 1, and that the initial conditions are Riemann initial conditions. If $E(u^\ell, u^r)$ is small enough, then u is invariant by the scaling $(x,t) \to (kx,kt)$, and, therefore, is the self-similar solution of Riemann problem.

Idea of the proof. We can see, from Theorem 1, that, for all time $\mathcal{Q}(u(t)) = 0$, and that $\mathcal{L}(u(t))$ is constant in time. This implies that u is a sequence of k-waves, according to definition (19) : for all time, the interaction potential is zero, so that the interior of the support of $M_k(u)$ is disjoint of the interior of the support of $M_j(u)$, for all $j \neq k$. Let $\mathcal{R}_k$ be a region where u is smooth ; there exists k such that

$$\ell_i\, u_x = 0, \text{ for all } (x,t) \text{ in } \mathcal{R}_k, \text{ and all } i \neq k.$$

From equation (1), we obtain

$$u_t + \lambda_k\, u_x = 0,$$

so that, in $\mathcal{R}_k$, the k-characteristics are straight lines.

Consider now a k-singularity belonging to the k-th family : as $\underline{\Lambda} = 0$, relations (47) and (48) imply that

$$(\ell_i u_x)^\ell = (\ell_i u_x)^r = 0, \ \forall\ i \neq k,$$

$$(\ell_k u_x)^\ell = (\ell_k u_x)^r = 0, \text{ if } k \in \text{GNL}.$$

The speed of the singularity satisfies the equation

$$\dot{s} = \lambda_k(u^\ell, u^r),$$

which we differentiate with respect to time

$$\ddot{s} = D_1\ \lambda_k(u^\ell, u^r)\ [\frac{\partial u^\ell}{\partial t} + \dot{s}\frac{\partial u^\ell}{\partial x}] + D_2\lambda_k(u^\ell, u^r)[\frac{\partial u^r}{\partial t} + \dot{s}\frac{\partial u^r}{\partial x}].$$

If $k \in$ GNL, it is immediate that $\ddot{s} = 0$, and the singularity is a straight line. If $k \in$ LD, this is still true, because we can write

$$\lambda_k(u^\ell, u^r) = \lambda_k(u^\ell) = \lambda_k(u^r),$$

so that

$$\ddot{s} = \sum_{j=1}^{N} D\lambda_k(u^\ell)\ r_j^\ell\ \{(\lambda_k - \lambda_j)\ \ell_j \frac{\partial u}{\partial x}\}^\ell,$$

and there is no contribution in the above sum of the k-th term. The singularity is still a straight line.

No singularities can meet, otherwise, there would be an atomic contribution in $\underline{\Lambda}$, and with some more geometrical reasoning, which is detailed in [14], the result follows.

This result must be compared to the result of DiPerna [3], which holds in a wider functional class, but only for a system of two equations, with a convex entropy function. Another comparable result is included in an article by Li Da-Qian and Yu Wen-Ci [10], who prove the uniqueness for the solution of Riemann problem, in the piecewise C^1 class, with a very different method from the one I employed here : their result relies on an assumption comparable to (54), but without ever stating it explicitly. I shall prove in next paragraph that the method described here can be generalized to piecewise Lipschitz continuous functions, with the help of a new trace theorem.

12. ESPRIT DE L'ESCALIER : A GENERALIZATION TO PIECEWISE LIPSCHITZ CONTINUOUS FUNCTIONS. The joke one remembers when leaving the party is part of the spirit of the party ; this is why I would not leave this generalization out of the paper.

When trying to extend the previous construction to the piecewise Lipschitz case, there is no particular difficulty in the smooth regions, nor at the points of S_2. The difficulty is to extend (47) and (48) which do not make obviously sense without a trace theorem. Therefore, we must be able to take traces of u_t and u_x along a singularity of u, if the singularity belongs to a genuinely non-linear family, or of $\ell_i u_x$ and $\ell_i u_t$ for $i \neq k$, otherwise.

The first result concerns the nature of singular curves of a piecewise Lipschitz continuous function which belongs to BV.

We have a lemma on the characteristic function of a set $\{x<s(t)\}$:

LEMMA 4. Let $E = \{(x,t) \ / \ x \leq s(t)\}$. Then, the following two assertions are equivalent :

(i) 1_E is locally of bounded variation ;

(ii) s is locally of bounded variation.

Proof. (Abridged). If f is piecewise constant on [0,1], the gradient of $u = 1_E$ estimates f in the following fashion :

$$\int_0^1 |df/dt| \ dt = \int_0^1 \int_{-\infty}^{\infty} |\partial u/\partial t| \ dx \ dt,$$

which is proved easily by an explicit computation.

Then, one approximates s by a piecewise constant s^h, and the rest is easy. ∎

With this kind of result, it is easy to see that u is piecewise Lipschitz continuous, and locally of bounded variation, and if its curves of discontinuity are parameterized by t, they are of bounded variation with respect to t. In particular, Rankine-Hugoniot condition makes sense, and therefore, is Lipschitz continuous, and, using the differential equation $\dot{s} = \lambda_k(u^\ell, u^r)$, s is continuously differentiable, and its second derivative is bounded.

Concerning the trace result, we remark first that if u is in $BV([0,1]^2)$, and satisfies (1) in the sense of distributions, then

$$f(u) \in BV([0,1]^2),$$

$$f(u)_x \in M^1([0,1]^2),$$

$$u_t \in M^1([0,1]^2),$$

and therefore,

$$u \in BV([0,1]_t ; M^1([0,1]),$$

which implies that u(t) has a limit in the sense of distributions as t decreases to t_0, or increases to t_0. If u is piecewise Lipschitz continuous, the diffuse part of its derivatives is bounded, and the atomic part will converge because the singularities are parameterized by continuous curves, and the strength of the singularity is continuous in time. Therefore, u(t) converges in BV equipped with the distance d. This remark is even more interesting, if one observes that it is possible to exchange the role of x and t, and to deform the curve on which the trace is taken. Let s parameterize a k-shock, and let us make the change of variable

$$y = x - s(t), \; t = \tau, \; u(x,t) = w(y,\tau).$$

Then, (1) becomes

$$w_\tau + G(w,\tau)_y = 0,$$

where

$$G(w,\tau) = f(w) - \dot{s}(\tau)\, w.$$

Thanks to the implicit functions theorem, and to the shock condition (49), G is invertible in a neighborhood of y = 0, $w = w^\ell(0,\tau)$, or y = 0, $w = w^r(0,\tau)$, and therefore, if

$$G(w,\tau) = z \quad w = H(z,\tau),$$

the following equation is satisfied

$$z_y + H(z,.)_\tau = 0.$$

According to the previous argument, z has BV traces on $y=0$, so that it is possible to take traces of u_t and u_x along the shock.

If k was the index of a linearly degenerate characteristic field, a simple refinement of the same argument would work : in this case, D_wG is of rank N-1 at $w = w^r$ or w^ℓ, $y = 0$, but we do not need to recover the trace of $\ell_k u_x$, which is the only one not provided by inverting G.

REFERENCES

[1] A. J. Chorin, Random choice solutions of hyperbolic systems, J. Comput. Phys. 22(1976)517-533.

[2] C. Dafermos, Generalized characteristics and the structure of solutions of hyperbolic conservation laws, Indiana Univ. Math. J. 26(1977)1097-1119.

[3] R. DiPerna, Uniqueness of the solutions of nonlinear hyperbolic systems of conservation laws, Indiana Math. J., 28(1979)137-188.

[4] A. Douglis, Some existence theorems for hyperbolic systems of partial differential equations in two independent variables, Comm. Pure Appl. Math. 5(1952)119-154.

[6] A. Harten, P.D. Lax, B. van Leer, On upstream differencing and Godunov type schemes for hyperbolic conservation laws, S.I.A.M. Review 25(1983)35-61.

[7] P. Hartman, A. Wintner, On hyperbolic partial differential equations, Amer. J. Math. 74(1952)834-864.

[8] F. John, Formation of singularities in one-dimensional nonlinear wave propagation, Comm. Pure Appl. Math. 27(1974)377-405.

[9] P.D. Lax, Hyperbolic systems of conservation laws, II Comm. Pure Appl. Math. 10(1957)537-567.

[10] Li Da-Qian and Yu Wen-Ci, Boundary value problems for the first order quasilinear hyperbolic systems and applications, J. Differential Equations, 41(1981)1-26.

[11] T. P. Liu, The Riemann problem for general systems of conservation laws, J. Differential Equations, 20(1976)369-388.

[12] T.P.Liu, The entropy condition and the admissibility of shocks, J. Math, Anal. Appl. 53(1976)78-88.

[13] T.P. Liu, Admissible solutions of hyperbolic conservation laws, Mem. A.M.S. 30 (1981).

[14] M. Schatzman, Continuous Glimm functionals and uniqueness of the solution of Riemann problem, Preprint #215, Center for Pure and Applied Mathematics, University of California, Berkeley , March 1984.

DEPARTEMENT DE MATHEMATIQUES
UNIVERSITE CLAUDE-BERNARD
69622 VILLEURBANNE CEDEX, FRANCE.

Lectures in Applied Mathematics
Volume 23, 1986

A PDE APPROACH TO SOME LARGE DEVIATIONS PROBLEMS

W.H. Fleming[1] and P.E. Souganidis[2]

ABSTRACT. We discuss some of the connections between large deviations and the theory of Hamilton-Jacobi PDE. We illustrate the effectiveness of viscosity solution methods by giving a new proof and extending a result of W.H. Fleming and C.-P. Tsai concerning the asymptotic formula for the minimum exit probability for a nearly deterministic process.

1. LARGE DEVIATIONS. The theory of large deviations is concerned with asymptotic formulas for exponentially small probabilities of events associated with stochastic processes. Such large deviations problems are typically formulated in terms of a family ξ_t^ε of processes depending on a small positive parameter ε. Let P^ε be the probability of some event depending on the sample paths of the process ξ_t^ε. If $-\varepsilon \log P^\varepsilon$ tends to a limit $I > 0$, then there is a large deviation. Usually, the limit I turns out to be characterized as the minimum in a certain associated optimization problem.

We are concerned here with large deviations problems of the kind treated by Freidlin-Ventsel [10], in which ξ_t^ε is a nearly

---

1980 Mathematics Subject Classification 60F10, 90D25, 35F20, 35L60.

[1]Partially supported by NSF under Grant No. MCS 8121940, by ONR under Grant No. N00014-83-K-0542 and by AFOSR under Grant No. AF-AFOSR 81-0116.

[2]Partially supported by ONR Grant No. N00014-83-K-0542 and by NSF under Grant No. DMS-8401725.

deterministic Markov diffusion in $\mathbb{R}^N$. Of particular interest are probabilities of events related to the first time τ^ε when ξ_t^ε reaches the boundary of a given region $D \subset \mathbb{R}^N$ (τ^ε is called the exit time). Particular examples considered in [10] are $P^\varepsilon = \Pr(\tau^\varepsilon \leq T)$, $P^\varepsilon = \Pr(\tau^\varepsilon > T)$, with T fixed, and $P^\varepsilon = \Pr(\xi_{\tau^\varepsilon}^\varepsilon \in \Gamma)$, where $\Gamma \subset \partial D$.

2. HAMILTON-JACOBI EQUATIONS AND LARGE DEVIATIONS. The theory of nonlinear, first order PDE of Hamilton-Jacobi type has been substantially developed with the introduction by M.G. Crandall and P.-L. Lions [2] of the class of viscosity solutions. This turns out to be the correct class of generalized solutions for such equations. M.G. Crandall, L.C. Evans and P.-L. Lions [1] provide a simpler introduction to the subject while the book by P.-L. Lions [11] and the review paper by M.G. Crandall and P.E. Souganidis [3] provide a view of the scope of the theory and references to much of the recent literature.

The results of Freidlin-Ventsel about exit (or nonexit) probabilities originally proved via probabilistic methods can also be obtained by PDE-viscosity solution methods [6],[8]. In outline, this is done in the following way. Consider, for instance, $P^\varepsilon = \Pr(\tau^\varepsilon \leq T)$. As a function of an initial time $s < T$ and an initial state $x = \xi_s^\varepsilon$, $P^\varepsilon(s,x)$ satisfies a linear, second order parabolic PDE (the backward PDE for the process ξ_t^ε). We make the logarithmic transformation $I^\varepsilon = -\varepsilon \log P^\varepsilon$. Then I^ε satisfies a nonlinear second order PDE, with a quadratic nonlinearity in the gradient DI^ε. For $\varepsilon = 0$ this equation reduces to a first order Hamilton-Jacobi equation with convex Hamiltonian. The boundary conditions for I^ε are 0 and $+\infty$. The technique is to show that, as $\varepsilon \to 0$, I^ε tends to a limit I, which must be a viscosity solution of the first order PDE. Then I is identified with the minimum in an associated calculus of variations problem.

Another case which we explain in more details in the next section is the one where ξ_t^ε is a controlled diffusion process and

P^ε is the minimum exit probability. In that case the nonlinearity in DI^ε is neither convex nor concave. Similar methods show that again $I^\varepsilon \to I$ as $\varepsilon \to 0$. The limit I turns out to be the lower value of an associated differential game. This was originally proved in a less general setting by W.H. Fleming and C.-P. Tsai [9] again via rather involved differential game theoretic and probabilistic arguments.

There are several advantages of the PDE-viscosity solution method (first used by L.C. Evans and H. Ishii [6] for convex cases) over the original probabilistic arguments. First, it presents a more or less unified approach to questions previously attached by different methods. Secondly, the techniques are almost totally analytic and much simpler than the probabilistic ones; the latter ones need either the rather difficult Freidlin-Ventcel estimates or an interpretation of I^ε as the value of a certain stochastic control or differential game theory problem. Finally, it allows us in the case of minimal exit probabilities to extend the previously known results.

3. MINIMUM EXIT PROBABILITIES AND DIFFERENTIAL GAMES. Let $\xi(\cdot)$ be an N-dimensional stochastic process with continuous sample paths defined for times $t \geq s$. Let $D \subset \mathbb{R}^N$ be open and bounded with smooth boundary ∂D. For initial time s and state $x = \xi(s) \in D$, let τ_{sx} denote the exit time from D (i.e., the first t such that $\xi(t) \in \partial D$). For fixed $T > s \geq 0$, $P(\tau_{sx} \leq T)$ is the exit probability.

We assume that $\xi(\cdot)$ is a controlled Markov diffusion process, satisfying in the Ito-sense the stochastic differential equation

$$d\xi(t) = b[\xi(t),y(t)]dt + \varepsilon^{\frac{1}{2}}\sigma[\xi(t)]dw(t) ,$$

where $y(t)$ is a control applied at time t, $\varepsilon > 0$ is a parameter, σ is an $N \times N$ matrix and $w(\cdot)$ is an N-dimensional brownian motion. We assume that $y(t) \in Y$, where $Y \subset \mathbb{R}^M$ is compact. Moreover, $b(\cdot,\cdot)$, $\sigma(\cdot)$ are Lipschitz; and the matrix

$a(x) = \sigma(x)\sigma'(x)$ has all eigenvalues $\geq c > 0$.

The control process $y(\cdot)$ is assumed to have the feedback form

$$y(t) = \underline{y}(t,\xi(t))$$

where $\underline{y}:[s,T] \times \mathbb{R}^N \to Y$ is a Borel measurable function. Let

$$q^\varepsilon_{\underline{y}}(s,x) = P(\tau_{s,x} \leq T).$$

Of course, the exit probability depends on ε and $\underline{y}$ in view of (3.1), namely $\tau_{sx} = \tau^{\varepsilon,\underline{y}}_{sx}$. The minimum exit probability is

$$q^\varepsilon = \min_{\underline{y}} q^\varepsilon_{\underline{y}} . \tag{3.2}$$

The function $q^\varepsilon(s,x)$ satisfies the dynamic programming equation

$$q^\varepsilon_s + \frac{\varepsilon}{2} \operatorname{tr} a(x) q^\varepsilon_{xx} + \min_{y \in Y}\{b(x,y) \cdot q^\varepsilon_x\} = 0, \tag{3.3}$$

in the cylinder $[0,T] \times D$, where q_x denotes the gradient vector and $\operatorname{tr} a q_{xx} = \sum_{i,j} a_{ij} q_{x_i x_j}$. The boundary conditions are

$$\begin{aligned} q^\varepsilon(s,x) &= 1 \quad \text{for} \quad s < T,\ x \in \partial D \\ q^\varepsilon(T,x) &= 0 \quad \text{for} \quad x \in D. \end{aligned} \tag{3.4}$$

In general, it is difficult to get effective information about q^ε and the optimal control law in this way. Instead, we seek an asymptotic formula for q^ε, valid for small $\varepsilon > 0$, of the form

$$-\lim_{\varepsilon \downarrow 0} \varepsilon \log q^\varepsilon = I, \tag{3.5}$$

where I turns out to be the lower value of a certain differential game. Equation (3.5) can be written as

$$q^\varepsilon = \exp\left\{- \frac{I+o(1)}{\varepsilon}\right\} ,$$

which is a weaker result than a WKB expansion

$$q^\varepsilon = \exp\left\{-\frac{I}{\varepsilon}\right\} \cdot \text{asymptotic series in powers of } \varepsilon. \tag{3.6}$$

A formal description of the game is as follows. There are two players, a maximizing player who chooses $y(t) \in Y$ and a minimizing player who chooses $z(t) \in \mathbb{R}^N$. The state $x(t)$ of the game at time t satisfies

$$x(t) = x + \int_s^t z(r)dr \ . \tag{3.7}$$

Let τ_x denote the exit time of $x(t)$ from D, and $\tau_x \wedge T = \min(\tau_x, T)$. Let

$$L(x,y,z) = \frac{1}{2}(b(x,y)-z)'a(x)^{-1}(b(x,y)-z) \ ,$$

$$\chi(x) = \begin{cases} 0, & x \in \partial D \\ +\infty, & x \in D. \end{cases}$$

The game payoff is

$$\int_s^{T\wedge\tau_x} L(x(t),y(t),z(t))dt + \chi(x(\tau_x \wedge T)) \ . \tag{3.8}$$

We consider the "lower" game in which (formally speaking) the minimizing player has the information advantage of knowing both $y(t)$ and $x(t)$ before $z(t)$ is chosen, while his opponent knows only $\phi(t)$ before choosing $y(t)$.

This formal description can be made precise in one of several possible ways, each of which involves concepts of game strategy. The Elliott-Kalton formulation [4],[5],[7],[12] is convenient here. Let $I = I(s,x)$ denote the lower value of the game, in the Elliott-Kalton sense. Let

$$H(x,p) = \max_{y\in Y} \min_{z\in\mathbb{R}^N} [L(x,y,z) + p\cdot z] \ . \tag{3.9}$$

The Isaacs (or dynamic programming equation) associated with this lower game is

$$I_s + H(x,DI) = 0 \tag{3.10}$$

THEOREM. ([8]).

(a) $I(s,x)$ <u>is the unique viscosity solution to (3.10) in the cylinder</u> $(0,T) \times D$ <u>with the boundary conditions</u>

$$\begin{aligned} &I(s,x) = 0 \quad \text{for} \quad 0 < s < T, \ x \in \partial D \\ &I(s,x) \to +\infty \ \text{as} \quad s\uparrow T, \ \text{for} \ x \in D. \end{aligned} \tag{3.11}$$

(b) Let $I^\varepsilon = -\varepsilon \log q^\varepsilon$. Then

$$\lim_{\varepsilon\to 0} I^\varepsilon = I.$$

As candidates for viscosity solutions we admit functions

$$I \in C^{0,1}([0,T'] \times \bar{D}), \ \forall T' < T, \tag{3.12}$$

where $C^{0,1}(\mathcal{O})$ is the space of continuous functions defined in $\mathcal{O}$ which are Lipschitz continuous with respect to x.

BIBLIOGRAPHY

1. Crandall, M.G., L.C. Evans and P.-L. Lions, "Some properties of viscosity solutions of Hamilton-Jacobi equations," Trans. AMS., 282 (1984), 487-502.

2. Crandall, M.G. and P.-L. Lions, "Viscosity solutions of Hamilton-Jacobi equations," Trans. AMS., 277 (1983), 1-42.

3. Crandall, M.G. and P.E. Souganidis, Developments in the Theory of Nonlinear First-Order Partial Differential Equations, Proceedings of International Symposium on Differential Equations, Birmingham, Alabama (1983), Knowles and Lewis, eds., North Holland.

4. Elliott, R.J. and N.J. Kalton, "Boundary value problems for nonlinear partial differential operators," J. Math. Anal. and Appl., 46 (1974), 228-241.

5. Evans, L.C. and H. Ishii, "Nonlinear first order PDE on bounded domains," to appear.

6. Evans, L.C. and H. Ishii, "A PDE approach to some asymptotic problems concerning random differential equations with small noise intensities," to appear.

7. Evans, L.C. and P.E. Souganidis, "Differential games and representation formulas for solutions of Hamilton-Jacobi-Isaacs equations," to appear in Indiana U. Math. J.

8. Fleming, W.H. and P.E. Souganidis, to appear.

9. Fleming, W.H. and C.-P. Tsai, "Optimal exit probabilities and differential games," Appl. Math. Optim., 7 (1981), 253-282.

10. Freidlin, M.I. and A.D. Wentzell, Random Perturbations of Dynamical Systems, Springer-Verlag, New York (1984).

11. Lions, P.-L., Generalized Solutions of Hamilton-Jacobi Equations, Pitman, Boston (1982).

12. Lions, P.-L. and P.E. Souganidis, "Differential games, optimal control and directional derivatives of viscosity solutions of Bellman's and Isaacs' equations," to appear in SIAM J. of Cont. and Opt.

LEFSCHETZ CENTER FOR DYNAMICAL SYSTEMS
DIVISION OF APPLIED MATHEMATICS
BROWN UNIVERSITY
PROVIDENCE, R.I. 02912

Lectures in Applied Mathematics
Volume 23, 1986

ON USING SEMIGROUPS AND THE TROTTER PRODUCT FORMULA TO SOLVE QUASI-LINEAR SYSTEMS

George H. Pimbley[1]

ABSTRACT. A quasi-linear system with linear coupling terms is treated. An approximate system for which the Trotter product of constituent semigroups converges is solved. Then it is shown that the approximate solutions converge to a solution of the original system, and that the resulting solution operator is a semigroup. Remarks are made about a nonlinearly coupled system.

1. INTRODUCTION. Until recently, semigroup solution of nonlinear initial value problems was successful mainly for single partial differential equations (PDE's), where accretiveness or monotonicity principles could be made to work, [Refs. 2,7]. In this paper we show how the Trotter product formula, [Ref. 1], can be used to find semigroups for coupled quasilinear systems of PDE's, for which monotonicity methods have not generally met with success. These Trotter product ideas were explained by J. Marsden, [Ref. 10], using a profound Banach manifold approach.

We carry out our method in the case of problems with linear coupling. This at least enables a complete treatment in a given Banach space. Later we make remarks about a case where the coupling is nonlinear.

1980 Mathematics Subject Classification. 34G05, 35F25, 35L65, 47D05.
[1]Supported by USDOE.

In the space $[L^1(R)]^2 = L^1(R) \times L^1(R)$, (where $R = (-\infty, \infty)$), let us study linearly coupled systems:

$$\rho_t + \phi(\rho)_x + \sigma u_x = 0 \quad , \qquad \sigma \geq 0,\ \phi(0) = \psi(0) = 0$$

$$u_t + \sigma\rho_x + \psi(u)_x = 0 \quad ; \qquad \phi' > 0,\ \psi' > 0$$

$$\underline{\text{I.C.}}:\quad \rho(x,0) = \rho_0(x),\ u(x,0) = u_0(x) \quad . \tag{1}$$

System (1) is a linear perturbation on a separated quasi-linear problem ($\sigma = 0$), that was treated in considerable generality by M. Crandall, [Ref. 2], using accretiveness principles in $L^1(R)$. This separated system can be written in vector form as follows:

$$\frac{dw}{dt} + A_1 w \ni 0 \quad ; \quad A_1 w \ni \begin{pmatrix} \phi(\rho)_x \\ \psi(u)_x \end{pmatrix} \quad , \quad w = \begin{pmatrix} \rho \\ u \end{pmatrix} \quad , \quad w(0) = w_0 \quad . \tag{2}$$

We note in (2) that nonlinear generators are considered to be set-valued. The perturbing system may be written:

$$\frac{dw}{dt} + \sigma A_2 w = 0 \quad ; \quad A_2 w = \begin{pmatrix} u_x \\ \rho_x \end{pmatrix} \quad , \quad w(0) = w_0 \quad , \quad \sigma > 0 \quad . \tag{3}$$

System (2) is nonlinear. Having no coupling terms, it can be solved globally, using accretiveness principles. This results in a generalized entropy solution (provided $w_0 \in [L^1(R)]^2 \cap [L^\infty(R)]^2$), given by a contraction semigroup: $w = S_1(t)w_0$. Problem (2) can also be solved locally, using the familiar method of characteristics; this carries the regularity of the initial data w_0 up to a finite breakdown time, after which only the distributional form of the solution is valid. System (3) is linear, but coupled, constituting a linear wave problem. The global solution of (3) is another semigroup: $w = S_2(t)w_0$, which is not contracting in $[L^1(R)]^2$, since it has Lipschitz constant 2. $S_2(t)$ is actually a group, since it extends to negative t.

It is known that the semigroup $S(t)$ corresponding to the sum $A + B$ of two non-commuting generators, each having the semigroups $S_A(t)$ respectively $S_B(t)$, is given by the Trotter product, [Refs. 1, 10, p. 51]:

$$S_{A+B}(t) = \lim_{n\to\infty} \left[S_A\left(\frac{t}{n}\right)S_B\left(\frac{t}{n}\right)\right]^n = S_A(t)*S_B(t) \tag{4}$$

provided S_A and S_B are such that (4) converges. We seek to make some application of this principle, involving the semigroups $S_1(t)$ and $S_2(t)$ generated in problems (2) and (3), in an attempt to solve problem (1). We rewrite problem (1), using the definitions of vectors and operators given in connection with problems (2) and (3):

$$\frac{dw}{dt} + A_1 w + \sigma A_2 w \ni 0 \quad , \qquad w(0) = w_0 \quad , \qquad \sigma > 0 \ . \tag{5}$$

2. AN APPROXIMATE PROBLEM. Sadly, there is a lack of theorems concerning the Trotter product (4) when either (or both) of the semigroups $S_A(t)$ or $S_B(t)$ is not contracting or quasi-contracting. In our application, the semigroup $S_1(t)$ is contracting in $[L^1(R)]^2$, but the semigroup $S_2(t)$ is not. The situation with $S_2(t)$ can be partially remedied, however, by substituting the Yosida approximation $A_{2\lambda} = A_2(I+\lambda A_2)^{-1}$ for its linear generator, [Ref. 12, p. 248, Eq. (7)].

The Yosida approximation $A_{2\lambda}$ of the generator A_2 in (3) is given concretely by:

$$A_{2\lambda}\begin{pmatrix}\rho\\u\end{pmatrix} = \begin{pmatrix} \frac{1}{2\lambda}\int_{-\infty}^{x} e^{-\frac{x-\hat{x}}{\lambda}}(\rho_x(\hat{x})+u_x(\hat{x}))d\hat{x} - \frac{1}{2\lambda}\int_{x}^{\infty} e^{-\frac{\hat{x}-x}{\lambda}}(\rho_x(\hat{x})-u_x(\hat{x}))d\hat{x} \\ \frac{1}{2\lambda}\int_{-\infty}^{x} e^{-\frac{x-\hat{x}}{\lambda}}(\rho_x(\hat{x})+u_x(\hat{x}))d\hat{x} + \frac{1}{2\lambda}\int_{x}^{\infty} e^{-\frac{\hat{x}-x}{\lambda}}(\rho_x(\hat{x})-u_x(\hat{x}))d\hat{x} \end{pmatrix} \tag{6}$$

wherein we see derivatives ρ_x, u_x being smoothed by δ-tending kernels. The linear operator $\sigma A_{2\lambda}$ is continuous on $[L^1(R)]^2$, with Lipschitz constant $\beta_\lambda = \frac{3\sigma}{\lambda}$. It generates the semigroup $S_{2\lambda}(t) = e^{-t\sigma A_{2\lambda}}$, which solves the problem:

$$\frac{dw}{dt} + \sigma A_{2\lambda} w = 0 \quad , \qquad w(0) = w_0 \quad . \tag{3_λ}$$

We note that $S_2(t)w_0 = \lim_{\lambda\to 0} S_{2\lambda}(t)w_0$. Using Gronwall's inequality in connection with problem (3_λ), it may be readily shown that

$$||S_{2\lambda}(t)w_1 - S_{2\lambda}(t)w_2|| \le e^{\beta_\lambda t}||w_1-w_2|| \ , \quad w_1,w_2 \in [L^1(R)]^2 \ . \tag{7}$$

Thus, $S_{2\lambda}(t)$ is quasi-contracting.

We propose now to study the convergence of the Trotter product:

$$S_\lambda(t)w_0 = w_\lambda = \lim_{n\to\infty}\left[S_{2\lambda}\left(\frac{t}{n}\right)S_1\left(\frac{t}{n}\right)\right]^n w_0 \quad , \tag{8}$$

which we shall use as a candidate for solving the approximate problem:

$$\frac{dw}{dt} + A_1 w + \sigma A_{2\lambda} w \ni 0 \quad , \qquad w(0) = w_0 \quad , \tag{5_λ}$$

involving operator (6). We remember in (8) that $S_1(t)$ was contracting in $[L^1(R)]^2$, while $S_{2\lambda}(t)$ is quasi-contracting, [c.f.(7)]. This indicates that the "Chernoff operator," $K_\lambda(t) = S_{2\lambda}(t)S_1(t)$, is quasi-contracting. Moreover, the generator $A_{2\lambda}$ of $S_{2\lambda}(t)$ is Lipschitz continuous. Thus, we can use J. Marsden's results, [Ref. 10, pp. 53-61, Th. 2.1; pp. 66-69], to show the convergence of (8) as $n \to \infty$.

Marsden has two main requirements on $K_\lambda(t)$. The first is that $K_\lambda(t)$ be quasi-contracting:

$$||K_\lambda(t)w_1-K_\lambda(t)w_2|| \leq e^{\beta_\lambda t}||w_1-w_2|| \ , \ w_1,w_2 \in [L^1(R)]^2 \ . \tag{9}$$

This is almost self-evident from (7) and the contractiveness of $S_1(t)$.

Using (9), we prove a result which substitutes for Marsden Lemma 2.2, but which goes farther in our case:

Lemma 2.2: Let $y \in W$, where $W \subset [L^1(R)]^2$ is a bounded closed region containing the null element θ. Then for $0 \leq t \leq T$, there exists the bounded region V^T, $W \subset V^T$, such that the sequence of iterates $\{K_\lambda\left(\frac{t}{n}\right)^n y\}$ of Chernoff's operator remains in V^T, $n = 1, 2, \ldots$

Proof: Since $K_\lambda(t)\theta = \theta$, and using (9), we write:

$$\begin{aligned} ||K_\lambda\left(\tfrac{t}{n}\right)^n y|| &= ||K_\lambda\left(\tfrac{t}{n}\right)^n y-K_\lambda\left(\tfrac{t}{n}\right)^n \theta|| \\ &\leq e^{\beta_\lambda \frac{t}{n}}||K_\lambda\left(\tfrac{t}{n}\right)^{n-1} y-K_\lambda\left(\tfrac{t}{n}\right)^{n-1} \theta|| \\ &\leq - - - - - - - - - - - - - - - - \\ &\leq e^{\beta_\lambda t}||y-\theta|| \leq e^{\beta_\lambda T}||y|| \ . \end{aligned}$$

Thus, given any initial element $y \in W$, and a time T, $K_\lambda\left(\frac{t}{n}\right)^n y$

remains within a radius $e^{\beta_\lambda T}||y||$, thus within some V^T. This ends the proof.

Lemma 2.2 here enables a global theory, while in Marsden's generality, there were only local results. The main item is that our manifold, $[L^1(R)]^2$, has a fixed point for $K_\lambda(t)$, namely θ.

Marsden's second requirement is that $K_\lambda(t)$ be an "approximate semigroup," in the sense that there exists a constant C_0 (which may depend on λ and y), such that as $s,t \to 0$,

$$||K_\lambda(t+s)y-K_\lambda(t)K_\lambda(s)y|| \leq C_0 st \quad , \qquad 0 \leq t \leq T \quad , \tag{10}$$

for $y \in V^T \cap [W^{2,1}(R)]^2$. (Here, $[W^{2,1}(R)]^2 = W^{2,1}(R) \times W^{2,1}(R)$ is the second Sobolev space.) For $K_\lambda(t) = S_{2\lambda}(t)S_1(t)$, property (10) is proved using the procedure of Marsden, [Ref. 10, p. 68], involving Taylor's theorem. The fact that $S_1(t)y$ carries the regularity of the initial data, for a small time, must be used in our case.

Morever, (10) is true, without the indicated regularity, for $y \in V^T$. If $y_j \to y$, $y_j \in V^T \cap [W^{2,1}(R)]^2$, $y \in V^T$, then the left side of (10) behaves continuously, and the only question involves C_0; does it blow up as $y_j \to y$? We have $K_\lambda(t + s) = S_{2\lambda}(t)S_{2\lambda}(s)$ $S_1(t)S_1(s)$ and $K_\lambda(t)K_\lambda(s) = S_{2\lambda}(t)S_1(t)S_{2\lambda}(s)S_1(s)$. Also by (7) and the contractiveness of $S_1(t)$, $S_{2\lambda}(t)-I$, and $S_1(t)-I$ are first order in time at $t = 0$ in terms of the $[L^1(R)]^2$ norm. No blowup of C_0 is indicated. Hence, (10) is true if $y \in V^T$.

There follow more Marsden lemmas, explicated for our case. The norm is of course that for $[L^1(R)]^2$. Constants may depend on λ and y.

<u>Lemma 2.3</u>: $||K_\lambda(t)y-K_\lambda\left(\frac{t}{\ell}\right)^\ell y|| \leq C_1 t^2$, $y \in V^T$, where C_1 is a constant, and $\ell = 1, 2, \ldots$

<u>Proof</u>: Using (10), we may write:

$$||K_\lambda(t)y - K_\lambda\left(\frac{t}{\ell}\right)^{\ell} y||$$

$$\leq \sum_{j=0}^{\ell-1} ||K_\lambda\left(\frac{t}{\ell}\right)^{j} K_\lambda\left(t-j\frac{t}{\ell}\right)y - K_\lambda\left(\frac{t}{\ell}\right)^{j} K_\lambda\left(\frac{t}{\ell}\right)K_\lambda\left(t-(j+1)\frac{t}{\ell}\right)y||$$

$$\leq \sum_{j=0}^{\ell-1} e^{j\beta_\lambda \frac{t}{\ell}} ||K_\lambda\left(t-j\frac{t}{\ell}\right)y - K_\lambda\left(\frac{t}{\ell}\right)K_\lambda\left(t-(j+1)\frac{t}{\ell}\right)y||$$

$$\leq \sum_{j=0}^{\ell-1} e^{j\beta_\lambda \frac{t}{\ell}} C_0\left(\frac{t}{\ell}\right)\left(t-(j+1)\frac{t}{\ell}\right) \leq C_0 e^{\beta_\lambda T} \frac{t^2}{\ell\left(\ell-\frac{1}{2}(\ell+1)\right)} \leq C_1 t^2$$

This ends the proof.

Lemma 2.5: For $m \geq n$ and $y \in W$ with $0 \leq t \leq T$, we have

$$||K\left(\frac{t}{n}\right)^{n} y - K\left(\frac{t}{m}\right)^{m} y|| < C_2 \frac{t^2}{n} ,$$

where C_2 is a constant.

Proof: First, we put $m = n\ell$ where ℓ is an integer. Then we may write:

$$||K_\lambda\left(\frac{t}{n}\right)^{n} y - K_\lambda\left(\frac{t}{n\ell}\right)^{n\ell} y||$$

$$\leq \sum_{j=0}^{n-1} ||K_\lambda\left(\frac{t}{n}\right)^{n-j} K_\lambda\left(\frac{t}{n\ell}\right)^{j\ell} y - K_\lambda\left(\frac{t}{n}\right)^{n-j-1} K_\lambda\left(\frac{t}{n\ell}\right)^{(j+1)\ell} y||$$

$$\leq \sum_{j=0}^{n-1} e^{\beta_\lambda t \frac{n-j-1}{n}} ||K_\lambda\left(\frac{t}{n}\right)K_\lambda\left(\frac{t}{n\ell}\right)^{j\ell} y - K_\lambda\left(\frac{t}{n\ell}\right)^{\ell} K_\lambda\left(\frac{t}{n\ell}\right)^{j\ell} y||$$

$$\leq e^{\beta_\lambda T} C_1 \frac{t^2}{n}\quad, \text{ using Lemmas 2.2 and 2.3.}$$

Next, for general integer m,

$$\begin{aligned}||K_\lambda\left(\frac{t}{n}\right)^n y - K_\lambda\left(\frac{t}{m}\right)^m y|| &\leq ||K_\lambda\left(\frac{t}{n}\right)^n y - K_\lambda\left(\frac{t}{nm}\right)^{mn} y|| \\ &+ ||K_\lambda\left(\frac{t}{m}\right)^m y - K_\lambda\left(\frac{t}{mn}\right)^{mn} y|| \\ &\leq e^{\beta_\lambda T} C_1 t^2\left(\frac{1}{n}+\frac{1}{m}\right) \leq 2C_1 e^{\beta_\lambda T}\frac{t^2}{n} = C_2\frac{t^2}{n}\ .\end{aligned}$$

This ends the proof.

Lemma 2.6: $\{K_\lambda\left(\frac{t}{n}\right)^n y\}$, $n = 1, 2, \ldots$, converges uniformly for $0 \leq t \leq T$, $y \in W$.

Proof: For $y \in W$, this result is obvious from Lemma 2.5 and the Cauchy criterion. This convergence is uniform in t over $0 \leq t \leq T$. Since $[L^1(R)]^2$ is complete and W is closed, we can define $S_\lambda(t)y = \lim_{n\to\infty} K\left(\frac{t}{n}\right)^n y$, $y \in W$. Also, we note that

$$\begin{aligned}||K_\lambda\left(\frac{t}{n}\right)^n y_1 - K_\lambda\left(\frac{t}{n}\right)^n y_2|| &\leq e^{\beta_\lambda \frac{t}{n}}||K_\lambda\left(\frac{t}{n}\right)^{n-1} y_1 - K_\lambda\left(\frac{t}{n}\right)^{n-1} y_2|| \\ &\leq e^{\beta_\lambda t}||\, y_1 - y_2||,\ y_1, y_2, \in W\quad,\end{aligned}$$

where we have used (9). Hence, in the limit as $n \to \infty$, we have

$$||S_\lambda(t)w_1 - S_\lambda(t)w_2|| \leq e^{\beta_\lambda t}||w_1 - w_2||,\ w_1, w_2 \in W \subset [L^1(R)]^2\quad . \tag{11}$$

This ends the proof.

Where needed in the next proof, which is due to J. Marsden, we can put $W = V^T$ in Lemma 2.2, and so generate a new $\widetilde{V}^T$.

$\{K_\lambda\left(\frac{t}{n}\right)^n y\}$ then remains in this new $\widetilde{V}^{\widetilde{T}}$, n = 1, 2, ... , provided that $y \varepsilon \widetilde{W} = V^T$, $0 \leq t \leq \widetilde{T}$. Thus, convergence is assured in Lemma 2.6.

Lemma 2.7: For $y \varepsilon W$, we have $S_\lambda(t+s)y = S_\lambda(t)S_\lambda(s)y$, $0 \leq t,s \leq T$.

Proof: First, we let t,s be rationally related, say $s = \frac{\ell t}{m}$, where ℓ and m are integers. Then

$$S_\lambda(t+s)y = \lim_{n\to\infty} K_\lambda\left(\frac{t+s}{n}\right)^n y = \lim_{n\to\infty} K_\lambda\left(\frac{t}{k}\right)^k K_\lambda\left(\frac{s}{k'}\right)^{k'} y \quad , \quad y \varepsilon W \quad , \qquad (12)$$

where $k = \frac{nm}{\ell+m}$, $k' = \frac{\ell n}{\ell+m}$

are chosen so that $\frac{t}{k} = \frac{s}{k'} = \frac{t+s}{n}$, and $k + k' = n$, and the limit in (12) is taken through multiples of $\ell + m$, to make k,k' integers. Such a choice of k,k' works because

$$\frac{t}{k} = \frac{t(\ell+m)}{nm} = \frac{s(\ell+m)}{n\ell} = \frac{s}{k'} = \frac{t}{n-k'} \Longrightarrow \frac{s}{k'} = \frac{t+s}{n} \quad .$$

Now with $y \varepsilon W$,

$$||K_\lambda\left(\frac{t}{k}\right)^k K_\lambda\left(\frac{s}{k'}\right)^{k'} y - S_\lambda(t)S_\lambda(s)y||$$

$$< ||K_\lambda\left(\frac{t}{k}\right)^k K_\lambda\left(\frac{s}{k'}\right)^{k'} y - K_\lambda\left(\frac{t}{k}\right)^k S_\lambda(s)y||$$

$$+ ||K_\lambda\left(\frac{t}{k}\right)^k S_\lambda(s)y - S_\lambda(t)S_\lambda(s)y||$$

$$\leq e^{\beta_\lambda t} ||K_\lambda\left(\frac{s}{k'}\right)^{k'} y - S_\lambda(s)y||$$

$$+ ||K_\lambda\left(\frac{t}{k}\right)^k S_\lambda(s)y - S_\lambda(t)S_\lambda(s)y|| \quad ,$$

and each of the last terms vanish as $k,k' \to \infty$. Hence, we get $S_\lambda(t+s)y = S_\lambda(t)S_\lambda(s)y$, $y \in W$, for rationally related t,s, which form a dense set in the interval $0 \le t, s \le T$. By continuity of $S_\lambda(t)y$ in t on bounded intervals (see proof of Lemma 2.6), $S_\lambda(t+s)y = S_\lambda(t)S_\lambda(s)y$ holds for all t,s with $0 \le t, s \le T$. This ends the proof.

The approximate problem (5_λ) is not expected to be solved by (8) in a classical sense. To show that (8) is a generalized entropy solution of (5_λ), we begin with a problem known to have a unique classical solution, namely

$$\frac{dw}{dt} + A_{1\mu}w + \sigma A_{2\lambda}w - \varepsilon\vec{\Delta}w = 0 \ , \ w(0) = w_0 \ , \ \vec{\Delta}w = \begin{pmatrix}\Delta\rho\\ \Delta u\end{pmatrix}, \qquad (5_{\lambda,\mu,\varepsilon})$$

where $w_0 \varepsilon [L^1(R)]^2 \cap [L^\infty(R)]^2$. This solution may be expressed as

$$w_{\lambda,\mu,\varepsilon}(t) = S_{2\lambda}(t)*S_{1\mu}(t)*S_\varepsilon(t)w_0, \qquad \lambda > 0,\ \mu > 0,\ \varepsilon > 0, \qquad (13)$$

i.e. a Trotter triple product of semigroups. Of these, $S_{2\lambda}(t)$ and $S_{1\mu}(t)$ have generators $A_{2\lambda}$ and $A_{1\mu}$ that are Lipschitz continuous and $S_\varepsilon(t)$ has the Gaussian kernel $(4\pi\varepsilon t)^{-\frac{1}{2}} \exp\{-\frac{|x-\tilde{x}|^2}{4\varepsilon t}\}$. Moreover $S_{2\lambda}(t)$ is quasi-contracting, while $S_{1\mu}(t)$ and $S_\varepsilon(t)$ are contracting. We have $\lim_{\varepsilon\to 0} S_\varepsilon(t)w_0 = w_0$, and $\lim_{\mu\to 0} S_{1\mu}(t)w_0 = S_1(t)w_0$, since $A_{1\mu}$ is the Yosida approximation of A_1, [Ref. 12, p. 447, 9]. Since each of these limits is in terms of the $[L^1(R)]^2$ norm, it can be shown that we may pass to the limit in expression (13) as $(\mu,\varepsilon) \to (0,0)$. The resulting composite limit is also in terms of the $[L^1(R)]^2$ norm.

Separating problem $(5_{\lambda,\mu,\varepsilon})$ into its two coupled component equations, multiplying the first equation through by $f(x,t)\Phi(\rho)$, and the second equation through by $g(x,t)\Psi(u)$, (where f,g are arbitrary positive, twice differentiable, compactly supported test functions, and Φ,Ψ are convex entropies), performing double integration of each equation over $0 \leq t \leq T$, $-\infty < x < \infty$, integrating by parts, and setting $\Phi(\rho) = |\rho-k_1|$, $\Psi(u) = |u-k_2|$, where k_1,k_2 are arbitrary constants (all in the manner of S. N. Kruzkov, [Ref. 8, p. 236]), we get the two entropy inequalities satisified by classical solutions of $(5_{\lambda,\mu,\varepsilon})$:

$$\int_0^T\int_{-\infty}^{\infty}\Big[|\rho_{\lambda,\mu,\varepsilon}-k_1|(f_t+\varepsilon f_{xx})+\mathrm{sign}(\rho_{\lambda,\mu,\varepsilon}-k_1)(-[A_{1\mu}(\rho_{\lambda,\mu,\varepsilon},u_{\lambda,\mu,\varepsilon})]_1 -[A_{2\lambda}(\rho_{\lambda,\mu,\varepsilon},u_{\lambda,\mu,\varepsilon})]_1)f\Big]dxdt\geq 0$$

$$\int_0^T\int_{-\infty}^{\infty}\Big[|u_{\lambda,\mu,\varepsilon}-k_2|(g_t+\varepsilon g_{xx})+\mathrm{sign}(u_{\lambda,\mu,\varepsilon}-k_2)(-[A_{1\mu}(\rho_{\lambda,\mu,\varepsilon},u_{\lambda,\mu,\varepsilon})]_2 -[A_{2\lambda}(\rho_{\lambda,\mu,\varepsilon},u_{\lambda,\mu,\varepsilon})]_2)g\Big]dxdt\geq 0. \tag{14}$$

In (14) $[A_{1\mu}(\rho,u)]_i$ represents the ith component of the Yosida approximation of the generator A_1, and $[A_{2\lambda}(\rho,u)]_i$ is the ith component of $A_{2\lambda}$ as given concretely by (6) (after integration by parts).

We may legitimately pass to the limit as $\mu \to 0$ and $\varepsilon \to 0$ in inequalities (14), using expression (13) for the classical solution of problem $(5_{\lambda,\mu,\varepsilon})$, since these limits are in terms of the norm of $[L^1(R)]^2$. These two inequalities are still satisfied in the limit, and the derivatives in the components of A_1 are then shifted to the test functions f,g, using integration by parts after the limit as $\mu \to 0$, $\varepsilon \to 0$ is taken. Thus, we conclude that (8) represents the generalized entropy solution of problem (5_λ).

This approach to the generalized solution of problem (5_λ) was suggested by Prof. J. A. Goldstein of Tulane University.

Thus, (5_λ) is solved in the distributional sense by the semigroup (8) (proved to converge in Lemma 2.6, and proved to be a semigroup in Lemma 2.7).

3. THE CONVERGENCE QUESTION. The semigroup $S_0(t)$ that would solve problem (5) ought now to be given by $w = \lim_{\lambda\to 0} w_\lambda$, where w_λ as specified in (8) is the weak solution of problem (5λ). We face the difficulty, however, that the Lipschitz constant in (7) blows up as $\lambda \to 0$, so that the semigroup $S_{2\lambda}(t)$ in (8) ceases to be quasi-contracting. We presently have no way of telling whether the Trotter product in (8) converges if we set $\lambda = 0$.

We avoid this problem by passing from Eq. (5_λ) to a linear vector equation of evolution:

$$\frac{dy}{dt} + A_1'(w_\lambda(t))y + \sigma A_{2\lambda}y = 0 \quad , \quad y(0) = w_0 \quad , \quad \lambda > 0 \quad , \tag{15}$$

where

$$A_1'(w_\lambda(t)) = \begin{pmatrix} \phi'(\rho_\lambda(x,t))\frac{\partial}{\partial x} & 0 \\ 0 & \Psi'(u_\lambda(x,t))\frac{\partial}{\partial x} \end{pmatrix} , \tag{16}$$

and $w_\lambda(t)$ is the known function given in (8) which solves problem (5_λ) weakly, with components $\rho_\lambda(x,t)$ and $u_\lambda(x,t)$. With identical initial data $y_\lambda(0) = w_\lambda(0) = w_0$, we expect that $y_\lambda(t) \equiv w_\lambda(t)$, where $y_\lambda(t)$ solves (15), since $A_1 w_\lambda = A_1'(w_\lambda(t))w_\lambda$. Thus, we study solutions of Eq. (15) as $\lambda \to 0$.

We define now the operators $A_\lambda(t) \equiv A_1'(w_\lambda(t))+\sigma A_{2\lambda}$, parameterized by t, $0 \le t \le T$, with the time-independent domain $[W^{1,1}(R)]^2$, and we write problem (15) simply as follows:

$$\frac{dy}{dt} + A_\lambda(t)y = 0 \quad , \qquad y(0) = w_o \quad , \qquad \lambda > 0 \quad . \tag{17}$$

Methods developed by T, Kato, relative to convergence of parameterized linear semigroups, [Ref. 5, pp. 497-509], and to linear evolution equations, [Ref. 6], are applied.

To use Kato's theory with Eq. (17), we must prove several items about the operators $A_\lambda(t)$ as a collection of generators in $[L^1(R)]^2$, parameterized by $t \in [0,T]$:

a) that $A_\lambda(t)$, for fixed t, generate a uniformly bounded semi-group, i.e., that $A_\lambda(t) \in G(M,0)$, where M is uniform in $t \in [0,T]$; (see [Ref. 6, p. 243] for notation).

b) that $\operatorname{s-lim}_{\alpha\to 0} [I+\alpha A_\lambda(t)]^{-1} = I$, uniformly in $\lambda > 0$, for each fixed $t \in [0,T]$.

c) that, uniformly in $t \in [0,T]$, there is a subsequence $\lambda_n \to 0$ such that $\operatorname{s-lim}_{n\to\infty} [\zeta I+A_{\lambda_n}(t)]^{-1}$ exists for $\zeta > 0$.

d) that $A_\lambda(t)$ satisfy

$$|||A_\lambda(t_1) - A_\lambda(t_2)||| \leq C_\lambda|t_1-t_2| , \qquad t_1,t_2 \in [0,T] ,$$

with the operator norm $|||\cdot|||$ being the norm for the space $B([W^{1,1}(R)]^2), [L^1(R)]^2)$.

e) that relative to some norm equivalent to the cartesian norm of the space $[L^1(R)]^2$, the collection of generators $\{A_\lambda(t)\}$, $0 \leq t \leq T$, is a "stable family," [Ref. 6, p. 244], i.e.

$$||\prod_{j=1}^{k} [\zeta I+A_\lambda(t_j)]^{-1}|| \leq M_o\zeta^{-k} \quad , \quad \zeta > 0, \quad M_o = \text{const}, \tag{18}$$

for every finite set $0 \leq t_1 \leq t_2 \ldots \leq t_k \leq T$, $k = 1, 2, \ldots$, and the product in (18) is time-ordered, with terms involving greater t_j standing progressively to the left in the product.

Items a) - d) can now be recast in terms of the equivalent norm determined in item e).

We describe briefly how properties a) to e) are valid for Eq. (15). Item a) is proved using Fourier transforms in the space $\mathcal{S}(R)$ of rapidly decreasing functions, [Ref. 12, pp. 146-

161], to solve the equation $\frac{dw}{ds} + A_\lambda(t)w = 0$, for fixed $t \in [0,T]$. Then operator calculus is used to bound uniformly the solution of the transformed problem, and this bound is shifted to the original equation in $[L^1(R)]^2$ by the continuity of the inverse transform in $\mathcal{S}(R)$. b) is more or less straightforward. The resolvent convergence in c) results from a weak* convergent subsequence selected from $\Phi'(w_\lambda) = (\phi'(\rho_\lambda(x,\cdot)), \Psi'(u_\lambda(x,\cdot)))$ as $\lambda \to 0$, where we assume that $|\phi'|$ and $|\Psi'|$ are bounded by a constant. This subsequence is actually uniformly weakly convergent over $t \in [0,T]$ by virtue of the equicontinuity of the semigroup $S_{2\lambda}(t)$ as a function of t, parameterized by $\lambda > 0$. This weak* convergence implies the uniform resolvent convergence. The Lipschitz continuity in d) results from the assumed smoothness of ϕ,Ψ, and a Lipschitz norm-continuity that we are able to establish for $w_\lambda(t)$. It is necessary to assume that $w_0 \in [L^\infty(R)]^2 \cap [L^1(R)]^2$ initially, and then to use a density argument. e) is proved using a procedure of Kato, [Ref. 6, p. 245], which includes a change to an equivalent norm; then steps a) to d) are readily expressible in this new norm.

With conditions a) to e) satisfied, we are led to modify an important result of Kato [Ref. 5, p. 505, Th. 2.17]. Accordingly, there exsists an operator collection $A(t) \in G(M,0)$, $t \in [0,T]$, such that if $U_{\lambda_n}(t,s)$ is the unique family of evolution operators pertaining to the equation: $\frac{dw}{dt} + A_{\lambda_n}(t)w = 0$, [Ref. 6, p. 246], then $U_{\lambda_n}(t,s) \overset{s}{\to} U(t,s)$ as $n \to \infty$, uniformly for $0 \leq s \leq t \leq T$. Here, $U(t,s)$ is the unique family of evolution operators for the equation: $\frac{dw}{dt} + A(t)w = 0$.

We use the word "pertaining" here. In general these families of evolution operators do not satisfy the equations to which they pertain in any classical sense, because of the lack of differentiability, [Ref. 6, p. 246]. They may be valid in a distributional sense.

Conditions a) to c) are similar to those of the original Kato theorem, [Ref. 5, p. 505]. Conditions d) and e) are required for the modification.

So we have for Eq. (15),

$$y_{\lambda_n}(t) \equiv w_{\lambda_n}(t) = U_{\lambda_n}(t,0)w_0 \to U(t,0)w_0 = w^*(t) \tag{19}$$

as $n \to \infty$, in norm. Moreover, by assuming Lipschitz continuity for the functions ϕ',Ψ', we can have, relative to the $[L^1(R)]^2$ norm,

$$(\phi'(\rho_{\lambda_n}(x,t)),\ \Psi'(u_{\lambda_n}(x,t))) \to (\phi'(\rho^*(x,t)),\ \Psi'(u^*(x,t))) \quad ,$$

where of course ρ^*,u^* are the components of w^* in (19), and $\phi'(\rho_{\lambda_n}(x,t)) - \phi'(0)$, etc., are in $[L^1(R)]^2$. We see then that

$$A_1'(w^*(t)) = \begin{pmatrix} \phi'(\rho^*(x,t))\frac{\partial}{\partial x}\cdot & 0 \\ 0 & \Psi'(u^*(x,t))\frac{\partial}{\partial x}\cdot \end{pmatrix} \tag{20}$$

is the limiting form of operator (16), that $A(t)=A_1'(w^*(t))+\sigma A_2$, and that, at least for $t=0$, with right time derivative,

$$\frac{dw^*}{dt} + A_1'(w^*(t))w^* + \sigma A_2 w^* = \frac{dw^*}{dt} + A_1 w^* + \sigma A_2 w^* = 0 \tag{21}$$

is valid, with $w^*(0) = w_0$. In the classical sense, no more can be said, because of differentiability problems. We must show that $w^*(t)$ is a generalized solution of Eq. (15) and of Eq. (5).

4. ON $S_0(t) = \lim_{\lambda_n \to 0} S_{\lambda_n}(t)$ BEING A SEMIGROUP. We know that for $\lambda > 0$, $S_\lambda(t+s) = S_\lambda(t)S_\lambda(s)$, (c.f. Lemma 2.7). We wonder if this

is true in the limit. For this it is required that $S_\lambda(t)$ be Lipschitz continuous uniformly in λ as $\lambda \to 0$, i.e. that:

$$||S_\lambda(t)w_1 - S_\lambda(t)w_2|| \leq M||w_1 - w_2|| \quad . \tag{22}$$

This relates to property a) in Section 3, but is somewhat deeper.

Using the Frechet derivative $S_\lambda'(t)$ of $S_\lambda(t)$, we can write:

$$\begin{aligned} ||S_\lambda(t)w_1 - S_\lambda(t)w_2|| &= \left|\left|\int_0^1 [S_\lambda'(t)w_\theta](w_1 - w_2)d\theta\right|\right| \\ &\leq \int_0^1 ||S_\lambda'(t)w_\theta||d\theta||w_1 - w_2|| \end{aligned} \tag{23}$$

where $w_\theta = (1-\theta)w_1 + \theta w_2$, and $S_\lambda'(t)\colon [L^1(R)]^2 \to B([L^1(R)]^2)$ has components h,k obtained by solving the linearized system:

$$\begin{aligned} & h_t + [\phi'(\rho_\theta(x))h]_x + \sigma[k_x]_\lambda = 0 \quad , \qquad w_\theta = (\rho_\theta, u_\theta) \\ & k_t + \sigma[h_x]_\lambda + [\Psi'(u_\theta(x))k]_x = 0, \; h(x,0) = \rho_0, \; k(x,0) = u_0 \end{aligned} \tag{24}$$

(see [Ref. 9, p. 297]). In (24), $[h_x]_\lambda, [k_x]_\lambda$ are the components of the Yosida approximation given in (6), (but with h_x, k_x substituted for ρ_x, u_x). System (24) is solved by using Fourier transforms in the space $\mathcal{S}(r)$ of rapidly decreasing functions, [Ref. 12, p. 146], to show that $||S_\lambda'(t)w_\theta|| \leq M$, $0 \leq \theta \leq 1$. (We dealt similarly with condition a) in Section 3 to show that $A_\lambda(t) \in G(M,0)$, $0 \leq t \leq T$.) Thus, (22) follows from (23). Given $\varepsilon > 0$, there exists $\delta > 0$, such that

$$||S_0(t+s)w_0-S_0(t)S_0(s)w_0||$$

$$\leq ||S_0(t+s)w_0-S_{\lambda_n}(t+s)w_0|| + ||S_{\lambda_n}(t)S_{\lambda_n}(s)w_0-S_0(t)S_0(s)w_0||$$

$$\leq ||S_0(t+s)w_0-S_{\lambda_n}(t+s)w_0|| + ||S_{\lambda_n}(t)S_{\lambda_n}(s)w_0-S_{\lambda_n}(t)S_0(s)w_0||$$

$$+ ||S_{\lambda_n}(t)S_0(s)w_0-S_0(t)S_0(s)w_0||$$

$$\leq ||S_0(t+s)w_0-S_{\lambda_n}(t+s)w_0|| \quad , \quad \hat{w}_0 = S_0(s)w_0 \quad ,$$

$$+ M||S_{\lambda_n}(s)w_0-S_0(s)w_0||+||S_{\lambda_n}(t)\hat{w}_0-S_0(t)\hat{w}_0|| < \varepsilon$$

provided $\lambda_n < \delta$. Hence, $w^*(t) = S_0(t)w_0$ where $S_0(t)$ has the semigroup property. That $S_0(t)$ is Lipschitz continuous is seen by inspection of (22) with $\lambda_n \to 0$. Moreover $||\hat{W}_0|| \leq M||w_0||$.

The entropy inequalities satisfied by weak solutions of approximate problem (5_λ) were obtained from (14) by going to the limit as $(\mu,\varepsilon) \to (0,0)$. This results in the following:

$$\int_0^T\int_{-\infty}^{\infty}\Big[|\rho_\lambda-k_1|f_t+\text{sign}(\rho_\lambda-k_1)((\phi(\rho_\lambda)-\phi(k_1))f_x-\sigma[A_{2\lambda}(\rho_\lambda,u_\lambda)]_1 f)\Big]dxdt \geq 0$$

$$\int_0^T\int_{-\infty}^{\infty}\Big[|u_\lambda-k_2|g_t+\text{sign}(u_\lambda-k_2)(-\sigma[A_{2\lambda}(\rho_\lambda,u_\lambda)]_2 g+(\Psi(u_\lambda)-\Psi(k_2))g_x)\Big]dxdt \geq 0 \tag{25}$$

where here, k_1,k_2 are the constants in (14), and (25) is satisfied, for arbitrary k_1,k_2, by the weak solution $\rho_\lambda(x,t)$, $u_\lambda(x,t)$

of problem (5_λ). Again, in (25), $[A_{2\lambda}(\rho,u)]_1$ and $[A_{2\lambda}(\rho,u)]_2$ are the components given by (6).

The process of letting the remaining parameter, i.e. $\lambda_n > 0$, vanish in (25) can also be handled. We have from (13)

$$w_{\lambda,0,0}(t) = [S_{2\lambda}(t)*S_1(t)]w_0 \quad , \quad w_0 \in [L^1(R)]^2 \cap [L^\infty(R)]^2 \quad , \tag{26}$$

and this process is represented in (19) in a norm equivalent to $[L^1(R)]^2$ convergence. Taking this limit in (25) shows that the functions $\rho^*(x,t) = \lim_{\lambda_n \to 0} \rho_{\lambda_n}(x,t)$ and $u^*(x,t) = \lim_{\lambda_n \to 0} u_{\lambda_n}(x,t)$ satisfy the following entropy inequalities, (for arbitrary k_1, k_2):

$$\int_0^T \int_{-\infty}^{\infty} \Big[|\rho^*-k_1| f_t + \Big(\text{sign}(\rho^*-k_1)(\phi(\rho^*)-\phi(k_1)) + \sigma \int_{x_0}^{x} \text{sign}(\rho^*(\hat{x})-k_1)u_x^*(\hat{x})d\hat{x}\Big) f_x\Big] dxdt \geq 0 \quad ,$$

$$\int_0^T \int_{-\infty}^{\infty} \Big[|u^*-k_2| g_t + \Big(\sigma \int_{x_0}^{x} \text{sign}(u^*(\hat{x})-k_2)\rho_x^*(\hat{x})d\hat{x} + \text{sign}(u^*-k_2)(\Psi(u^*)-\Psi(k_2))\Big) g_x\Big] dxdt \geq 0 \quad , \tag{27}$$

noting that the process of taking the limit as $\lambda_n \to 0$ in (6) is $[L^1(R)]^2$ normed convergence.

This shows that the semigroup expression $S_0(t)w_0$, $w_0 \in [L^1(R)]^2 \cap [L^\infty(R)]^2$ gives the weak entropy solution of problem (5), and therefore of problem (1).

5. NONLINEARLY COUPLED SYSTEM. We ask if a similar approach might serve in a case of nonlinear coupling. Let us keep the same conditions on the functions ϕ,Ψ, (c.f. (1)), and consider the system:

$$\begin{aligned} &\rho_t+\phi(\rho)_x+\sigma u_x=0 \;, && \underline{\text{I.C.}}:\ \rho(x,0)=\rho_0(x), u(x,0)=u_0(x) \;, \\ &u_t+\sigma\Upsilon(\rho)_x+\Psi(u)_x=0 \;, && \Upsilon(0)=0, \Upsilon''(0)=0, \rho\Upsilon''(\rho)>0, \rho\neq 0, \Upsilon'>0 \end{aligned} \tag{28}$$

We get a form of Eq. (5) wherein $A_2 w \ni \begin{pmatrix} u_x \\ \Upsilon(\rho)_x \end{pmatrix}$ is nonlinear. Separated system (2) is the same, but perturbing system (3) is now difficult since it involves the nonlinear operator A_2. This perturbing problem has been studied by R. DiPerna, [Ref. 3, pp. 55-57], using vanishing viscosity and compensated compactness. A global weak entropy solution is obtained, but the uniqueness properties discovered to date do not yet indicate a semigroup.

A conjecture has been made, (not by the author), that the following property <u>may</u> hold between any pair of solutions $w_1(t)$ and $w_2(t)$ of this DiPerna equation:

$$||w_1(t)-w_2(t)|| \leq k(t)||w_1(0)-w_2(0)|| \quad , \quad k(t) > 1 \; . \tag{29}$$

<u>If</u> this were true, and <u>if</u> $k(t)$ were bounded on $[0,T]$, then we might be justified in considering the solution operator a uniformly Lipschitz semigroup $S_2(t)$, [Ref. 4, pp. 617, 618]

There would be the problem of arranging that $S_1(t)$ and $S_2(t)$ are simultaneously quasi-contracting on some space or manifold. A theorem of Marsden, [Ref. 10, Th. 4.1] would be used. After taking the Yosida approximation of A_1, and obtaining the contraction $S_{1\lambda}(t)$, a non-Banach space metric d could be found such that $d(S_2(t)w_1, S_2(t)w_2) \leq e^{\beta_1 t} d(w_1,w_2)$, and which is also such that $d(S_{1\lambda}(t)w_1, S_{1\lambda}(t)w_2) \leq e^{\beta_2 t} d(w_1,w_2)$, [Ref. 10, pp. 67-69]. In this way, we could have a quasi-contracting Chernoff function

$K_\lambda(t)$. This may be in line with some thinking of B. Temple, [Ref. 11], on the feasibility of Banach space norms for general conservation laws.

The author wishes to record here his appreciation to Professors J. E. Marsden and J. A. Goldstein for suggestions concerning this work.

BIBLIOGRAPHY

1. P. R. Chernoff, "Product Formulas, Nonlinear Semigroups, and Addition of Unbounded Operators," AMS Memoirs, No. 140 (1974).

2. M. G. Crandell, "The Semigroup Approach to First Order Quasi-linear Equations in Several Space Variables," Israel J. of Math 12 (1972), pp. 108-132.

3. R. J. DiPerna, "Convergence of Approximate Solutions to Conservation Laws," Arch. Rat. Mech. & Anal., Vol. 82, No. 1 (1983), pp. 27-70.

4. E. Hille and R. Phillips, Functional Analysis and Semi-Groups, Amer. Math. Soc. Coll. Publ., Vol. 31, 1957.

5. T. Kato, Perturbation Theory for Linear Operators, 2nd ed., Springer-Verlag, 1980.

6. T. Kato, "Linear Evolution Equations of Hyperbolic Type," J. Fac. Sci., Tokyo U., Sect. I, Vol. 17 (1970), pp. 241-258.

7. B. Keyfitz, "Solutions with Shocks: An Example of an L_1-Contractive Semigroup," Comm. P and A Math, Vol. XXXIV (1971), pp. 125-132.

8. S. N. Kruzkov, "First Order Quasilinear Equations in Several Independent Variables," Math. Sbornik, Engl. transl., Vol. 10 (1973), No. 2, pp. 217-243.

9. L. Lusternik and V. Sobolev, Elements of Functional Analysis, 2nd ed., Hindustan Publ. Co. - John Wiley, Delhi-New York, 1974.

10. J. Marsden, "On Product Formulas for Nonlinear Semigroups," J. of Funct. Anal., Vol. 13 (1973), pp. 51-72.

11. B. Temple, "No L_1 - Contractive Metrics for Systems of Conservation Laws," preprint, March 1983.

12. K. Yosida, Functional Analysis, 6th ed., Springer-Verlag, 1980.

LOS ALAMOS NATIONAL LABORATORY
LOS ALAMOS, NEW MEXICO 87545

ABCDEFGHIJ–AP–89876